INSTABILITIES IN LUMINOUS EARLY TYPE STARS

ASTROPHYSICS AND SPACE SCIENCE LIBRARY

A SERIES OF BOOKS ON THE RECENT DEVELOPMENTS
OF SPACE SCIENCE AND OF GENERAL GEOPHYSICS AND ASTROPHYSICS
PUBLISHED IN CONNECTION WITH THE JOURNAL
SPACE SCIENCE REVIEWS

VOLUME 136
PROCEEDINGS

INSTABILITIES IN LUMINOUS EARLY TYPE STARS

PROCEEDINGS OF A WORKSHOP
IN HONOUR OF PROFESSOR CEES DE JAGER
ON THE OCCASION OF HIS 65TH BIRTHDAY
HELD IN LUNTEREN, THE NETHERLANDS, 21–24 APRIL 1986

Edited by

HENNY J. G. L. M. LAMERS

Sonnenborgh Observatory and SRON Laboratory for Space Research, Utrecht, The Netherlands

and

CAMIEL W. H. DE LOORE

Astrophysical Institute, Vrije Universiteit Brussel and Rijksuniversitair Centrum Antwerpen, University of Antwerp, Belgium

D. REIDEL PUBLISHING COMPANY

A MEMBER OF THE KLUWER ACADEMIC PUBLISHERS GROUP

DORDRECHT / BOSTON / LANCASTER / TOKYO

Library of Congress Cataloging in Publication Data

Instabilities in luminous early type stars.

(Astrophysics and space science library; v. 136)
Includes indexes.
1. Shell stars—Congresses. 2. Wolf-Rayet stars—Congresses. 3. Jager, C. de (Cornelis de), 1921– . I. Jager, C. de (Cornelis de), 1921– . II. Lamers, Henny J. G. L. M., 1941– . III. Loore, Camiel W. H. de. IV. Series.
QB843.S53I57 1987 523.8 87–9664
ISBN-13: 978-94-010-8232-7 e-ISBN: 978-94-009-3901-1
DOI 10.1007/978-94-009-3901-1

Published by D. Reidel Publishing Company,
P.O. Box 17, 3300 AA Dordrecht, Holland.

Sold and distributed in the U.S.A. and Canada
by Kluwer Academic Publishers,
101 Philip Drive, Assinippi Park, Norwell, MA 02061, U.S.A.

In all other countries, sold and distributed
by Kluwer Academic Publishers Group,
P.O. Box 322, 3300 AH Dordrecht, Holland.

PUBLISHER'S NOTE

We herewith take the opportunity to express our sincere gratitude and admiration to Cees de Jager who for many years has been the driving force and architect of the Reidel Astronomy programme.

Since the early sixties Cees de Jager has invested a lot of time and energy in this programme and has advised the company on directions to go and indicated gaps in the literature that needed to be filled. This was all done for the sole motive of serving the field of Astronomy.

It is no exaggeration to state that his activities have resulted in a great many publications that have enriched the literature and stimulated research in this exciting field.

Although officially retired, Cees seems now to be busier than ever. We hope that our association will last for many more years. All of us at Reidel wish Cees de Jager a happy and active retirement.

J. F. Hattink

TABLE OF CONTENTS

SECTION I: INVITED REVIEWS AND GENERAL DISCUSSIONS

SECTION II: POSTER PAPERS

PREFACE

On April 28 1986 Cornelis de Jager reached the age of 65 years. On April 30 he officially retired from the University of Utrecht where he has held a Chair for Stellar Astrophysics, later changed into Space Physics, since 1958.

Cees de Jager, as he prefers to be called by his friends, has had an active and successful life in science. His interest in astronomy was raised by his father under the clear skies of Celebes (Indonesia). He started a study in physics and astronomy as a student of the late M. Minnaert in Utrecht during World War II. When in 1943 the occupying forces recruited students who did not want to sign the declaration of loyalty for their war-efforts, Cees and three fellow students went into hiding at the Observatory in Utrecht. During this very "quiet" period van de Hulst developed the theory of the 21 cm radiation of neutral hydrogen and de Jager started the observations of variable stars in the pitch dark nights of a country at war. The study of Beta Cepheids rapidly awoke his interest which was kept throughout the years. In 1958 he organized an international campaign to observe 12 Lac spectroscopically and photometrically, which was a great success.

His main interest was initially directed at the Sun. After a thesis on "The Hydrogen Lines in the Solar Spectrum" he wrote an extensive review article on solar physics for Handbuch der Physik. When the first opportunities for observations from rockets above the earth atmosphere became available in the early sixties, he realized their importance for solar and stellar research and started to develop rocket X-ray instruments. This led to the foundation of the Laboratory for Space Research in Utrecht in 1961 of which he became the first director. At about the same time (1963) he succeeded Minnaert as a director of the Observatory. Later (1968) he founded the Astrophysical Institute in Brussels.

De Jager is an excellent organizer. With a clear vision of the important role of astronomy for the development of science and with a keen eye for technical possibilities on one hand and a broad insight in astrophysical problems on the other, he has played an important role in many international scientific organizations: ESRO (now ESA), IAU (General Secretary from 1970 to 1973), COSPAR (President from 1972 to 1978 and from 1982 to 1986), ICSU (President from 1978 to 1980) and many others. His affable and honest character combined with his tenacity in defending his ideas has made him a successful advocate for international astronomy.

Cees de Jager is one of the rare astronomers who reach a second youth in their scientific career after the age of 50. After 1977, when his burden of directorates was lightened, he enthusiastically returned to research in particular to his lifelong pet-subject: stellar instabilities. With a group of colleagues he discovered the phenomenon of "shock-driven mass loss" from ß Ceph stars. He also started a very ambitious project to write a book about "The Brightest Stars" which was published by Reidel in 1980. Since then he is full time studying the hypergiants in order to understand the processes in their teneous atmospheres and their variability. His investigations of the motions in these atmospheres led him to the suggestion that turbulence sets an upperlimit to the luminosity of very bright stars. This predicted turbulent upperlimit is now called the "de Jager limit". His large room at the Laboratory for Space Research is always a center of activity where he is surrounded by five or six students each one studying one of his pet hypergiants. He shows no sign of diminishing interest or curiosity for stellar instability.

The first ideas for a meeting in honour of Cees de Jager was born in 1984 at a meeting of the Dutch Studygroup of Extended Stellar Atmospheres (SUA). A preliminary organizing committee, consisting of A. van Genderen, H. Lamers and P.S. Thé, was formed. Very early on it was decided to choose the subject of "Instabilities in Luminous Early Type Stars" because that was the main interest of de Jager in recent years. Moreover it was felt that the meeting should be shaped as a "workshop" with a small number of invited participants, in order to stress the ongoing interest and active research of de Jager in this field.

The Scientific Organizing Committee, consisting of I. Appenzeller (Heidelberg), P. Conti (Boulder), R. Humphreys (Utrecht), C. de Loore (Brussels), H. Lamers (Utrecht, chairman), P.S. Thé (Amsterdam), and A. van Genderen (Leiden), met at the IAU Symposium nr 116 in Greece in May 1985 and shaped the scientific program, suggested invited review speakers as well as participants to be invited. The format of the meeting consisted of 12 sessions of a review and an extended general discussion. New results could be presented in posters and mentioned in the general discussion. This format turned out to be very successful in generating lively discussions between theoreticians and observers. This led a worried observer to ask: "Did we make the wrong observations?" and an impatient theoretician exclaim "You observers should have more patience!". The discussions are included in these proceedings.

The Local Organizing Committee consisting of K. van der Hucht (chairman) and J. Vogel has done an excellent job in selecting the very nice accommodation of "De Blije Werelt" located in the woods of Lunteren, and smoothly and efficiently running the local organization including a 20 mile bicycle trip through the woods of the National Park to the van Gogh paintings in the Kröller-Müller museum. The LOC was supported at the workshop by the students J. Coté, P. Mulder, A. Pieters and J. Wollaert. The discussions were recorded by means of hand-out forms by R. Waters.

We like to thank all those organizations who contributed generously, either financially or otherwise. In particular we acknowledge financial contributions from: SRON Space Research Laboratory in Utrecht; Sonnenborgh Observatory in Utrecht; the University of Utrecht; the Dutch Royal Academy of Sciences; the Leids Kerkhoven Bosscha Fonds; COSPAR; ESA and Fokker.

We are grateful to the Reidel Publishing Company, in particular G.

Kiers, for the publication of these proceedings with a colour portrait of Cees de Jager. This portrait was painted by W.C. van de Hulst, a brother of the astronomer, and presented on the occasion of Cees' retirement. The discussions were skilfully typed by Louise Cramer and Celia Roovers. The photographs were taken by H. Nieuwenhuijzen.

Most of all we want to thank all the participants for their contributions. The main goal of this workshop has certainly been achieved: "To teach Cees de Jager many new and fascinating facts about instabilities in luminous early type stars, which will be useful for him in his ongoing scientific career."

The participants in front of the Conference Center "De Blije Werelt" ("the happy world") in Lunteren.

LIST OF PARTICIPANTS

I. Appenzeller, Landessternwarte, Königstuhl, Heidelberg, B.R. Deutschland.
D. Baade, ST-ECF, ESO, Garching bei München, B.R.Deutschland.
L. Bianchi, Osservatorio Astronomico di Tirone, PINO TORINESE (TO), Italia.
R. Bonnet, ESA Headquarters, Paris, France.
G. Burki, Observatoire de Genève, Sauverny, Suisse.
A. Cassatella, ESA-IUE Tracking Station, Villafranca del Castillo, Madrid, Spain.
J. Cassinelli, Astronomy Dept., University of Wisconsin, Madison WI 53706, U.S.A.
J. Castor, L.Livermore Natl. Lab., University of California, Livermore, CA, U.S.A.
P. Conti, JILA, University of Colorado, Boulder, CO, U.S.A.
J. Coté, SRON, Ruimteonderzoek Utrecht, Utrecht, Nederland.
K. Davidson, Dept. of Astronomy, Univ. of Minnesota, Minneapolis, U.S.A.
D. Dawanas, ITB Observatorium Bosscha, Lembang, Jawa Barat, Indonesia.
J.-P. De Cuyper, Etterbeek, België.
A. van Genderen, Sterrewacht Leiden, Leiden, Nederland.
T. de Graauw, SRON Ruimteonderzoek Groningen, Groningen, Nederland.
A. Greve, IRAM, Université de Grenoble, St-Martin-d'Heres, France.
M. de Groot, Armagh Observatory, Armagh, Northern, Ireland, U.K.
G. Habets, Sterrenkundig Instituut Anton Pannekoek, Amsterdam, Nederland.
G. Hammerschlag, Sterrenkundig Instituut Anton Pannekoek, Amsterdam, Nederland.
A. Hearn, Sterrewacht Sonnenborgh, Utrecht, Nederland.
H. Henrichs, JILA, University of Colorado, Boulder, CO, U.S.A.
E. van den Heuvel, Sterrenkundig Instituut Anton Pannekoek, Amsterdam, Nederland.
L. Houziaux, Institut d'Astrophysique, Université de Liège, Cointe-Ougrée, Belgique.
K. van der Hucht, SRON Ruimteonderzoek Utrecht, Utrecht, Nederland.
R. Humphreys, Dept. of Astronomy, Univ. of Minnesota, Minneapolis, MN, U.S.A.
C. de Jager, Sterrewacht Sonnenborgh and SRON-ROU, Utrecht, Nederland.
M. Jerzykiewicz, Astronomical Institute, Wroclaw, Poland.
P. Korevaar, Sterrewacht Sonnenborgh, Utrecht, Nederland.
H. Lamers, Sterrewacht Sonnenborgh and SRON-ROU, Utrecht, Nederland.
C. Leitherer, Landessternwarte, Königstuhl, Heidelberg, B.R.Deutschland.
C. de Loore, Astrofysisch Instituut, VUB, Brussel, België
A. Maeder, Observatoire de Genève, Sauverny, Suisse
A. Moffat, Départ. de Physique, Univ. de Montréal, Montréal, Canada.
P. Mulder, SRON Ruimteonderzoek Utrecht, Utrecht, Nederland.
E. Müller, Basel, Suisse.
H. Nieuwenhuijzen, Sterrewacht Sonnenborgh, Utrecht, Nederland.
A. Noels, Institut d'Astrophysique, Univ. de Liège, Cointe-Ougrée, Belgique.
Y. Osaki, Department of Astronomy, Univ. of Tokyo, Tokyo, Japan.

S. Owocki, Center for Astrophys. and Space Science, UC San Diego, La Jolia, CA, U.S.A.
A. Piters, SRON Ruimteonderzoek Utrecht, Utrecht, Nederland.
F. Praderie, Observatoire de Meudon, Meudon, France.
M. Raharto, Sterrewacht Leiden, Leiden, Nederland.
J. Rountree-Lesh, Bethesda, MD, U.S.A.
G. Rufener, Observatoire de Genève, Sauverny, Suisse.
G. Rybicki, Harvard-Smithsonion Center for Astrophysics, Cambridge, MA, U.S.A.
R. Schulte-Ladbeck, Landessternwarte, Köningstuhl, Heidelberg, B.R. Deutschland.
L. Smith, Dept. of Physics and Astronomy, UCL, London, U.K.
J. Smolinski, Centrum Astronomiczne im M. Kopernika, Torun, Poland.
O. Stahl, European Southern Observatory, Garching bei München, B.R. Deutschland.
D. Stickland, Rutherford Appleton Laboratory, SAD, Chilton, Didcot, U.K.
P. Thé, Sterrenkudig Instituut Anton Pannekoek, Amsterdam, Nederland.
H. Tjin A Djie, Sterrenkundig Instituut Anton Pannekoek, Amsterdam, Nederland.
P. Ulmschneider, Institut für Theor. Astrophys. Heidelberg, B.R. Deutschland.
R. Viotti, Instituto Astrofysica Spaziale CNR, Frascati, Italia.
J.-M. Vreux, Institut d'Astrophysique, Univ. de Liège, Cointe-Ougrée, Belgique.
C. Waelkens, Astronomisch Instituut, Univ. van Leuven, Heverlee, België.
W. Wamsteker, ESA-IUE Observatory VILSPA, Madrid, Spain.
L. Waters, SRON Ruimteonderzoek Utrecht, Utrecht, Nederland.
P. Wesselius, SRON Ruimteonderzoek Groningen, Groningen, Nederland.
P. Williams, The Royal Observatory, Edinburgh, Schotland, U.K.
A. Willis, Dept. of Physics and Astronomy, UCL, London, U.K.
H. van Woerden, Kapteyn Laboratorium, Groningen, Nederland.
B. Wolf, Landessternwarte, Königstuhl, Heidelberg, B.R. Deutschland.
J. Wollaert, SRON Ruimteonderzoek Utrecht, Utrecht, Nederland.
J.-P. Zahn, Observatoire de Toulouse, Toulouse, France.

Section I

Invited Reviews and General Discussions

The final meeting of the Scientific Organizing Committee on the evening before the workshop. Peter Conti, Immo Appenzeller, Arnoud van Genderen, Bert de Loore, Pik-Sin Thé, Roberta Humphreys, Henny Lamers and Cees de Jager (consultant).

THE UPPER HR DIAGRAM - AN OBSERVATIONAL OVERVIEW

ROBERTA M. HUMPHREYS
University of Minnesota
Department of Astronomy
116 Church Street, S.E.
Minneapolis, MN 55455

The most massive and most luminous stars have always intrigued astronomers with considerable speculation about the upper limits to stellar masses and luminosities. Only about 20 years ago the nominal upper limit to stellar masses was supposed to be about 60 $M_{\odot}$ (Ledoux 1941, Schwarzschild and Harm 1959) due to vibrational instability. In a more massive star the pulsation amplitude was expected to grow until its outer layers were ejected reducing the mass or even destroying the star. Theory seemed to preclude the existence of very massive stars. But a few stars appeared to be very massive such as Plaskett's star, and the brightest supergiants in the Large and Small Magellanic Clouds observed by Feast et al. (1960). They found that many stars in the Clouds have visual luminosities that require masses close to 100 $M_{\odot}$. And by 1970, η Car (Davidson 1971) had been recognized as a very massive star. At about the same time, theorists (Appenzeller 1970, Ziebarth 1970) found that the vibrational instability was limited and the upper mass limit was closer to 100 $M_{\odot}$. Non-catastrophic mass loss might occur but massive stars were theoretically possible.

During the past decade or so, both the observational and theoretical studies of massive stars and their evolution have progressed very rapidly and in new directions. Observationally, this development was spurred by 1) the recognition of the importance of mass loss in stars >20 or 30 $M_{\odot}$ across the entire HR diagram thanks largely to results from IUE and infrared observations (see reviews by Hutchings 1978, Barlow 1978), 2) modern discussions of the HR diagrams for the luminous stars in our Galaxy and the Magellanic Clouds (Humphreys 1978, Humphreys and Davidson 1979), plus observations of the brightest and most massive stars in other nearby galaxies, and 3) the addition of mass loss and internal mixing which produced physically more realistic models and better agreement with the observations (see recent review by Chiosi and Maeder 1986).

The observed upper HR diagram and the characteristics of some of its most luminous stars provided the empirical evidence for an upper luminosity boundary likely determined by the relative stability or instability of the photospheres of the evolved most massive stars. The observed HR diagram for massive stars for our solar neighborhood and

H. J. G. L. M. Lamers and C. W. H. de Loore (eds.), Instabilities in Luminous Early Type Stars, 3–22.

other nearby galaxies is the basis for our current scenarios of massive star evolution and for much of the motivation behind our interest in the role of instabilities on their evolution.

In this opening paper I will review the upper HR diagram and the observational uncertainties in its determination. I will also discuss the upper luminosity boundary defined by the most luminous hot and cool stars, and the observational evidence for high mass loss rates and instabilities in these stars' atmospheres.

1. THE OBSERVED HR DIAGRAM - ITS DERIVATION AND UNCERTAINTIES

The HR diagram is normally plotted in one of two ways; M_V versus spectral type or color and M_{Bol} versus log T_{eff}. The first is based only on observed parameters while the second depends on calibrations to produce a diagram for comparison with the theoretical models and evolutionary tracks.

We rely very heavily on these HR diagrams and the locations on the HRD of different groups of stars for our understanding of massive star evolution and the role of atmospheric instabilities and resulting mass loss on that evolution. Before discussing the main features of the HRD and their implications I will briefly summarize the calibrations and uncertainties on which it depends.

1.1. Observed Parameters - Spectral Type and Photometry

There are of course personal errors in classification which are difficult to assess, but my comparisons of published types suggest that an uncertainty of $\pm$ half a type and $\pm$ half a luminosity class are typical; among the supergiants this is usually at most the difference between Ia and Iab for example. The errors in the photometry, at least for stars in our galaxy and the Magellanic Clouds, is usually quite small. The effective temperature and bolometric correction depends on the spectral type, but its uncertainties are <10% and are small compared to differences among the different effective temperature and bolometric correction scales.

1.2. Calibrated Parameters

The determination of effective temperatures for most hot stars and supergiants depends mostly on an analysis of the absorption line spectrum plus a model atmosphere with some empirical data from energy distributions and angular diameters. For the hot stars there have basically been two temperature scales, the 'hot' scale advocated by Conti (1973) based on non-LTE models by Auer and Mihalas (1972) and the 'cool' scale proposed by Underhill (1980, 1983) based on the stellar continua. The hot scale has been modified downwards somewhat by the recent non-LTE analysis by Kudritzki and his collaborators (Kudritzki 1980, Kudritzki et al. 1983, Simon et al. 1983). The now modified hot scale has in my opinion consensus support and is based on a firm foundation in stellar atmospheres. Use of the cool scale (see review by Böhm-Vitense 1981) produces an HR diagram in which the hottest stars are shifted to lower temperatures and luminosities leaving a large gap near the upper main sequence which must be explained; for example, by supposing most of the O-type stars must be hidden (see Garmany et al. 1982).

The fundamental data on effective temperature for stars later than A-type is also based on model atmospheres. There is less uncertainty about their calibration. The temperature scale for the cool supergiants, types K and M, is now much better determined with increased use of infrared data (see Lee 1970, Ridgway et al. 1980).

With increased ultraviolet and infrared observations the energy distributions and thus total luminosities of many more stars are available. The bolometric corrections (B.C.) for hot stars largely depend on the adopted effective temperature scale, thus uncertainties in the temperatures propagate to the bolometric corrections. The bolometric corrections for the evolved intermediate type supergiants are small and any uncertainties do not cause large changes in M_{Bol}. Bolometric luminosities for the red supergiants are large and for a long time were considered very uncertain, but again the infrared observations have greatly improved the accuracy of their bolometric corrections (see Elias, Frogel and Humphreys 1985). Indeed, given that a star is an M supergiant, the bolometric correction and the bolometric luminosity is known to an accuracy of ± 0.1 mag. With JHK photometry the luminosities of the red supergiants are now very well defined.

These are the fundamental calibrations required to produce the M_{Bol} versus log T_{eff} diagram; however, there are additional sources of errors that arise from the need to correct the data for interstellar extinction and to derive the absolute luminosities from a distance. Although there is evidence for variations in the IS extinction law in the ultraviolet in our galaxy and others, most of the available data supports a nearly uniform extinction law in the blue-visual to near infrared wavelength region.

The intrinsic color data for stars of all types have been summarized by many investigators (Johnson 1966, FitzGerald 1970, Schmidt-Kaler 1983). In a recent paper, Conti et al. (1986) have questioned the previously published U-B colors for the O-stars, based on their observations in the LMC. For both the blue and intermediate-type luminous stars, though, the extinction corrections are usually based on the (B-V) colors and for the red supergiant the infrared observations allow us to determine their luminosities at a wavelength where extinction can be essentially ignored.

The distance is perhaps the single most important pice of information in determining the HR diagram for luminous stars. Nearly all of these stars are too distant for direct measurement of their distances and thus their luminosities. In our galaxy, their visual and bolometric absolute luminosities are derived from membership in clusters and associations with known distances (e.g., Humphreys 1978, Humphreys and McElroy 1984). A fundamental reference for this procedure is Blaauw (1965). Walborn (1972, 1973) has provided a revised calibration for O and early B-type stars, and recent summaries by Schmidt-Kaler (1983) and Humphreys and McElroy (1984) give luminosity calibrations for larger data sets. These calibrations typically have a standard deviation of

0.5 mag for an individual luminosity and standard errors of the mean of 0.1 to 0.2 mag. For a star in a stellar aggregate, the luminosities are better determined with typical errors of ±0.25 mag due to uncertainties in the sitance moduli. For studies of stellar populations in other galaxies we must know the distance, and for the most part, we rely on published distances from RR Lyrae stars and the Cepheid period-luminosity relation. The uncertainties for Local Group galaxies are ±0.2 to 0.3 mag.

In Table 1 I have summarized the uncertainties in the observational data and the calibrations used to produce the HR diagram.

Table 1
Uncertainties in Observed and Calibrated Parameters
in an Individual Star

Sp. Type, UBV } ±10% − A_V ± 10% = V_o ± .15–.20 mag

T_{eff} (OB) ±5–10% (hot scale) B.C. (OB) ±.25 mag
(AFG) ±5–10% (AFG) ±.10 mag
(KM) ±5–10% (KM) ±.10 mag (IR)

Distances ±.25 mag → M_V ±.30 mag → M_{Bol} (OB) ±.4 mag
(AFG) ±.3 mag
(KM) ±.3 mag

The uncertainties are typically ±.3 to .4 mag in luminosity and ≈10% in effective temperature for an individual star, and the errors in the observations should be random not systematic. It should be emphasized that these errors apply to the location of a single star on the HR diagram and not to the locus of a whole group such as the hot stars, the red supergiants, or the location of the upper luminosity boundary. Because these are based on many stars whose parameters and distances come from different sources, they are better determined, probably to about half the above uncertainties.

The mass loss rate, while not an input parameter for the observed HR diagram, has figured very significantly in the evolutionary models, in the comparison of theory and observations, and especially for understanding the instabilities in these luminous stars. As a result of observations in both the ultraviolet and the infrared we now have mass loss occurring to some degree in all stars in the upper part of the HR diagram. Reviews by Hutchings (1978), Barlow (1978, 1981), and Zuckerman (1980) summarize the situation for the luminous hot and cool stars, and recent papers by Hagen et al. (1981), Lambert et al. (1981), and Kunasz and Morrison (1982, 1983) discuss the mass loss rates by intermediate type supergiants. Based on the published mass loss rates by different investigators and with different techniques I estimate the uncertainty in the mass loss rate for an individual star as a factor of two.

With this background, let's look at the observed HR diagram, its implications for the evolution of the luminous stars and the role of instabilities in their evolution.

The HR diagram, M_{Bol} versus Log T_{eff}, for the luminous stars in 91 associations and clusters in the solar neighborhood of our Galaxy is shown in Figure 1 in Humphreys and McElroy (1984). Even though the stars are essentially restricted to a region only about 3 kpc in radius, centered on the Sun, this diagram is representative of the stellar population in the spiral arms in the outer parts of our Galaxy. The most significant features of this diagram are:

1) A group of very hot, luminous stars with M_{Bol} = -10 to -11 mag, with an upper envelope of declining luminosity with decreasing temperature,
2) a lack of cooler counterparts, and
3) an upper luminosity limit for the cooler stars near M_{Bol} = -9.5 to -10 mag.

These features define the upper luminosity boundary in the HR diagram.

The HR diagram for the luminous stars in the Large Cloud (Fig. 3 in Humphreys and McElroy) looks very much like that for the solar neighborhood of our galaxy revealing very similar distributions of stellar temperatures and luminosities in the two galaxies and the same upper envelope of stellar luminosities (Humphreys and Davidson 1979). In the Small Cloud (Fig. 4 in Humphreys and McElroy) the hottest stars are both less luminous and fewer in number (Humphreys 1983). There is now an O3 star (Walborn 1986) in the SMC but on the basis of published data there are still no stars with initial masses greater than 80-85 $M_\odot$ in the Small Cloud. There are fewer of the brighgest stars at all of the early spectral types in the SMC. This is largely a statistical effect due to the smaller sample size in the Small Cloud. It is not due to chemical composition differences.

Even with these stochastic differences, the large scale features of the HR diagram for the SMC are essentially the same as for the Galaxy and the LMC, especially the upper luminosity boundary for the cooler supergiants.

Incompleteness of the data sample is very apparent in the HR diagrams for the Magellanic Clouds. It is clear that the data set for the hot stars becomes incomplete for luminosities below $M_{Bol} \simeq -8$ mag corresponding to ≈ 30 $M_\odot$. Luminosity functions for these data sets show marked turnover in numbers fainter than M_{Bol} = -8 mag (see Fig. 7 in Humphreys and McElroy 1984). This is the classic signature of incompleteness. The galactic data shows the same trend so we suspect that is also incomplete at the fainter luminosities. The degree of incompleteness is harder to evaluate at the higher masses, and luminosities, but the decrease in numbers for the highest luminosities may be due to incompleteness. Conti and collaborators have surveys in progress for both Magellanic Clouds to search for the O-type stars which will produce more complete samples of the hottest stars in these galaxies. That does not necessarily mean, however, that incompleteness does not exist among other groups of stars in the HR diagram.

In this symposium we are interested in the most luminous stars and this very likely means stars with initial masses above 30 or even 40 $M_\odot$, what I call the upper HR diagram. The upper HR diagrams for the Galaxy, the Clouds, and M33 are consistent with an upper envelope of declining luminosity with decreasing temperature which becomes nearly constant at M_{Bol} = -9.5 to -10 mag for the cooler evolved supergiants.

2. THE UPPER LUMINOSITY BOUNDARY AND THE STABILITY LIMIT OF THE PHOTOSPHERES OF THE MOST MASSIVE STARS

The temperature dependence of the luminosity boundary for the most luminous, hot stars suggests that it defines a critical location on the HR diagram, one that is mass dependent.

Humphreys and Davidson (1979) first drew attention to this empirical luminosity boundary and to the lack of evolved very massive stars. We suggested that this observed boundary was due to a stability limit encountered as the most massive stars (>50 $M_\odot$) evolve away from the main sequence and accompanied by enhanced mass loss. For a brief review of the observations and theoretical considerations see Humphreys and Davidson (1984).

This critical boundary in the HR diagram is marked by some well known very luminous stars, η Car, P Cyg, S Dor and the Hubble-Sandage variables in M31 and M33. These stars are unstable with high mass loss rates (10^{-5} to 10^{-4} $M_\odot$/yr) which are typically ten times higher than the 'normal' stars of comparable temperatures and luminosities. They may have even higher losses in occasional dramatic ejections. Temperatures, luminosities, and mass loss rates of some of these luminous blue variables (LBV's) in the Galaxy and the LMC have been determined from IUE observations (Wolf et al. 1980, 1981a,b, Shore and Sanduleak 1984, Lamers et al. 1983, Stahl et al. 1983, Leitherer et al. 1985), and these stars are shown on the schematic HR diagram in Figure 1. The LBV's are presumably the most massive stars near the stability limit of their photospheres. When the star encounters the instability it loses enough mass to become temporarily 'stable' again until the next ejection. During visual maximum, at the time of the shell ejection phase, the spectra of the LBV's may resemble the cooler photospheres of A or even F-type supergiants even though the star is quite hot (20,000-30,000°K). Leitherer et al. (1985) have described the development of the 'false photosphere' in S Dor quite thoroughly. Var C in M33 recently brightened by two magnitudes and a very recent spectrum resembled an F5 supergiant, perhaps the coolest shell that has been observed in an LBV. The LBV's maintain essentially constant luminosity between maximum and minimum light (R71 and S Dor, Appenzeller and Wolf 1981; Var 83 in M33, Humphreys et al. 1984; and R127, Stahl et al. 1983). These variations in 'photospheric' temperature are shown as dashed lines in Figure 1.

An alternative interpretation of the LBV's, originally suggested by Bath (1979) and advocated by Gallagher and co-workers (Gallagher et al. 1981, Kenyon and Gallagher 1985), is that they are really accretion disks around less massive stars (20-30 $M_\odot$) in binary systems. But the accretion disk model cannot account for all of the observations, such as high temperatures (25-30,000°K) at minimum light.

The exact cause of the instability in these luminous, hot stars is not known; several possibilities have been reviewed by Stothers and Chin (1983). Appenzeller (1986) and Lamers (1986) have independently suggested radiation pressure from many weak lines and a turbulent pressure gradient has been proposed by de Jager (1980, 1984) and Maeder (1983). Several reviewers at this symposium will be discussing the

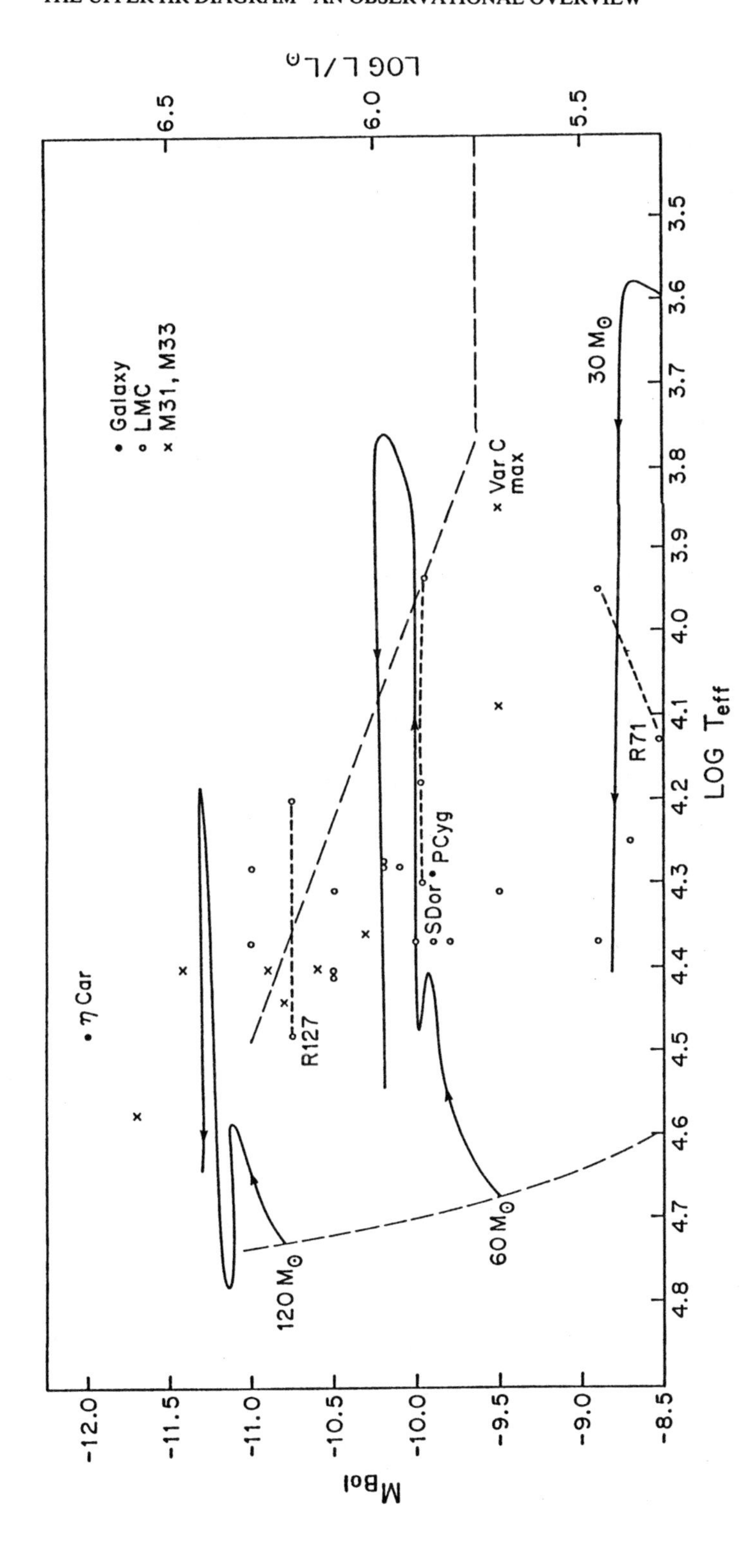

Fig. 1 - The schematic HR diagram for stars more luminous than $M_{Bol} = -8.5$ mag. The location of the LBV's in the Galaxy, LMC, SMC, M31 and M33 are shown. The dashed line is the empirical upper luminosity boundary.

different mechanism for instabilities in the LBV's as well as other luminous hot stars. The most massive stars ($\geq$60 $M_\odot$) very likely do not evolve to cooler temperatures because of the instabilities which develop in their atmospheres as they evolve away from the main sequence. The LBV's are a relatively short stage (few $\times 10^5$ yrs) in which these hot stars shed mass at high rates in transition to the WR stage.

de Jager (1984) has published a very informative HR diagram showing the locus of mass loss rates. Stars near the luminosity boundary have very high mass loss rates, especially the cooler supergiants with mass loss rates 10 to 100 times those of the hot stars of the same luminosity. Many of the hot stars along the luminosity boundary ($M_{Bol} < -9.5$ or -10 mag), especially the LBV's, will be discussed in some detail during this workshop; therefore I want to focus on some interesting phenomena occurring in the cooler hypergiants and questions concerning the evolution of the slightly less luminous, less massive stars (40 < M_i < 60 $M_\odot$). Most of the LBV's have luminosities more than $M_{Bol} \simeq -10$ mag but there are a few which are significantly fainter. They have presumably come from stars of lower initial mass (40-50 $M_\odot$) and have evolved through a cool supergiant stage, thus their evolution has been quite different from the more luminous LBV's, and the instabilities and mass loss in the cooler hypergiants may be very relevant to the origin of these LBV's and even some WR stars.

Figure 2 is a schematic HR diagram showing the location of some very luminous evolved supergiants plus the OH/IR supergiants, the LBV's, and WR stars. I will use this diagram to discuss some questions concerning the lower temperature stars that lie along the observed luminosity boundary.

First of all I want to point out some A-type hypergiants, specifically, Cyg OB212 and HD33579 in the LMC, which are the visually brightest stars in their galaxies. Their mass loss rates are not exceptionally high ($\approx 10^{-5}$), yet with initial masses of 50-60 $M_\odot$ they must be approaching the Eddington limit for their masses. Will they reach the region of the red supergiants or evolve back to higher temperatures becoming LBV's? To get some idea, let's look at the slightly less luminous cooler hypergiants.

The F, G, and K-type hypergiants in the Galaxy and LMC essentially define the upper limit for cool stars. Many of them are clearly unstable as evidenced by variations in light and in their spectra, high mass loss rates (few $\times 10^{-4}$ $M_\odot$/yr, Hagen et al. 1981) and by the presence of circumstellar dust shells around many. ρ Cas is well known for its shell episode (1946-47) in which it decreased 1.5 mag and had the spectrum of an M star. Many of these very luminous G supergiants show the 10μ silicate feature due to circumstellar dust (Humphreys et al. 1971). These stars are all close to the limits of their stability, and de Jager's turbulence pressure gradient is a plausible cause of the instabilities in the cool hypergiant. I think it likely that these stars are evolving toward the red supergiant region.

The most luminous M supergiants are well represented by stars like μ Cep, HD143183, and KW Sgr, but there is a group of very luminous late-type stars which are later in spectral type than the normal M supergiants. They have very large infrared excesses, many with

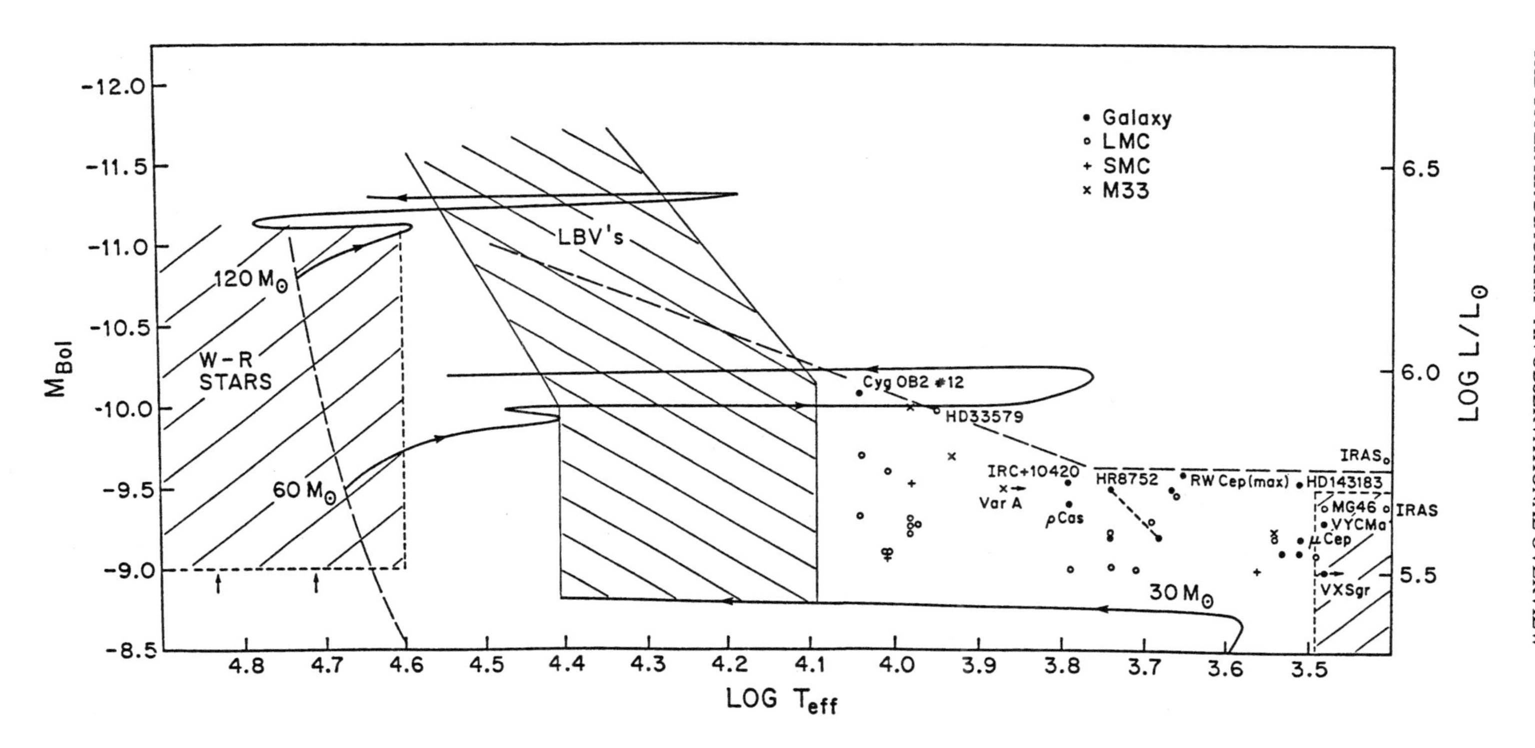

Fig. 2 - The schematic HR diagram showing the location of the cool hypergiants, the supergiant OH/IR sources, the LBV's and WR stars.

optically thick dust shells, and are also OH masers. These are known as the supergiant OH/IR sources. They are likely the most evolved M supergiants; M supergiants that have lost sufficient mass that their dust shells are now optically thick. The mass loss from these OH/IR sources may be very high 10^{-4} to 10^{-3} $M_{\odot}$/yr. A few with optically thin shells are visibly bright, like VY CMa (M3-5eIa), VX Sgr (M4-8eI), and S Per (M4eIa). Two highly obscured, highly luminous late-type stars (Elias, Frogel and Schwering 1986) discovered by IRAS presumably belong to this group; OH emission has just been detected from one (Wood et al. 1986). What will eventually become of these supergiant OH/IR sources? My colleague Terry Jones and I are studying the possibility that they may evolve to WR stars, analogous to their less massive counterparts which become the central stars of planetary nebulae.

IRC+10420 may be a good candidate for a star in transition from red supergiant to WR star. IRC+10420 has the spectrum of a very luminous F supergiant (F8Ia+) plus a very large IR excess from a circumstellar dust shell (Humphreys et al. 1972). It is also the earliest (warmest) known OH/IR source (Giguere et al. 1976). Recent OH observations show the 1665 MHz feature is weakening while the 1612 MHz feature is growing (Lewis et al. 1986). This is what we would expect if the dust shell were dissipating. Interestingly it has also been getting visually brighter (Gottleib and Liller 1978). If this trend continues a very plausible model for IRC+10420 will be a post M supergiant-OH/IR star blowing off its cocoon of dust and gas as it evolves to the left to warmer temperatures on the HR diagram.

I finally want to mention Var A in M33, a very enigmatic object whose history is difficult to interpret. Var A is one of the original Hubble-Sandage variables, it showed a slow increase in brightness over more than fifty years from ≃18 to 15.7 (mpg) mag at maximum in 1951 when it had an early F-type supergiant spectrum. It then rapidly declined by about 4 mag in less than a year (Rosino and Bianchini 1973). It has been faint and red ever since. A recent spectrogram shows very strong TiO bands (>M5 in type). It also has a strong infrared radiation which resembles some of the OH/IR stars. It is also very luminous, M_V (and M_{Bol}) $\simeq -9.5$ mag, at maximum.

Is Var A's normal state an F or an M supergiant? Is its M star spectrum produced in a cool shell? Var A in many ways resembles ρ Cas; although, Var A has presumably had its M star spectrum for more than 30 years and has a circumstellar dust shell which ρ Cas lacks.

I have singled out some individual cool hypergiants which are right at the upper luminosity boundary. They are obviously unstable as evidenced by their light and spectral variability, high mass loss rates and circumstellar shells. The luminous stars that define this luminosity limit represent important stages with instabilities that determine their evolution and eventual fate perhaps as unstable luminous hot stars.

I conclude this review by presenting some evolutionary scenarios for the massive stars that pose several questions.

1. $M_i \geq 60\ M_{\odot}$ O → Of → LBV → WR

This is now the commonly accepted model for the most massive stars. It is very unlikely that stars this massive can evolve to the red supergiant region.

2. $M_i \simeq 50\text{-}60\ M_\odot$ O → blue super - ? { LBV → WR ; OH/IR → LBV → WR }

It is also doubtful whether stars this massive can evolve to the red supergiant region, but if they did the mass loss rates in that region (see de Jager 1984) would be so high they would not be visible; they would be shrouded in dust as OH/IR sources.

3. $M_i \simeq 40\text{-}50\ M_\odot$ MS → blue super → yellow super → red super → OH/IR = ? { LBV → WR ; SN }

Do the most luminous red supergiants become WR stars perhaps via the OH/IR stage during which they lose a lot of mass? IRC+10420 may be an example of such a star.

4. $M_i \simeq <40\text{-}30\ M_\odot$ MS → blue super → yellow super → red super → yellow super and Cepheid → blue super (He-rich) → yellow super and Cepheid → red super → OH/IR → SN

This scenario has a blue loop in the HRD. The He-rich blue stars refer to a group of peculiar A-type stars in the Clouds (Humphreys 1983) which may be He-rich.

Some final questions:

1) What is the cause of the instability in the LBV's and what causes their more violent eruptive stages as observed in η Car and P Cyg?

2) Have the less luminous LBV's (M_{Bol} = -9 to -10 mag) been cooler supergiants?

3) Do supergiant OH/IR sources become WR stars?

I expect many of the papers at this symposium will bring us closer to answering some of these questions.

REFERENCES

Appenzeller, I., 1970, A. & A., 5, 355.
Appenzeller, I., 1986, in IAU Symposium #116, Luminous Stars and Associations in Galaxies.
Appenzeller, I. and Wolf, B., 1981, in ESO Workshop on the Most Massive Stars.
Auer, L. and Mihalas, D., 1972, Ap.J. Suppl., 24, 193.
Barlow, M.J., 1978, in IAU Symposium #83, Mass Loss and Evolution of O-Type Stars, p. 119.
Barlow, M.J., 1981, in IAU Symposium #99, Wolf-Rayet Stars, p. 149.
Bath, G.T., 1979, Nature, 282, 274.

Blaauw, A., 1965, in Stars and Stellar Systems, Vol. 3, Basic Astronomical Data, p. 383.
Böhm-Vitense, E., 1981, Ann. Rev. of Astron. and Astrophys., Vol. 19, 295.
Chiosi, C. and Maeder, A., 1986, Ann. Rev. of Astron. and Astrophys., in press.
Conti, P.S., 1973, Ap.J., 179, 181.
Conti, P.S., Garmany, C.D. and Massey, P., 1986, A.J., in press.
Davidson, K., 1971, MNRAS, 154, 415.
de Jager, C., 1980, The Brightest Stars.
de Jager, C., 1984, A. & A., 138, 246.
Elias, J.H., Frogel, J.A. and Humphreys, R.M., 1985, Ap.J. Suppl., 57, 91.
Elias, J.H., Frogel, J.A. and Schwering, P.B.W., 1986, Ap.J., 302, 675.
FitzGerald, M.P., 1970, A. & A., 4, 234.
Gallagher, J.S., Kenyon, S.J. and Hege, E.K., 1981, Ap.J., 249, 83.
Garmany, C.D., Conti, P.S. and Chiosi, C., 1982, Ap.J., 263, 277.
Giguere, P.T., Woolf, N.J. and Webber, J.C., 1976, Ap.J. (Letters), 207, L195.
Gottlieb, E.W. and Liller, W., 1978, Ap.J., 225, 488.
Hagen, W., Humphreys, R.M. and Stencel, R.E., 1981, Pub.A.S.P., 93, 567.
Humphreys, R.M., 1978, Ap.J. Suppl., 38, 389.
Humphreys, R.M., Blaha, C., D'Odorico, S., Gull, T.R. and Benvenuti, P., 1984, Ap.J., 278, 124.
Humphreys, R.M. and Davidson, K., 1979, Ap.J., 232, 409.
Humphreys, R.M. and Davidson, K., 1984, Scienc3, 223, 243.
Humphreys, R.M. and McElroy, D.B., 1984, Ap.J., 284, 565.
Humphreys, R.M., Strecker, D.W., Murdock, T.L. and Low, F.J., 1973, Ap.J. (Letters), 179, L49.
Humphreys, R.M., Strecker, D.W. and Ney, E.P., 1971, Ap.J. (Letters), 167, L35.
Hutchings, J.B., 1978, in IAU Symposium #83, Mass Loss and Evolution of O-Type Stars, p. 3.
Johnson, H.L., 1966, Ann. Rev. Astron. Astrophys., Vol. 4, 183.
Kenyon, S.J. and Gallagher, J.S., 1985, Ap.J., 290, 542.
Kudritzki, R.P., 1980, A. & A., 85, 174.
Kudritzki, R.P., Simon, K.P. and Hamann, W.-R., 1983, A. & A., 118, 245.
Kunasz, P.B. and Morrison, N.D., 1982, Ap.J., 263, 226.
Kunasz, P.B., Morrison, N.D. and Spressant, B., 1983, Ap.J., 266, 739.
Lambert, D.L., Hinkle, K.H. and Hall, D.N.B., 1981, Ap.J., 248, 638.
Lamers, H.J.G.L.M., 1986, in IAU Symposium #116, Luminous Stars and Associations in Galaxies.
Lamers, H.J.G.L.M., de Groot, M. and Cassatella, A., 1983, A. & A., 128, 299.
Ledoux, P., 1941, Ap.J., 94, 537.
Lee, T.A., 1970, Ap.J., 162, 217.
Leitherer, C., Appenzeller, I., Keare, G., Lamers, H.J.G.L.M., Stahl, O., Waters, L.B.F.M. and Wolf, B., 1985, A. & A., 153, 168.
Lewis, B.M., Terzian, Y. and Eder, J., 1986, Ap.J. (Letters), 302, L23.
Maeder, A., 1983, A. & A., 120, 113.
Ridgeway, S.T., Joyce, R.R., White, N.M. and Wing, R.F., 1980, Ap.J., 235, 126.

Rosino, L. and Bianchini, A., 1973, A. & A., 22, 453.
Schmidt-Kaler, Th., 1983, in Landolt-Bornstein, New Series, Group 6, Vol. 2, Part B.
Schwarzschild, M. and Harm R., 1959, Ap.J., 129, 637.
Shore, S.N. and Sanduleak, N., 1984, Ap.J. Suppl., 55, 1.
Simon, K.P., Jonas, G., Kudritzki, R.P. and Rahe, J., 1983, A. & A., 125, 34.
Stahl, O., Wolf, B., Klare, G., Cassatella, A., Krautter, J., Persi, P. and Ferrari-Toniolo, M., 1983, A. & A., 127, 49.
Stothers, R. and Chin, C.-W., 1983, Ap.J., 264, 583.
Underhill, A.B., 1980, Ap.J., 239, 220.
Underhill, A.B., 1981, Ap.J., 244, 963.
Walborn, N.R., 1972, A.J., 72, 312.
Walborn, N.R., 1973, A.J., 78, 1067.
Walborn, N.R. and Baades, J.C., 1986, Ap.J. (Letters), 304, L17.
Wolf, B., Appenzeller, I. and Cassatella, A., 1980, A. & A., 88, 15.
Wolf, B., Appenzeller, I. and Stahl, O., 1981b, A. & A., 103, 94.
Wolf, B., Stahl, O., de Groot, M.J.H. and Sterken, C., 1981a, A. & A., 99, 351.
Wood, P.R., Bessell, M.S. and Whiteoak, J.B., 1986, Ap.J. (Letters), in press.
Ziebarth, K., 1970, Ap.J., 162, 947.
Zuckerman, B., 1980, Ann. Rev. Astron. Astrophys., Vol. 18, 263.

Roberta Humphreys:
singing or giving a review?

DISCUSSION

HEARN: In the slide of the HR diagram of stars of our own galaxy, the upper limit seems to have a minimum and then goes back up again. Is this a significant effect?
The shape of the upper limit is of course important in determining the constraints on possible theoretical explanations.

HUMPHREYS: The minimum you refer to is the region of the luminous A-type supergiants. In the HR diagram for the LMC there are several stars in this same region. I suspect the lack of these luminous A-type supergiants in our galaxy is due to observational selection.

VIOTTI: About variable A in M33 I think that there is a simple explanation of its behaviour. It could be a binary system with a luminous M supergiant and a hotter companion which is now masked by a dust envelope. The formation of this dust envelope may explain the deep fading which is applied only to one component of the system.
A similar situation has occurred in the galactic symbiotic system PU Vulpeculae which underwent a big fading in 1980 from V = 8-9 to V = 13 lasting about one year, and followed by the disappearance of the F-type spectrum and appearance (but this is only a contrast effect) of an M type spectrum.
This hypothesis on Var A is putting forward a more general problem. Maybe many of these objects are binaries or multiple systems. How much would your HR diagram be affected by this? I think that this is a very important problem if one wants to understand the physics going on in the atmospheres of these objects.

HUMPHREYS: I have also considered the possibility that Var A could be a binary. I have planned CCD imaging in UBVR to look for a warmer companion. However, I actually doubt if it is a binary because it will be very difficult to explain a binary system with a very luminous F type supergiant and an M supergiant on the basis of evolution and the time scales required.

MAEDER: The correspondence between the observed and theoretical upper star limit in the HRD depends very much on the assumed mass loss rates $\dot{M}$ in the LBV stars. For low mass loss rates, the evolution proceeds rapidly up to the red supergiant stage; on the contrary for very high mass loss rates, the evolution immediately turns to the blue in the HR diagram. Now, a good correspondence between the observed and theoretical limit is obtained for an average mass loss rate value of about 10^{-3} $M_{\odot}$/yr in LBV stars. And it is nice that this well corresponds to the observed mass loss rates given by observations (cf. Lamers, in IAU Symposium 116).

HUMPHREYS: Yes. There is presently a good agreement between the observed limit from observations in the HR diagram and the expectations from the evolutionary tracks with higher mass loss in the upper HR diagram.

MOFFAT: In the case of R136a, Weigelt's speckle interferometry yields no star inside R136a which is optically more luminous than any star outside R136a in the 30 Dor cluster. Hence if the limit is 100 $M_{\odot}$ for individual resolved 30 Dor stars, it must be the same to the stars in R136a.

HUMPHREYS: I commented that the most luminous, most massive stars may have initial masses close to 200 $M_{\odot}$, i.e. R136a, based on comments by Nolan Walborn at IAU Symposium # 116.
There is now evidence that η Car (Weigelt, 1986) is also multiple bringing its mass down to 100 $M_{\odot}$. η Car will be discussed in detail by Davidson later in this symposium.

DE JAGER: Are all early-type stars near the Humphreys-Davidson limit Luminous Blue Variables, and are all LBV's situated near the HD limit?

HUMPHREYS: No, the limit is defined by the distribution of normal stars in the HRD. Yes, many of the LBV's are on or near the empirical limit in the HRD.

DE JAGER: In that connection I draw attention to AG Carina, often classified as a P Cyg star, but having a luminosity well below the HD limit, with a rate of mass loss comparable to the values found at the limit. Life would be much easier when the bolometric luminosity of AG Car would be larger than the commonly assigned value.

HUMPHREYS: That is one of the reasons I drew attention to the less luminous LBV's in the HRD and why I have suggested that they may have been through a previous cool supergiant stage. If their luminosities are an indicator of initial mass then presumably the evolution of these less luminous LBV's has been different from those with $M_i \geqslant 60$ $M_{\odot}$.

LAMERS: The problem of the absolute luminosity of LBV's which are apparently below the luminosity upper limit is a very important one because it implies that stars can overcome the LBV instability and still evolve into a red supergiant.
The critical test for this is AG Carina which is far below the Humphreys-Davidson limit. It should be realized, however, that the distance of this star is very uncertain since it is located in a direction where we see the spiral arm tangentially, so the star may be more distant then presently assumed. There are some WR stars in that direction which are more distant than the presently adopted distance of AG Car.

DE JAGER: That sounds promising, but in order to get the star close enough to the instability limit its distance should be three to five times larger than presently assumed.

WOLF: 1) Another case of a less luminous S Dor variable is R71 of the LMC for which we know the absolute magnitude rather well. This star shows all the characteristics of the S Dor variables and is definitely located below the boundary limit.

2) I would like to draw your attention to another interesting group of stars located at the Humphreys-Davidson instability limit, the Be-supergiants of the LMC, which have been discussed recently by our group. These stars have many spectral characteristics in common with S Dor variables, are surrounded by dust, but have not shown any variability.

HUMPHREYS: Yes, I agree about R71, it is one of the less luminous LBV's. I tend to group these luminous early-type supergiants with the emmission line spectra in the general category with the LBV's even though they are not now observed to be variable. Many of these stars may be relatively quiescent for long periods. Some of the H-S variables in M31 and M33 are like that, for example Var C in M33.

OWOCKI: What are observed mass loss rates for those cooler Hypergiants? Is the episodic mass loss enough to reduce the stellar mass? From the observational point of view, is stellar mass loss dominated by continuous or episodic mass loss?

HUMPHREYS: 1) The cooler hypergiants, particularly those near the upper luminosity boundary, have mass loss rates of 10^{-5} to 10^{-4} $M_\odot$/yr.
2) Yes, during the more violent episodes of mass ejection such as occurred in η Car the mass loss may be 10^{-3} $M_\odot$/yr or higher, depending on whose estimates one believes.
3) If the true mass loss during the violent episodes is as high as 10^{-3} $M_\odot$/yr in η Car-like ejections and 10^{-4} $M_\odot$/yr in LBV's at maximum light (the shell ejection phase) as opposed to a continuous loss of 10^{-5} $M_\odot$/yr, then obviously the more violent episodes and shell ejections will dominate (see Lamers, IAU Symp. 116).

NOËLS: Would it be possible to imagine that there could be, in the process of star formation, a slight dependence on the heavy elements content so that the less massive stars observed in the SMC would be a real effect and not only a statistical effect?

HUMPHREYS: The observations suggest that the lower luminosity of the brightest stars and their fewer numbers in the SMC are primarily due to the smaller size of the population sample of this smaller, lower mass galaxy. The same effect is observed in the other irregular galaxies of our Local Group. These differences are not obviously dependent on metallicity.
I suspect that the IMF's are essentially the same but the star formation rate for massive stars may be lower in these smaller galaxies.

CONTI: You seemed to imply during your talk that red hypergiants, above the Humphreys-Davidson limit, but as yet unidentified, may exist. Would you comment on this?
Is variable A (M33) which is now an M supergiant, above the limit?

HUMPHREYS: If they did exist they would very likely be surrounded by optically thick circumstellar dust clouds because of the high mass loss rates that they should have. Less luminous red supergiants already have

$\dot{M}$ of a few times 10^{-4}. Thus I would not expect these stars to be visible even if they existed. They could presumably be detected in the infrared. Although our surveys in the optical are essentially restricted to a few kiloparsecs around the sun, the surveys for OH/IR sources cover the entire galaxy. No supergiant OH/IR sources have been detected above $M_{bol} \simeq -9.5$ mag.; therefore, I do not think they exist, at least not many.
No, variable A is not above the limit. At visual maximum it had $M_V \simeq -9.5$ mag which for an F-type spectrum is essentially the bolometric luminosity. We will determine its bolometric luminosity as an M star when we have the infrared data. But I do not expect it to be significantly different.

APPENZELLER: If the 'Less Luminous Blue Variables' are indeed in an evolutionary stage following a OH/IR source state, would you expect to see still some FIR radiation from the remains of the dust ejected earlier and could the presence of such dust be used to prove your scenario?

HUMPHREYS: I think it would depend on the time scales involved. I suspect that there is enough time for the previous red supergiant (OH/IR star) to shed its circumstellar shell in transition back to the blue side of the HRD. I see no reason why an LBV could not produce its own dust shell with its high mass loss rates.

WILLIAMS: The long term fading and reddening of HR 5171 you mentioned cannot be due to growth of its circumstellar dust shell as the dust shell is optically thin. The star itself must be changing.

VAN DEN HEUVEL: I have two questions:
1) Do you think - taking observational selection effects into account - that the diagrams with the distribution of massive luminous stars in the HRD are the same in the galaxy, LMC and SMC, despite the differences in metal abundance?
2) If they are indeed similar, would this then not suggest that things like mass loss rates and evolution are quite independent of Z? This would seem to suggest then that the mass loss is not (physically) driven by atmospheric processes, like radiation pressure in lines in the atmosphere (as atmospherically-driven mass loss is expected to depend rather strongly on Z), but rather by something having to do with the deep interior structure and evolution of the star.

HUMPHREYS: 1) The HRD's for the LMC and solar neighbourhood are very much alike. There certainly are differences between the LMC and SMC, and Milky Way and SMC: the lack of the most luminous O stars in the SMC, for example. This effect however has been well documentated in many smaller, less massive galaxies and is usually attributed to the smaller population sample. Metallicity differences certainly exist but they are the major determinant of the observed population of massive stars. What we observe appears to depend more on stochastic effects and the star formation rate.
2) I would not say that they are entirely independent of Z but it is not

the major driver, at least not over the range of chemical abundances represented by the Galaxy and Magellanic Clouds.

CONTI: There seems to be no demonstrated dependence of $\dot{M}$ on 'Z' for O type stars (See Garmany + Conti, Ap. J, 1985). The SMC stars seem to have the same $\dot{M}$ as similar luminosity galactic stars, even with 'Z' lower by a factor 6 (by number). The Humphreys-Davidson limit probably does not depend on the past $\dot{M}$ history but rather on an instability at that location in the HRD.

MOFFAT: If one were to make a complete photometric survey of all stars in the LMC or SMC down to a given apparent magnitude would we see the same distribution and upper limit as you show based on spectral (selective?) surveys? (Is anyone doing this?)

HUMPHREYS: I would think so for the observed upper luminosity boundary. The only possibility is for some very peculiar star to be discovered that was suffering unusually high interstellar or circumstellar reddening.
Conti and collaborators are surveying both Magellanic Clouds for OB stars but their observations do not go further than m_V = 14 magn.

STICKLAND: On the HR diagrams we see the evolutionary tracks for various masses. What direct evidence do we have for the high masses at the upper end of the diagram?

HUMPHREYS: The most massive known stars from binary orbits are Plaskett's stars for which the components apparently have masses of 60 $M_{\odot}$ or slightly more. Higher masses for the hot stars are implied by the high luminosities.

CONTI: Plaskett's star still is the highest mass binary system, at some 60 $M_{\odot}$. There are stars more luminous and of earlier spectral type, thus more massive. The high mass you find for ι Ori seems inconsistent with our present understanding of the masses in the upper HRD. Presumably there will be some solution of the problem you raise, perhaps in the distance.

LAMERS: I have a question to the evolutionists. If a massive star does not lose enough mass it will evolve to the red supergiant phase. If the star loses too much mass it will evolve from the B supergiant phase directly into a WR star. My question is: is there a mass loss rate which can keep a star for a period of about 10^4 years close to the instability limit as a LBV B-type supergiant? Is this mass loss rate very critical? If so, it would be difficult to explain the number of LBV's.

MAEDER: An average mass loss rate for the LBV stage of the order of 10^{-3} $M_{\odot}$ y^{-1} keeps the star close to the instability limit for 10^4 years. Now, if you would reduce the mass loss rate by a factor of two, you would increase the lifetime close to the limit by about a factor of two, since the definite departure to the blue side only occurs when the helium core mass fraction has a given critical value. In the example just given here, the star would take a time twice larger to reach the critical value of the core mass fraction.

DE LOORE: The time spent by a luminous star in the Hubble-Sandage variable (or LBV) region depends on the mass loss rate, this is evident. What is more, from the evolutionary viewpoint it is possible to match the observations, at least on an order of magnitude basis. I can show you a practical example. I have calculated the evolution of a 50 $M_{\odot}$ star, with a normal mass loss rate, as observed (order of a few times 10^{-6} $M_{\odot}$ yr^{-1}) until the star enters the quiescent P Cygni phase (after 6.6 10^{6} yrs), and then imposing a mass loss rate of 2.10^{-4} $M_{\odot}$ yr^{-1}. Some 10^{4} yrs later the star reverses its evolutionary sense, and returns to the left in the HRD. The star remains 37.000 yrs in the instability region, and evolves later into a WR star.

WOLF: In connection with the life-time of S Dor variables around the instability limit I would like to draw your attention to our poster on R127. Quite recently we found this star surrounded by an AG Car-like nebula. The kinematic age of this nebula is about 20 000 years.

DE JAGER: A rate of mass loss of 10^{-4} $M_{\odot}$/yr fits very well to the value observed near the instability limit. Hence we came to the interesting conclusion that just the 'average' rate of mass loss of hypergiants is sufficient to explain evolutionary the absence of stars beyond the HD limit, and that the addition of a significant contribution by episodical mass loss may not be needed.
I use this opportunity to mention that the diagram on $\dot{M}$ in the Hertzsprung Russell diagram shown by you is a few years old and that we have now more data, including a larger number of stars (250). A preprint is distributed at this meeting.

DAVIDSON: Perhaps we should avoid quoting specific mass-loss rates near the instability limit. Various hints seem to indicate that the limit behaves rather like a valve, which opens as the star's surface approaches some critical line in the L, T_{eff} diagram. Episodic cases like Eta Carinae may represent imperfections in the 'valve' causing unsteady behaviour. My point is this: Have any evolutionary calculations been done with artificial 'valves' of this type? Just as in a machine, the precise quantitative details of the artificial 'valve' might be unimportant. These remarks are entirely empirical, and are not based on any theoretical pre-suppositions, I hope!

DE LOORE: The way I treated the problem in the case I just mentioned is somewhat like a simulation of a valve, I mean, as mass loss during the O-type stage, I adopted an average of 10^{-6} $M_{\odot}$ yr^{-1}; when entering the quiescent Hubble-Sandage variable strip, I raised this to 2.10^{-4} in 10 steps of 100 year, and than I left the star at this regime for some time. This could then represent an episodic mass loss of a couple of solar masses in 100 years, at least for the overall evolutionary picture, since for the nuclear reactions in the interior the mass loss itself has not very much importance. A real treatment as a 'valve mechanism', i.e. an extreme mass loss on a very short time scale, is not possible by the kind of evolutionary codes we use for the 'overall' evolutionary pattern, (the evolutionary codes are made for the treatment of stable, spherical, equilibrium models, and are able to handle mass loss, when time is allowed for the model to restore equilibrium) and should require an hydrodynamical code.

MAEDER: I want to emphasize the importance of surface chemistry as an observational test of stellar evolution, in addition to the HR diagram and number counts. Due to mass loss and possible overshooting, the surface abundances may show evidences of CNO processed elements. The C/N and O/N ratios are especially sensitive: they may change by a factor of 10^2 from cosmic abundances to CNO equilibrium values. A cartography of the expected C/N and O/N ratios in the HR diagram is now made (Astron. Astrophys. in press), which can be used as a comparison basis for observations. For example, supergiants on bluewards evolutionary tracks show much lower C/N and O/N ratios than those on the first redwards crossing of the HRD.

VIOTTI: Let me make a more general comment about the interchange between observers and theoreticians.
There are two categories of observers: those interested in objects and-those interested in projects. (The denominations were suggested by P.M. Williams, Edinburgh.) The former ones are studying individual objects, such as P Cyg and Eta Carinae, collecting observations over several years, in as many wavelength ranges as possible. The latter ones are developing big projects including several stars and making impressive HR diagrams such as those shown by Roberta Humphreys.
Theoreticians are interested in the physical processes occurring in the stellar interior and atmospheres and on the evolutionary tracks. I think that in general the first kind of observations (on objects) is more suitable for the study of the physics, rather than the latter one. In addition I would like to recall that astrophysicists are commonly reluctant to use the standard statistical procedures to analyse their observations. I am sure that from the wide use of these procedures, such as the cluster method, we shall gain a lot from the observational data and derive pictures better suitable for theoretical modelling.
One should also try not to generalize the results from an individual object to the whole category, or conversely ask why a single object like AG Car does not fit with the instability limit, provided that we do not know the exact value of its luminosity from one hand, and where to place that limit on the other hand because of the rather poor statistics and errors on luminosity.

OBSERVATIONS AND INTERPRETATIONS OF STELLAR PULSATIONS

G. Burki
Observatoire de Genève
1290 SAUVERNY, Switzerland

ABSTRACT. The three most useful methods for the identification of the stellar pulsation modes are described: firstly, the method based on the values of the period P and/or pulsation quantity Q; secondly, the method using the profile variations of the photospheric lines; thirdly, the method using the amplitudes and phases of the light, colour and radial velocity curves. Some of the most significant examples are given for each method, taken among the Cepheid, RR Lyrae, β Cephei and δ Scuti stars.

For two classes the complex variability of stars is described in detail: the 53 Persei stars and the supergiants. In these two classes non-radial modes have been identified and the complete mode identification has still to be largely improved.

1. MODE IDENTIFICATION

The various methods in use for the mode identification of stellar pulsations have been reviewed by Lesh (1980) and Percy (1980). Among these methods the most useful are certainly those which are based on i) the period and pulsation quantity values; ii) the line profile variations; iii) the amplitudes and phases of the light, colour and velocity curves.

1.1. Pulsation Quantity Q and Period P

The pulsation quantity Q is defined by

$$Q = P\ (\ \rho\ /\ \rho_{\odot})^{1/2} \qquad (1)$$

or

$$\log Q = \log P + 0.5 \log (M/M_{\odot}) - 1.5 \log (R/R_{\odot}) \qquad (2)$$

For a star of known radius and mass the knowledge of P implies the value of Q and thus it is possible to determine in which mode and overtone the pulsation occurs by comparison with the theoretical values of P and Q.

H. J. G. L. M. Lamers and C. W. H. de Loore (eds.), Instabilities in Luminous Early Type Stars, 23–38.

1.1.1. Classical cepheids. In the case of the classical cepheids with period shorter than 10 days the agreement is fairly good between the observational log R - log P relation, based on the results of the Baade-Wesselink method (Burki, 1985), and the theoretical relation based on cepheid models (Fernie, 1984). Thus, for these radial pulsators it is possible to derive the following relation, valid for the fundamental mode (Burki and Meylan, 1986):

$$Q = 0.0238\ M^{0.5}\ R^{-0.1} \tag{3}$$

The log R - log P diagram can be used to select the overtone pulsators: at a given log R value these stars have a lower log P value with respect to the standard relation. The known cases are SU Cas, probably pulsating in the 1st overtone (Gieren, 1976), V473 Lyr (HR 7308), which seems to be a 2nd (or higher ?) overtone pulsator (Burki et al., 1987) and eventually EU Tau (Burki, 1985).

For the double mode cepheids the period ratio P_{short}/P_{long} is in the range 0.70-0.71, a reliable indication that the two modes are the fundamental and the 1st overtone. There is only one known exception, i.e. CO Aur. Its two periods are 1.783 d and 1.429 d, the period ratio is 0.80, which means that this star is a double mode cepheid pulsating in the 1st and 2nd overtones (Antonello et al., 1986; Babel et al., 1987).

1.1.2. RR Lyrae stars. These stars pulsate either in the fundamental or in the 1st overtone radial modes. Double mode pulsators are known in the field (AQ Leo) and in some globular clusters (M3, M15, M68) as well as in the Draco Galaxy. The period ratio is in the range 0.74-0.75, confirming the pulsation in the fundamental mode P_0 and 1st overtone P_1. From the knowledge of the period ratio P_1/P_0 it is possible to derive the mass of the double mode star. As shown by Jorgensen and Petersen (1967), the best way to do this is to use the P_1/P_0 vs. log P_0 diagram, calibrated in terms of stellar mass from model calculations. Cox (1984) derived masses from 0.55 to 0.65 $M_{\odot}$ for the known double mode RR Lyrae.

1.1.3. β Cephei stars. As listed by Lesh and Aizenman (1978), roughly half of the known β Cephei exhibit two or more closely spaced periods. Thus, non-radial modes are certainly present in these stars. However, as shown by Percy (1980), in no case is a radial mode known to be absent. In addition, according to the calculations by Lesh and Aizenman (1974), the value of the pulsation quantity Q is generally in favour of the 1st overtone for these radial modes.

1.2. Line Profile Variations

The variation of the photospheric line profiles became very useful for the mode identification when photoelectric high-resolution spectrometers came into use. The pioneer work was done observationally by Petrie (1954) on the extreme β Cephei star BW Vul and theoretically by Osaki (1971) who calculated the line profile variations for various cases of non-radial pulsations.

In a large series of papers started in 1978, M. Smith and his collaborators have identified the pulsation modes in several β Cephei, δ Scuti and 53 Persei stars by fitting the observed line profiles to those calculated by the modelisation of the pulsating stellar surface. The parameters of the fit are essentially the strength of the velocity fields (axial rotation, macroturbulence and pulsation), the parameters of the pulsation in terms of spherical harmonic components (spherical harmonic order l, azimuthal number m) and the inclination of the line of sight. For example, in the case of the star δ Scuti, Campos and Smith (1980) derived two modes of pulsation: a radial mode of amplitude 6 km s^{-1} and a non-radial component of amplitude 3 km s^{-1}. In the case of the β Cephei star 12 Lac, Smith (1980) obtained a four-mode solution consistent with the periods determined by Jerzykiewicz (1978): a radial mode of amplitude 13 km s^{-1} and three non-radial modes (l = 2, m = 0, -1, -2) of amplitude 40 km s^{-1}.

A high-order non-radial oscillation was identified by Vogt and Penrod (1983) in the O-type star ζ Oph. The quality of the observed line profiles is very good so that the mode l = 8, m = -8 can be postulated unambiguously. Other nice examples have been studied by D. Baade (see elsewhere in this volume). For instance, he has determined a two-mode oscillation l = 10, m = +10 and l = 2, m = +2 in the case of the Be star μ Cen (Baade, 1984).

Recently Balona (1986) has proposed a simpler method which merely requires the first two or three moments of the line profile. The modes are determined by the periodograms of these moments. Thus the complete profile modelling is not required any more. This method could be an interesting simplification of the line profile analysis; however its powerfulness must be confirmed on real stellar cases.

1.3. Amplitudes and Phases of the Light, Colour and Radial Velocity Curves.

Dziembowski (1977) derived formulae for the light and radial velocity variations of a non-radially oscillating star in the linear approximation. Following this idea, Balona and Stobie (1979) described a method to identify the mode of pulsation from the knowledge of the light, colour and velocity variations. One basic parameter is the phase difference $\Delta\phi$ between the light and colour curves: for the stars in the cepheid instability strip, $\Delta\phi$ is negative in the case l = 0 (radial pulsation), positive for even l values and zero for odd l values.

In the case of the star δ Scuti, Balona and Stobie (1981) identified two oscillation modes of periods 0.194 d. and 0.187 d. From the $\Delta\phi$ values, it is assumed that the first mode is radial (l = 0) and the second one is a quadruple oscillation (l = 2). The closeness of the two periods rules out the possibility that the radial oscillation is in the fundamental tone. Note that the results of this analysis are in perfect agreement with those of Campos and Smith (1980) on the same star but based on the line profile variations (see section 1.2).

The mode identification is almost achieved in the case of the cepheids and RR Lyrae stars. It is in progress in the case of δ Scuti and β Cephei stars and in the coming years their pulsation modes will probably be definitely known, in particular because of the shortness of their periods which is yet an observational advantage (the observing runs are generally short and the modern spectrometers have a high time-resolution).

As to the identification of the pulsation modes of the supergiants and 53 Persei stars, much progress is yet needed. Their periods are longer than 1 d (up to about 1000 d for the supergiants) and the light curves frequently change from one cycle to another, the pulsation being multiperiodic or/and unstable. Consequently, the possibilities for completely covering an entire pulsation cycle of these stars from ground based observations are: i) for the long periods, the monitoring of circumpolar stars; ii) for the short periods, organizing multisite intercontinental monitoring or observation during winter from the polar regions. Certainly, an efficient solution to these observational difficulties would be space observations by satellite.

2. VARIABILITY OF THE 53 PERSEI STARS

53 Persei, a B4 IV star, is the prototype of a class of slowly rotating B stars showing line profile variations (Smith, 1977). In 53 Per Smith et al. (1984) have identified two non-radial modes $l = +3$, $m = -3$ and $l = +3$, $m = -2$. The periods are 1.68 and 2.14 d, much larger than the fundamental radial period. Thus these two oscillations are g-modes of high overtones.

It is highly probable that this group of spectroscopic variables is to be identified with the group of B-type photometric variables called slow variables (Le Contel et al., 1981; Burki, 1983) or mid-B variables by Waelkens and Rufener (1985, called hereafter WR). The systematic photometric survey of early-type stars carried out by WR allowed a good definition of this group of variables: spectral type B3 to B8, luminosity class V to III. According to WR, the photometric variations can be characterized as follows:
1) There is a clearly defined period in the range 1-3 days.
2) Colour variations are observed, in phase with the light variations. The amplitude in U-B is one order of magnitude larger than the other colour amplitudes.
3) The amplitudes vary from one cycle to another, but the colour-to-light ratio remains practically constant.

This photometric behaviour is illustrated by Fig. 1 (WR). The mean amplitudes differed significantly in 1981 and 1983. In addition, the scatter around the mean light curves remains too large, so there is also a cycle-to-cycle amplitude variation.

The results of WR are in agreement with the identification of g modes of high overtones suggested by Smith et al. (1984). It is consequently very likely that the displacements of the pulsating layers are mainly confined to the outer stellar envelope. WR also suggests that the excitation of these non-radial g modes could be driven by the second Helium ionization zone located near the surface in these stars and, following Ledoux (1967), that the periods of the observed g modes could be essentially determined by the stellar rotation period.

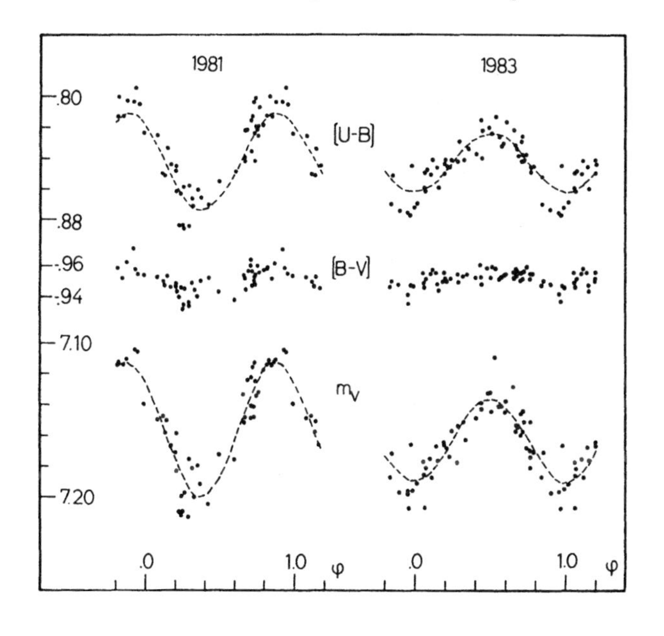

Figure 1.
Phase diagram of the 53 Persei star HD 160124 (Waelkens and Rufener, 1985)

3. VARIABILITY OF THE SUPERGIANT STARS

3.1. Quasi-Periods and Pulsation Modes

All supergiants exhibit light variations (Maeder and Rufener, 1972) whose expected amplitudes have been cartographed in the the HR diagram by Maeder (1980). According to many observers, these light variations are neither strictly periodic nor multi-periodic. In fact, an irregular component is frequently detected in addition to the generally quasi-periodic behaviour (e.g. Abt, 1957; Sterken, 1977; Rufener et al., 1978; van Genderen, 1979; Percy and Welch, 1983; Arellano Ferro, 1985).

Nevertheless a (quasi-period) - luminosity - colour relation (PLC relation can be defined for the supergiants if the quasi-period is defined as "the most probable characteristic time of the variation" or "the characteristic time of the variation exhibiting the largest amplitude". Fig. 2 shows the shape of this PLC relation in the HR diagram as given by Burki (1978). By using recent evolutionary models with mass loss to estimate the stellar masses, Maeder (1980) gives the following PLC relation for the supergiants:

$$\log P = -0.346\ M_{bol} - 3 \log T_{eff} + 10.60 \qquad (4)$$

This form of the PLC relation is thus empirical for the periods, effective temperatures, bolometric magnitudes and theoretical for the masses.

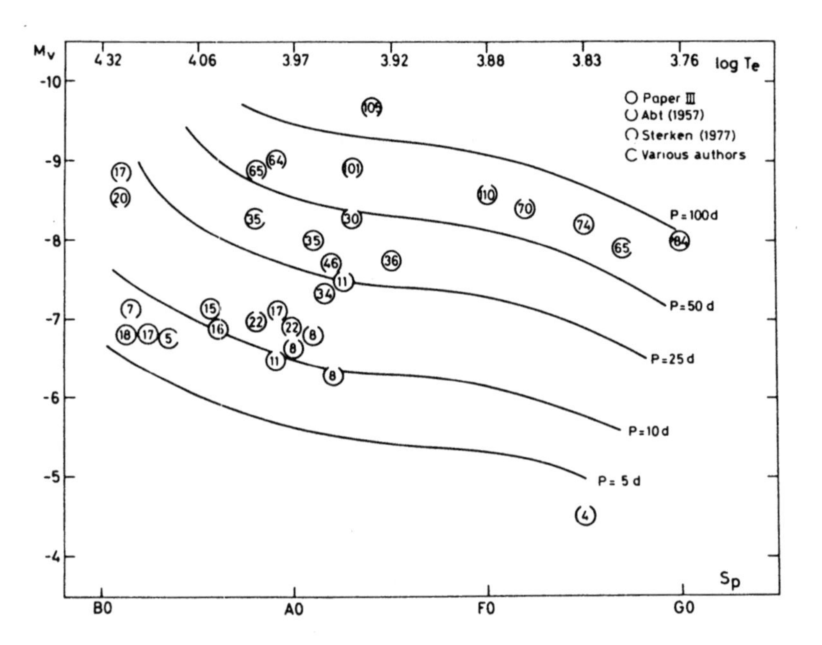

Figure 2. Calibration of the upper left part of the HR diagram in terms of isoperiod lines (Burki, 1978)

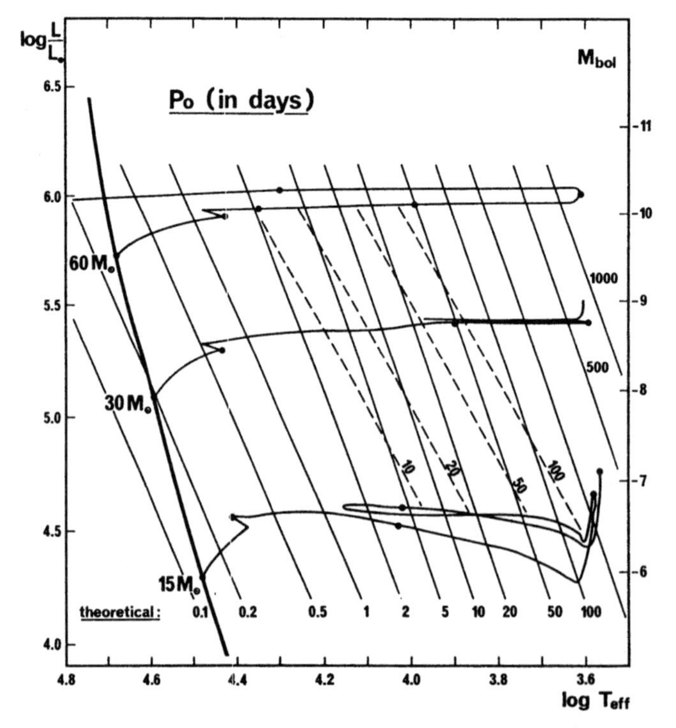

Figure 3. Theoretical isoperiod lines (fundamental radial mode) in the HR diagram (Lovy et al., 1984)

Lovy et al. (1984) have determined the period and pulsation quantity Q values for the radial modes (fundamental, 1st and 2nd overtones) of supergiants by using stellar models with medium mass loss rates chosen in the grids computed by Maeder (1981). They found it necessary to split the HR diagram into two zones according to T_{eff}. The corresponding theoretical PLC relations for the fundamental radial mode are:

$$\log P_o = -0.232 \, M_{bol} - 2.850 \log T_{eff} + 9.294 \quad (5)$$

for $\log T_{eff} > 4.3$ and:

$$\log P_o = -0.275 \, M_{bol} - 3.918 \log T_{eff} + 14.543 \quad (6)$$

for $\log T_{eff} < 4.1$. These two relations are plotted on Fig. 3 (Lovy et al., 1984) where the essentially empirical relation (4) is as well represented. As we see, the observed periods P_{obs} are generally too long with respect to the theoretical fundamental radial periods P_o, especially at higher temperatures. According to Lovy et al. this can be due to various non-exclusive causes: i) Many stars are in a post-red supergiant stage, moving bluewards in the HR diagram (in their calibrations, P_o stands for stars moving rightwards). ii) The stars are pulsating in non-radial g modes rather than in radial modes. iii) The mass loss rate affecting the stellar evolution is higher than the values adopted in the models.

The recent observational studies of supergiants confirm these various results. Arellano Ferro (1985) derived the pulsational mode of five yellow supergiants from the Q values. He found that ϵ Aur, V509 Cas and ρ Cas are likely non-radial pulsators, 89 Her seems to have a non-pure radial pulsator behaviour and HD 161796 can be a radial pulsator having exhibited period switches between fundamental and 1st overtone radial modes, as already suggested by Fernie (1983). Remember that these last two stars are supergiants located at high galactic latitude. As first proposed by Burki et al. (1980), they are probably not normal, massive, supergiants. Stars of this type are now classified as UU Herculis stars (Sasselov, 1984).

Similar conclusions have also been drawn by Percy and Welch (1983) on the basis of a photometric analysis of 16 early-type supergiants (09-A3). All but two of the stars are found to be variable. The ratio P_{obs}/P_o (quasi-period over theoretical radial fundamental period) varies from 2-10 in the earliest types to 0.2 - 2 in the later types. On this basis, they also conclude that these supergiants are most likely non-radial g mode pulsators.

The case of V810 Cen could be unique if the results of the period identification corresponds in fact to the physical reality. The first period determination is due to Eichendorf and Reipurth (1979) who suggested that this G0Ia star is a long period cepheid-like star with P $\simeq$ 125 d. Using 6 years of photometric monitoring by Eichendorf and Reipurth (1979), Dean (1980) and Geneva observers in La Silla, Burki (1984) determined a possible light curve with five periods (see (Fig. 4):

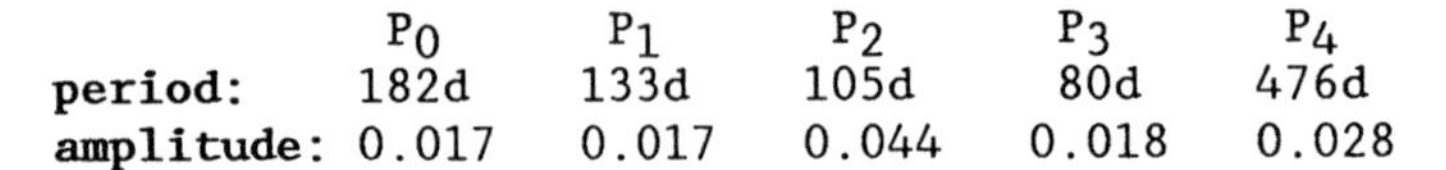

	P_0	P_1	P_2	P_3	P_4
period:	182d	133d	105d	80d	476d
amplitude:	0.017	0.017	0.044	0.018	0.028

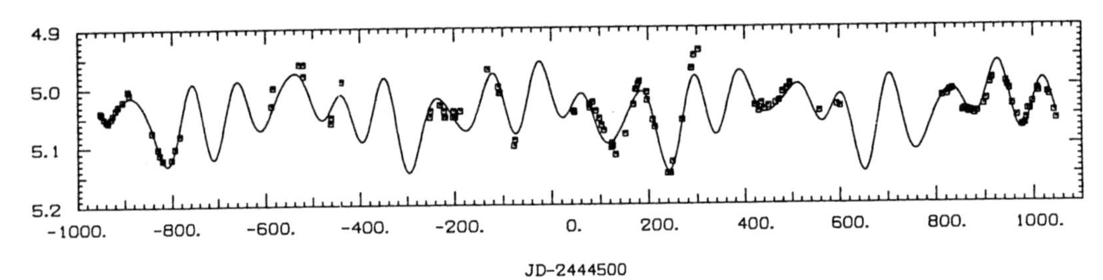

Figure 4. Light curve of V810 Cen

The residual standard deviation is 0.014. The interesting point is that the period ratios are $P_1/P_0 = 0.73$, $P_2/P_0 = 0.58$ and $P_3/P_0 = 0.44$, values very similar to those predicted in the case of the radial pulsation of the standard model: 0.74, 0.57, 0.46. This strongly suggests that P_0, P_1, P_2 and P_3 correspond to the fundamental and three first overtones, a result which would be absolutely unique if it is confirmed in the future. The 5th period P_4 could be related to the orbital motion of the companion, a B1 III star detected by Parsons (1981).

3.2. δ variations and convulsive pulsation

Van Genderen (1985, 1986 for the most recent papers) in an analysis of supergiant variability also observed that the main quasi-periods are generally longer than the predicted theoretical P_0 values. In addition, van Genderen (1985) discussed an important characteristic of the supergiant variability. Superimposed to the quasi-periodic variations, probably due to non-radial or/and radial oscillations, he put into evidence a second type of non-periodic variations, the so-called " δ variations". The rather detailed photometric analysis of these variations led van Genderen to the suggestion that at least in the O-type supergiants, they are caused by a "convulsive type of pulsation", i.e. the boundary between the conception of pulsation and that of the possible presence of large convective elements moving up and down in the stellar atmosphere becomes more undefined.

These δ variations can also be connected to various characteristics of the supergiant variability, such as:

i) in the hypergiant ζ^1 Sco, classified B1 Ia$^+$, Burki et al. (1982) have detected absorption components of variable intensity in the UV resonance lines observed with the IUE satellite. These absorption features become stronger when the photometric variations are larger, suggesting that part of the variability is connected to a discontinuous loss of stellar matter (see Fig. 5).

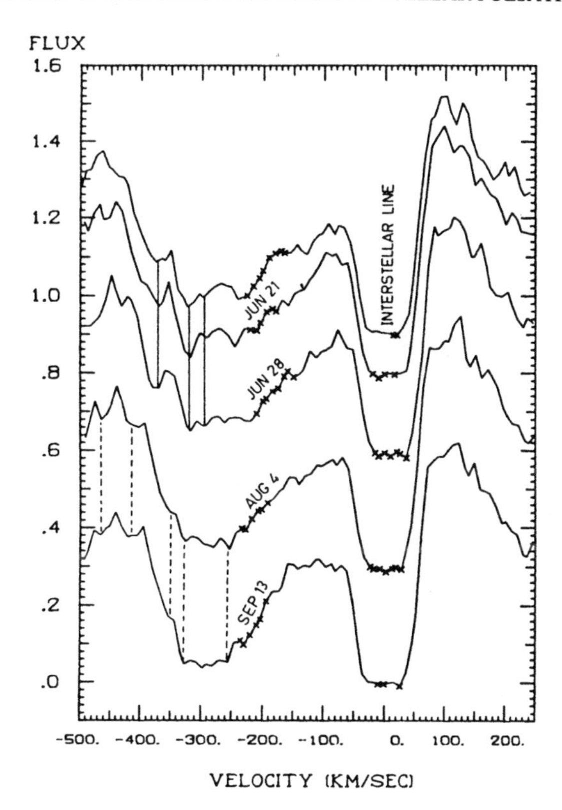

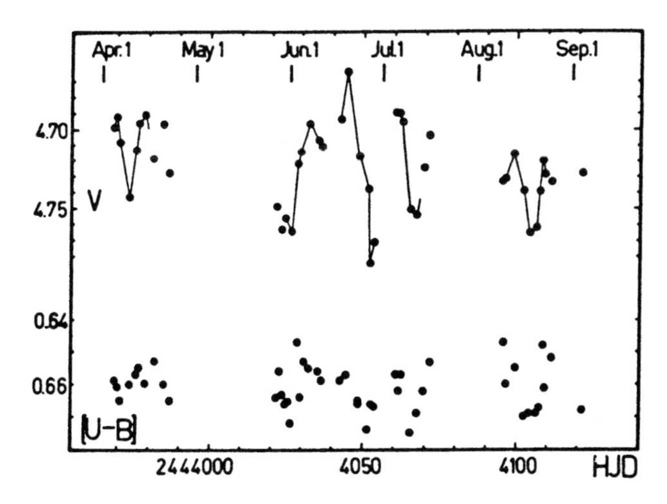

Figure 5. Simultaneous monitoring of ζ^1Sco in UV spectroscopy (IUE satellite) and visual photometry (Geneva system). Top: variation of the absorption components on the MgII line at 2795 Å. Bottom: photometric variability at the same epoch.

ii) De Jager et al. (1984) proved that the motion field of the lower photosphere of α Cyg (A2 Ia) can be described by a field of outward progressing sound waves which turn into shock waves in the photosphere. According to de Jager et al., this picture is in agreement with the view that the atmospheric motions are caused by large convective elements or by non-radial oscillations.
iii) According to de Jager (1980), from the values of the turbulent velocity components, "it is possible to draw a conclusion as to the extent of the areas of the star that move coherently. It is then found that the moving elements in the least-bright supergiants have an average size of approximately 0.05 stellar radius. For bright supergiants this is 0.25 stellar radius and for extreme supergiants approximately 0.50 stellar radius. Formally such variations are described by non-radial pulsations. However, it seems very unlikely that there is any correlation between motions in different parts of the star. For that reason it appears more probable that individual elements of the atmosphere respond to disturbances with their own characteristic pulsation frequency."

REFERENCES

Abt, H.A., 1957, Astrophys. J. **126**, 138
Antonello, E., Mantegazza, L., Poretti, E., 1986, Astron. Astrophys. **159**, 269
Arellano Ferro, A., 1985, Mon. Not. Roy. Astron. Soc. **216**, 571
Baade, D., 1984, Astron. Astrophys. **135**, 101
Babel, J., Burki, G., Mayor, M., 1987, Astron. Astrophys., to be submitted
Balona, L.A., 1986, Mon. Not. Roy. Astron. Soc. **219**, 111
Balona, L.A., Stobie, R.S., 1979, Mon. Not. Roy. Astron. Soc. **189**, 649
Balona, L.A., Stobie, R.S., 1981, Mon. Not. Roy. Astron. Soc. **194**, 125
Burki, G., 1978, Astron. Astrophys. **65**, 357
Burki, G., 1983, Astron. Astrophys. **121**, 211
Burki, G., 1984, Microvariability in the HR Diagram, in Space Research Prospects in stellar activity and variability, Eds. A. Mangeney and F. Praderie, Observatoire de Meudon, p. 69
Burki, G., 1985, Radius determinations for nine short-period cepheids, in Cepheids: Theory and Observations, IAU coll. no. 82, Ed. B.F. Madore, Cambridge Univ. Press, Cambridge, p. 34
Burki, G., Heck, A., Bianchi, L., Cassatella, A., 1982, Astron. Astrophys. **107**, 205
Burki, G., Meylan, G., 1986, Astron. Astrophys. **159**, 261
Burki, G., Schmidt, E.G., Arellano Ferro, A., Fernie, J.D., Sasselov, D., Simon, N.R., Percy, J.R., Szabados, L., 1987, Astron. Astrophys., in press
Campos, A.J., Smith, M.A., 1980, Astrophys. J. **238**, 667
Cox, A.N., 1984, Constraints on stellar evolution from pulsations, in Observational Tests of the Stellar Evolution Theory, IAU Symp. no. 105, Eds. A. Maeder and A. Renzini, D. Reidel, Dordrecht, p. 421
Dean, J.F., 1980, Inf. Bull. Variable stars 1892

Dziembowski, W., 1977, Acta Astron. **27**, 203
Eichendorf, W., Reipurth, B., 1979, Astron. Astrophys. **77**, 227
Fernie, J.D., 1983, Astrophys. J. **265**, 999
Fernie, J.D., 1984, Astrophys. J. **282**, 641
von Genderen, A.M., 1979, Astron. Astrophys. Suppl. **38**, 381
van Genderen, A.M., 1985, Astron. Astrophys. **151**, 349
van Genderen, A.M., 1986, Astron. Astrophys. **157**, 163
Gieren, W., 1976, Astron. Astrophys. **47**, 211
Le Contel, J.M., Sareyan, J.P. Valtier, J.C., 1981, Intrinsic Variability in B stars, in Workshop on Pulsating B stars, Eds. G.E.V.O.N. and C. Sterken, Observatoire de Nice, p. 45
Ledoux, P., 1967, Oscillation Theories of Magnetic variable stars, in The magnetic and related stars, Ed. R.C. Cameron, Mono Book Corporation, Baltimore, p. 65
de Jager, C., 1980, ESA Journal **4**, 123
de Jager, C., Mulder, P.S., Kondo, Y., 1984, Astron. Astrophys. **141**, 304
Jerzykiewicz, M., 1978, Acta Astron. **28**, 465
Jorgensen, H.E., Petersen, J.O., 1967, Z. f. Astrophysik **67**, 377
Lesh, J.R., 1981, Mode identification in pulsating B stars, in Workshop on pulsating B stars, Eds. G.E.V.O.N. and C. Sterken, Observatoire de Nice, p. 301
Lesh, J.R., Aizenman, M.L., 1974, Astron. Astrophys. **34**, 203
Lesh, J.R., Aizenman, M.L., 1978, Ann. Rev. Astron. Astrophys. **16**, 215
Lovy, D., Maeder, A., Noëls, A., Gabriel, M., 1984, Astron. Astrophys. **133**, 307
Maeder, A., 1980, Astron. Astrophys. **90**, 311
Maeder, A., 1981, Astron. Astrophys. **102**, 401
Maeder, A., Rufener, F., 1972, Astron. Astrophys. **20**, 437
Osaki, Y., 1971, Publ. Astron. Soc. Japan **23**, 485
Parsons, S.B., 1981, Astrophys. J. **245**, 201
Percy, J.R., 1980, Space Sc. Rev. **27**, 313
Percy, J.R., Welch, D.L., 1983, Publ. Astron. Soc. Pacific **95**, 491
Petrie, R.M., 1954, Publ. Dom. Astrophys. Obs. **10**, 39
Rufener, F., Maeder, A., Burki, G., 1978, Astron. Astrophys. Suppl. **31**, 179
Sasselov, D.D., 1984, Astrophys. Space Sci. **101**, 161
Smith, M.A., 1977, Astrophys. J. **215**, 574
Smith, M.A., 1980, Astrophys. J. **240**, 149
Smith, M.A., Fitch, W.S., Africano, J.L., Goodrich, B.D., Halbedel, W., Palmer, L.H., Henry, G.W., 1984, Astrophys. J. **282**, 226
Sterken, C., 1977, Astron. Astrophys. **57**, 361
Vogt, S.S., Penrod, G.D., 1983, Astrophys. J. **275**, 661
Waelkens, C., Rufener, F., 1985, Astron. Astrophys. **152**, 6

DISCUSSION

PRADERIE: You illustrated the mode identification method using the Q constant in the case of Cepheids. Which mass of the Cepheids is used in your log R, log L diagram to compare observed modes and computed ones? Is the conclusion as to the pulsation in fundamental, first and second harmonics dependent of this mass, and is the problem of mass resolved in these stars?

BURKI: The log R-log P diagram I used (Burki, G., Meylan, G.: 1986, Astron. Astrophys., in press) is based on purely observational data since the radii are derived by the Baade-Wesselink method. Thus, it does not require any assumption on the Cepheid masses. On the other hand, the so-called 'mass discrepancy problem' seems to be resolved if we exclude the binary Cepheids, at least in the case of Cepheids with P <10d. (See Burki, G.: 1984, Astron. Astrophys.)

PRADERIE: It may be that one of your conclusions is optimistic: you said that more identification might be achieved soon in δ Scuti and Cepheids. Even in these well studied cases, I wonder if excitation mechanisms are so well known, and if the advent of more homogeneous, high photometric precision, long term observations revealing eventually very low amplituded modes are not also necessary here, as for 53 Per or sugergiants?

BURKI: Only radial modes are observed until now in Cepheid stars and there is no indication for additional non-radial modes. In δ Scuti stars, both radial and non-radial modes have been detected. These stars being in the Cepheid instability strip, the mode identification is relatively more simple than in the case of ß Cepheid stars for example, and seems to be only a problem of observing time dedicated to this purpose.

JERZYKIEWIZ: In your talk you mentioned the well-known ß Cephei variable BW Vul in two different contexts. First you placed it among the ß Cephei stars which were found to be radial pulsators according to the light to colour amplitude ratio argument. Later you quoted O Dell's polarisation observations, which he has interpreted in terms of quadrupole oscillations. Kubiak (whom you did not mention) has also concluded from a careful line profile analysis that the star is a non-radial pulsator. Could you comment on these mode identification discrepancies.

BURKI: The same kind of discrepancy is relatively common among the ß Cephei stars. It may be the signature of multi-mode pulsators or/and due to the fact that mode identification methods are not yet enough accurate.

ROUNTREE-LESH: I think it is very optimistic to suggest that the problem of mode identification in the ß Cephei stars will be solved in the near future through observations of the line profile variations. Interpretations of the exciting observations still lead to very disparate results. Moreover, not all ß Cephei stars exhibit profile variations.

ß Canis Majoris is an example of a bright, well-observed, double-mode variable in which the line profiles are quite stable. More theoretical work is needed to discover the instability mechanism, so that we can predict which modes will be excited.

HENRICHS: You emphasized that all supergiants behave differently. How would you characterize the variability of any specific supergiant? In other words how variable is the variability? On what typical timescales do for instance the periods and/or amplitudes change?

BURKI: The characteristic time of variation can change from cycle to cycle (see e.g. Burki et al.: 1982, Astron. Astrophys. 107, 205; van Genderen, A.M. et al.: 1985, Astron. Astrophys. 153, 163)

DE JAGER: It is significant to speak of 'semi-periods' (P_s) or 'quasi-periods', a notion introduced by the Geneva group. They found the P_s-values for super- and hypergiants to range between a week and a few hundred days, dependent on the spectrum (shortest for early-type stars). The quasi-periods are maintained for a few quasi-pulsations only, and then tend to change. Translated in mathematical terms this means that a Fourier analysis of the brightness- or V_{rad}-variations would yield one or a few significant terms over a relatively short period of time (say 5 to $10P_s$) but when extended over a larger period the signifcance of the components would decrease.

HENRICHS: Is there any information known about simultaneous Hα behaviour during amplitude and/or period change in supergiants?

BURKI: I don't know. But note that Hα variations are related to the mass loss variations rather than to variations in the photospheric layers.

LAMERS: You mentioned that about 40% of the supergiants are in the fundamental mode and 60% have smaller frequencies. Is there any systematic difference between those stars which pulsate in the fundamental mode and the other ones?

BURKI: P_{obs} is larger than P_o (fundamental radial) in a majority of early-type supergiants (e.g. Percy, J.R., and Welch, P.L.: 1983, P.A.S.P., 95, 491; van Genderen, A.M., 1985, Astron. Astrophys., 151, 349). For the yellow supergiants, P_{obs} is in agreement with P_o in a majority of cases.

APPENZELLER: If the observed 'quasi-periods' were the fundamental mode pulsations, it seems surprising that these pulsations appear and disappear so fast. From the usual analysis of stellar pulsations one gets decay times for fundamental mode pulsations of such stars of at least years (and up to 10^3 years). How can such pulsations damp out within a few periods?

BURKI: The variations never stop, but well defined cycles are not continuously observed.

CASTOR: The time scale for leaving a finite amplitude pulsation mode, when periodic pulsation in that mode is unstable, is affected strongly by non-linear effects, and can be much less than the linear decay time. Mode-switching times in Cepheids can be a few times less than the linear decay times. The time scale discrepancy remains large, however.

ZAHN: Has it been possible to check the phase coherence of the four components which show a stable amplitude in 12 Lac?

JERZYKIEWICZ: The four oscillations are coherent over each observing season separately. However, the question whether they are coherent over the whole time interval shown in the figure has not been answered yet. One difficulty in trying to get an answer is that the periods are short - of the order of five hours - and therefore there are ambiguities in the cycle counts between observing seasons separated by several years. The other difficulty is related to the mean errors of the seasonal phases of the oscillations. The formal errors which one gets from the least square analyses are very small. If one believes that these errors are correct, one has to conclude that the oscillations are probably not strictly coherent. If, however, the errors were in fact somewhat larger, strict coherence would not be inconsistent with the data.

NOËLS: I would like to suggest that there could be an instability mechanism in supergiant variables, which are H-burning shell models. This would be a vibrational instability of g^+ modes trapped in the H-burning shell, especially if there is a intermediate connective zone outside the shell which could prevent the amplitude of the mode to increase too quickly outside the trapping region. Of course, the damping in the external layers would be a problem. Another problem is that the Brunt-Vaïsalä frequency is so high that the low order g^+ mode would have periods too short to be compatible with the observations.

OSAKI: I would like to point out two problems for the model proposed by Dr. Noëls. 1) The frequencies of non-radial g-modes trapped in the nuclear burning shell will be much higher than that of the radial fundamental mode in supergiant stars because the Brunt-Vaïsalä frequency in the burning shell will be very high in supergiants.
2) If those g-modes reappear near the surface, the dissipation in the outer zone will be significant and it might kill the nuclear energized instability.

OSAKI: A comment for the possibility of nuclear energized instability of low-frequency high overtone g-modes in supergiants mentioned by Dr. Noëls. I think that possibility will be very unlikely because we expect very heavy radiative dissipation for such modes due to the existence of a large number of modes in the radial direction.

MOFFAT: Among the supergiants, the lack of strict periods and the increase in amplitude towards the Humphreys-Davidson luminosity limit (where mass loss rates are observed to increase) makes it at least thinkable that inhomogeneities in the winds may be an important cause of the variations. How sure can we be that non-radial pulsations are the cause of the observed variations as opposed to other effects (e.g. blobs accelerated in the winds)?

BURKI: The observed light variations come mainly from the photopshere. Thus, pulsation and/or erratic non-coherent motions are the most probable explanations.

DE JAGER: The observed pulsations of super- and hypergiants are definitely photospheric since the radial velocities are as a rule measured with photospheric lines. A classic example is Alpha Cygni of which the V_r-measurements were made in the thirties, and analysed in 1976 by Lucy, who found sixteen, mostly significant components with periods between a week and hundred days. Such an example repeated with modern spectrographic methods, during a number of years with a dedicated instrument, preferably from space, may strongly push forward the research of hypergiant pulsations.

MAEDER: Concerning the reality of non-radial pulsation in supergiants, which was just questionned, we must remember that there are, as shown by Dr. Burki, several methods of diagnostics of the pulsation modes and the results of these different methods seem to converge in favour of non-radial pulsation in early supergiants.

WESSELIUS: I would like to make a plea for a dedicated satellite which could measure photometric variables with a high accuracy over long time-scales. As an example of its possibility, look at the ANS-UV instrument flying in 1974-1976: a 30-cm telescope with a few percent efficiency. In 1 year operation a repeatability of better than 1% was reached for stars down to $m_V = 8^m$–9^m. The calibration procedure was simple: observe ε and ν Dor every day.

V.D. HEUVEL: I thought that Hipparcos was going to give this, as a by product, for hundreds of thousands of stars. So, do you still need a separate satellite for this?

WESSELIUS: A project connected with the star mappers of Hipparcos, called Tycho, will observe the photometric and astrometric properties of a few hundred thousand stars. Each star will be observed 10-20 times on time scales determined by the HIPPARCOS operation, thus not by the observer. The photometric repeatability expected will be 1% for bright ($< 8^m$) stars to 2-3% for 10^m–11^m stars.

LAMERS: Maybe ESA's director of science, Dr. R. Bonnet, can comment on the possibilities of having a dedicated satellite for photometric variability studies.

BONNET: If the scientific justification of this satellite is based on solid theoretical and instrumental grounds it would have equal chances with the numerous competitors. However, the chances of this project appear to me higher if it were to be installed on the co-orbiting platform of the Columbus Space Station or on a Eureca platform. This appears to be a fairly simple project at first sight and, I agree with Wesselius, could be cheap and should remain such. Then its chance will be increased. I can therefore only recommend that the interested scientific community defines the scientific rationale for the project and starts building up the pressure necessary for getting the full support it deserves.

ROUNTREE-LESH: Another argument for a dedicated photometric satellite is the need for a survey of all the early-type stars - especially those near the Main Sequence - to determine the percentage of variables. Estimates of this quantity range from about 10% to 100% - in other words, from a narrow 'instability strip' in the HR diagram to universal variability. This has important implications for the theories of stellar evolution and stability. Although the High Speed Photometer on the Hubble Space Telescope is technically capable of doing this job, the chances of getting enough observing time for such a project are very low.

PRADERIE: A satellite project devoted to probing stellar interiors by means of long term microvariability and activity observations was precisely presented to ESA less than one month ago. However, it was not retained for on assessment study, and now we have indeed as nearest possible horizon the Eureca and/or the Space Station Programs as said by R. Bonnet.

I stress that with a careful instrumentation, one will reach the photon noise limit in photometry, and therefore improve the capacity of detecting light fluctuations by say 3 orders of magnitude relative to what is obtained from the ground. For this, one needs to monitor individual stars for long observing runs; the Hipparcos satellite will not provide this capacity.

VREUX: My feeling is that such a proposal would benefit of a stronger emphasis on the interest of astroseismology in different parts of the HR diagram. I also have the impression that it would probably have a good chance if it were submitted when a call will be issued for platform experiments.

THEORY OF NON-RADIAL PULSATIONS IN MASSIVE EARLY-TYPE STARS

Yoji Osaki
Department of Astronomy, Faculty of Science
University of Tokyo
Bunkyo-ku, Tokyo
Japan

ABSTRACT. The present theory of non-radial pulsations in massive early-type stars is reviewed. We first describe the basic property of non-radial pulsations in stars. Variation in modal-property with stellar evolution is then discussed with the help of the so-called propagation diagram. Problems of vibrational instabilities, effects of rotation, possible consequence of existence of non-axisymmetric waves on the stellar atmosphere are briefly discussed.

1. INTRODUCTION

Theory and observations of non-radial pulsations (hereafter abbreviated as NRP) in stars have been developed greatly in recent ten or fifteen years. This new development basically owes to the discovery of pulsations and pulsation-related phenomena in many stars, which were hitherto regarded to be non-pulsating stars. They include our sun itself, white dwarfs, Ap stars, and early-type stars. The most important characteristics of pulsations in these newly discovered variables are that their pulsations are often multi-periodic so that many of stellar eigenmode oscillations are involved and that most of their pulsations are usually non-radial pulsations (NRP). In fact, this development has opened a new important field of research called 'helio- and asteroseismology', which has been proved so succussful in the case of the sun.

It may be interesting to note here that the first observational evidence of NRP in stars was presented by Ledoux (1951) for pulsations of early-type variable stars of Beta Cephei type. Although it is now thought that the main pulsation of Beta Cephei stars is radial, NRPs are believed to be involved in some of Beta Cephei stars which show the multi-periodic beating. In recent years, many of early-type stars (with spectral types O and B) have been found to be intrinsically variable either by photometry (Waelkens and Rufener 1985) or by spectroscopy (Smith and Penrod 1985). Relatively long observed periods (longer than that of the radial fundamental) and the character of line-profile variations in these stars indicate that variabilities in these early-type stars are most likely caused by NRPs. Further interest in NRP has

H. J. G. L. M. Lamers and C. W. H. de Loore (eds.), Instabilities in Luminous Early Type Stars, 39–54.

recently been aroused by the discovery of possible connection between NRPs and stellar mass loss in some O and B stars (see, e.g., Abbott et al. 1986).

Since observational materials on NRPs in massive early-type stars will be presented by other reviewers (see, parallel talks by Burki and Baade in the proceedings), this review is restricted to theoretical aspects of NRPs in masive early-type stars. More detailed account of theory of stellar NRPs in general may also be found in monographs by Unno et al. (1979) and by Cox (1980).

2. WHAT ARE NRPs?

There exist two kinds of modes in free oscillation of a gaseous star. The first one is the radial pulsation in which a star executes periodic oscillation of expansion and contraction around its equilibrium state while keeping its spherical shape. This type of oscillations has extensively been studied in connection with classical pulsating variables such as the Cepheid and the RR Lyrae stars. What concerns us here is the second type of oscillations, called nonradial pulsations (NRPs). They are the more general type of stellar oscillations in which a star oscillates in non-spherical form. A normal mode of NRP for the displacement vector $\underset{\sim}{\xi}$ is described in the spherical polar coordinates (r, θ, ϕ) as

$$\underset{\sim}{\xi} = \left(\xi_r, \xi_h \frac{\partial}{\partial\theta}, \xi_h \frac{1}{\sin\theta}\frac{\partial}{\partial\phi}\right) Y_\ell^{m}(\theta, \phi)\, e^{i\omega t} \tag{1}$$

where

$$Y_\ell^{m}(\theta, \phi) = P_\ell^{m}(\cos\theta)\, e^{im\phi} \tag{2}$$

is the spherical harmonics, $P_\ell^m(\cos\theta)$ being the associated Legendre function. More specifically, the radial component of the displacement vector is written as

$$\xi_r(\underset{\sim}{r}, t) = \xi_{n\ell}(r)\, P_\ell^{m}(\cos\theta)\, e^{i(\omega t + m\phi)} \tag{3}$$

where $\xi_{n\ell}(r)$ represents the radial dependence of the eigenfunction. An eigenmode of NRP and its eigenfrequency ω are specified by three quantum numbers (n, ℓ, m). The quantum number n, corresponding to the principle quantum number in quantum mechanics, specifies the physical character of modes (such as p-, g-, and f-modes) and the number of nodal surfaces in the radial direction. The quantum numbers ℓ and m describe the horizontal dependence of eigenfunctions over the spherical surface, in which m represents the number of the nodal lines in longitude and $\ell - |m|$ that in latitude.

The eigenfrequency $\omega_{n\ell m}$ is degenerate in m for a non-rotating star, but the degeneracy is lifted for a slowly rotating star, giving rise to equi-distant $(2\ell + 1)$ - multiplet:

$$\omega_{n\ell m} = \omega_{n\ell}^{0} - m\,(1 - C_{n\ell})\,\Omega \tag{4}$$

where $\omega_{n\ell}^{0}$ stands for an eigenfrequency in non-rotating star and Ω denotes the angular frequency of rotation, and $C_{n\ell}$ is a numerical constant (usually less than unity) which depends on the particular mode and particular stellar model. A mode with $m \neq 0$ represents traveling-wave mode around the equator, a mode with $m < 0$ being the prograde mode in which waves travel in the same direction as rotation while that with $m > 0$ being the retrograde mode.

Two different kinds of restoring forces (i.e., the pressure force and the gravity force or buoyancy force) operate in NRPs and there exist, therefore, two different kinds of modes: pressure modes (p-modes) and gravity modes (g-modes). The p-modes form a sequence of increasing eigenfrequency with the order of modes, n, while the g-modes form a sequence of decreasing frequency. Besides that, there exists an eigenmode called the f-mode whose eigenfrequency lies between those of p_1- and g_1-modes and whose eigenfunction has no-node in the radial direction. This simple classification of NRP modes by Cowling (1941) applies only for simple stellar models such as the zero-age main-sequence star but it fails for more complicated stellar models, as will be discussed in the next section.

Nonradial eigenfunctions given in equation (1) are correct only in a non-rotating star. When a star is rotating, its nonradial oscillations are affected by rotation, particularly for NRP modes whose eigenfrequencies are comparable to the rotational frequency. Thus, g-modes are more strongly affected by rotation than p-modes. Besides that, there appear new modes called Rossby modes or r-modes (Papaloizou and Pringle 1978). They are basically horizontal eddy motion on the sphere and their velocity vector **V** is described in the limit of slow rotation as

$$V_{Rossby} = B \left(0, \frac{1}{\sin\theta}\frac{\partial}{\partial\phi}, -\frac{\partial}{\partial\theta}\right) P_{\ell}^{m}(\cos\theta)\, e^{i(\omega t + m\phi)} \tag{5}$$

where B is the velocity amplitude of the r-mode. The angular frequency of a r-mode in the inertial frame of reference is given by

$$\omega = m\Omega \left(-1 + \frac{2}{\ell(\ell+1)}\right) . \tag{6}$$

Thus, r-modes are prograde in the inertial frame but retrograde in the corotating frame of the star.

In a fully rotating star, eigenfunctions of NRP modes can never be described by a single spherical harmonics either in the case of the spheroidal p- or g-mode of equation (1) or in the case of the toroidal r-mode of equation (5), but they are generally described as a combination of both (Berthomieu et al. 1978, Saio 1982, Lee and Saio 1986):

$$\begin{aligned} \underset{\sim}{\xi} &= \underset{\sim}{\xi}(r, \theta)\, e^{i(\omega t + m\phi)} \\ &= \sum_{\ell \geq |m|} \left(\xi_r,\ S_\ell \frac{\partial}{\partial\theta} + T_\ell \frac{\partial}{\sin\theta\,\partial\phi},\ S_\ell \frac{\partial}{\sin\theta\,\partial\phi} - T_\ell \frac{\partial}{\partial\theta}\right) \\ &\qquad P_\ell^{m}(\cos\theta)\, e^{i(\omega t + m\phi)} \end{aligned} \tag{7}$$

where S_ℓ, T_ℓ are expansion coefficients of spheroidal and toroidal components, respectively. Nevertheless, one can still distinguish NRP modes into three types from their physical character. They are p-modes slightly modified by rotation, g-modes modified by rotation (sometime called inertial gravity modes), and r-modes modified by stratification.

The strongest observational evidence for NRP in massive early-type stars is line-profile variations seen in many O and B stars, which have successfully been explained in terms of velocity fields produced by NRPs on the surface of a rotating star. The profile modeling has usually been done based on the spheroidal NRP p- and g-modes given by equation (1). However, Osaki (1986b) has recently suggested a possibility of r-modes in some of line-profile variations. Further work is evidently necessary to distinguish among possible NRP modes. To do so, full eigenfunctions of equation (7) should be used in future for the line-profile modeling.

3. NRPs IN MASSIVE EARLY-TYPE STARS

The basic equations governing linear adiabatic NRPs in a non-rotating star are a system of four first-order ordinary differential equations for the radial dependence of the eigenfunctions. The equilibrium structure of a given stellar model comes in coefficients of these differential equations. These equations together with appropriate boundary conditions at the center and at the surface form a proper eigenvalue problem, and they can be solved numerically for a given model. Results of calculations for massive early-type stars may be found, for instance, in Osaki (1975) for the evolutionary sequence of a 10 solar mass star in the main-sequence stage.

To understand the basic nature of NRP modes, the most useful is the so-called propagation diagram. In this diagram, we exhibit spatial variations of two critical frequencies: the Brunt-Väisälä frequency, N, for gravity-wave propagation and the Lamb frequency, L_ℓ, for acoustic-wave propagation. They are, respectively, defined by

$$N^2 = g\left(\frac{1}{\Gamma_1}\frac{d\ln P}{dr} - \frac{d\ln\rho}{dr}\right) \tag{8}$$

and

$$L_\ell^{\,2} = \frac{\ell(\ell+1)}{r^2}\,C_s^{\,2} \tag{9}$$

where $C_s = \sqrt{\Gamma_1 P/\rho}$ is the sound velocity, ℓ is the spherical harmonic degree, and other symbols have their usual meaning. Figure 1 illustrates a schematic propagation diagram for a simple stellar model such as the massive, zero-age main-sequence star. A wave with a given frequency ω and at a given depth r in the stellar interior behaves locally as the pressure (or acoustic) wave if $\omega^2 > L_\ell^{\,2}$ and $\omega^2 > N^2$, and as the gravity wave if $\omega^2 < L_\ell^{\,2}$ and $\omega^2 < N^2$, while it is non-propagating (called evanescent wave) if $N^2 < \omega^2 < L_\ell^{\,2}$ or $L_\ell^{\,2} < \omega^2 < N^2$. This diagram

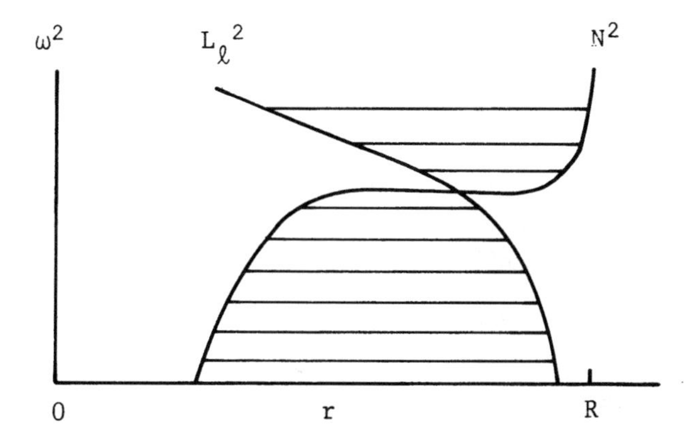

Fig. 1. Propagation diagram, for the massive, zero-age main-sequnce star.

shows in what part of the stellar interior a wave with a given frequency ω has locally a propagating or non-propagating character. The propagation diagram may be comparable to the potential energy diagram of the Schrödinger wave equation in quantum mechanics in such a way that a propagation zone corresponds to a potential well and a non-propagation zone to a potential wall. An eigenmode is a standing wave trapped in a potential well bounded at both ends by potential walls, satisfying a proper phase condition.

Let us now discuss NRP modes in massive stars. As the schematic propagation diagram in Figure 1 indicates, the stellar model for the zero-age main-sequence star (ZAMS) is simple as a non-radial oscillator. Its NRP modes can be easily classified as p-modes and g-modes. However, as the star evolves away from ZAMS, NRP modes become very much complicated. Figure 2 illustrates the hydrogen profile (the upper

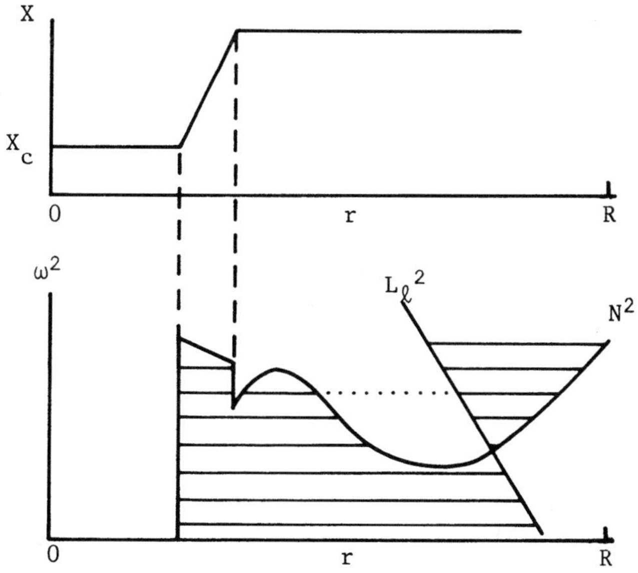

Fig. 2. Propagation diagram for an evolved model

panel) and the corresponding propagation diagram (the lower panel) for an evolved model near the end of the core hydrogen-burning main-sequence. The most important feature in this model is the trapezoidal profile of the N^2-curve just outside the convective core. This local peak in the N^2-curve is formed by the μ-gradient zone left behind by the receding convective core. As seen from Figure 2, those NRP modes with frequencies ω slightly less than the peak value of the N-curve have then a mixed-mode character in which they behave like gravity waves in the deep interior but they behave like acoustic waves in the envelope.

Figure 3 illustrates variation in eigenfrequencies ω (in dimension-

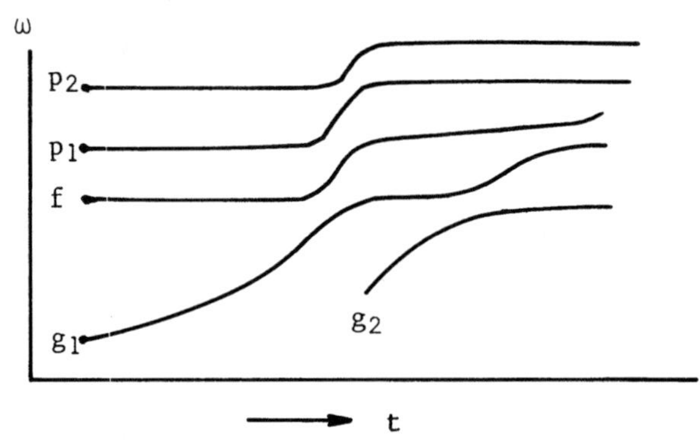

Fig. 3. Variation in eigenfrequencies with evolution of a massive star.

less units) with evolution of a massive star. As the star evolves away from ZAMS, frequencies of g-modes increase rapidly because of the increase in the Brunt-Väisälä frequency N around the μ-gradient zone. On the other eigenfrequencies of the envelope f- and p-modes essentially remain constant (in dimensionless units). Then, the frequencies of the f-mode and the g_1-mode draw near and they are going to cross each other. However, they never cross, but they pull away from each other after the two modes interact strongly at the crossing. This phenomenon was called "avoided crossing" by Aizenman et al. (1977). The two modes exchange their modal character at the avoided crossing, and the f-mode looks like the interior g-mode and the g_1-mode looks like the envelope f-mode after the avoided crossing. As the star evolves, the interior g-mode collides successively with p_1- modes, p_2-modes and so on. This phenomenon may be understood as the coupling between the two potential wells: the interior gravity-wave potential well and the envelope pressure-wave potential well.

When the star evolves further away from the main-sequence and evolves into the giant stage, the interior peak of the Brunt-Väisälä frequency N increases further more because of highly centrally condensed nature of the model. Then, even high frequency envelope p-modes behave like gravity wave of extremely short wavelength in the deep interior, as first noticed by Dziembowski (1971). Thus, radial modes and non-radial p-modes are basically different in giant and supergiant stars. However, there exists another interesting phenomenon called "wave trapping". We have so far emphasized the coupling between the two potential wells in NRP modes. However, if the degree ℓ of NRP modes is high, this coupling becomes weak and the two potential wells may be regarded to be almost independent. In such a case, eigenfunctions of NRP modes are essentially confined either in the envelope p-zone or in the interior g-zone, and they can be classified as the envelope p-modes and the interior g-modes (Shibahashi and Osaki 1976a), although their frequency ranges overlap each other.

4. EXCITATION MECHANISMS OF NRPs IN MASSIVE STARS

Various excitation mechanisms of both radial and non-radial pulsations in massive early-type stars have been proposed in connection with the Beta Cephei variables and line-profile variable stars. Since their review has been presented before by the present author (Osaki 1982, 1986a), it is not repeated. Here only a few additional comments are given.

Two destabilizing mechanisms (i.e., the so-called ε-mechanism and κ-mechanism) are known which may drive radial pulsations. The same mechanisms can in principle work for NRP modes as well. As for the ε-mechanism, it is not effective in most cases because the amplitudes of eigenfunctions are rather small in the nuclear burning region. However there exists a possibility of vibrational instability due to the ε-mechanism for non-radial g-modes trapped in the hydrogen-burning shell. In fact, Shibahashi and Osaki (1976b) and Noëls and Scuflaire (1987) have found that some of g-modes of intermediate harmonic degree $\ell \sim 10$ are vibrationally unstable in massive early-type stars. However, those g-modes are so effectively trapped in the nuclear burning shell that their amplitudes at the stellar surface are invisibly small, and they are unlikely to be responsible to variability observed in massive early-type stars.

Let us discuss briefly on vibrational stability of NRPs in luminous giant and supergiant stars. It has been shown by Dziembowski (1971) that NRP modes of low spherical harmonic degree ℓ (i.e., $\ell \lesssim 4$) are strongly damped in giant and supergiant stars because of enhanced radiative dissipation in the g-zone in the deep interior. It is thus very difficult to excite NRP modes of low degree ℓ in giant and supergiant stars that have a highly centrally condensed structure. However, non-radial p-modes are found to be vibrationally unstable in the Cepheid stars due to the κ-mechanism in the hydrogen and helium ionization zones, if their spherical harmonic degree ℓ is sufficiently high, i.e., $\ell \gtrsim 6 - 7$ (Osaki 1977). This is because these modes are so well trapped in the envelope that the interior dissipation is negligible. It is further found by Dziembowski (1977) and by Shibahashi and Osaki (1981) that the vibrational instability against high ℓ f-modes extends far to the blue beyond the blue edge of the (radial) Cepheid instability strip. Shibahashi and Osaki (1981) have accidentally found in their numerical calculations some violently unstable NRP modes in very luminous supergiant stars with $L \gtrsim 10^5 L_\odot$, in which the imaginary part of the complex eigenfrequency is comparable to its real part. It is presumed that these violently unstable modes are originated from some kind of coupling between the oscillatory non-radial f-mode and convectively unstable g$^-$

-mode under an extremely non-adiabatic condition, although exact physical and mathematical nature of these modes still remain to be clarified.

So far, our discussions have been restricted to non-rotating stars. If rotation of stars is taken into account, a new possibility of excitation of NRP waves arises. One of such possibilities is oscillatory convection or overstable convection in a rotating star. Rotation can exert restoring force on otherwise unstable convective g^--modes and it gives them a necessary wave character. Osaki (1974) has proposed, as a possible excitation mechanism of Beta Cephei pulsations, the resonant excitation of a global NRP mode by oscillatory convection in the rapidly spinning convective core of a massive star. Lee and Saio (1986) have recently calculated NRP eigenmodes in a rotating star having the convective core, and they have shown that there exist some overstable NRP modes, whose eigenfunctions have large amplitude both in the convective core and in the outer layers, thus confirming this resonant coupling mechanism. Further work will be worthwhile in extending this to fully non-adiabatic calculations.

5. EFFECTS OF NRP WAVES ON THE STELLAR ATMOSPHERE

Another interesting aspect of NRPs in stars is that NRP waves can transport energy and angular momentum from one part of the star to another and thus they could be an important source of non-radiative (mechanical) energy and angular momentum to drive stellar winds in massive stars (see, Abbott 1986). The stellar wind driven by shocks generated by NRPs in B and Be stars has been discussed by Willson (1986), while angular momentum trasport by non-axisymmetric NRP waves has been discussed by Osaki (1986a). Ando (1986) has recently proposed, as a possible driving mechanism of episodic mass loss in Be stars, the quasi-periodic variation (called the vacillation) of rotation profiles in the stellar atmosphere, which is produced by interaction between rotation (mean flow) and NRP waves. This phenomenon has been known in the meteorology as the quasi-biennial oscillation of the zonal wind in the tropical stratosphere. Further work along this line may be found fruitful.

REFERENCES

Abbot, D. C., Garmany, C. D., Hansen, C. J., Henricks, H. F., and Pesnell, W. D. 1986, Publ. Astron. Soc. Pacific, **98**, 29.

Aizenman, M. L., Smeyers, P. and Weigert, A. 1977, Astron. Astrophys., **58**, 41.

Ando, H. 1986, Astron. Astrophys., in press.

Berthomieu, G., Gonczi, G., Graff, Ph., Provost, J., and Rocca, A. 1978, Astron. Astrophys., **70** , 597.

Cowling, T. G. 1941, Mon. Not. R. astr. Soc., **101**, 367.

Cox, J. P. 1980, Theory of Stellar Pulsation (Princeton: Princeton University Press).

Dziembowski, W. 1971, Acta Astron., **21**, 289.

Dziembowski, W. 1977, Acta Astron., **27**, 1.
Lee, U. and Saio, H. 1986, Mon. Not. R. astr. Soc., in press.
Ledoux, P. 1951, Astrophys. J., **114**, 373.
Noels, A. and Scuflaire, R. 1987, in these proceedings.
Osaki, Y. 1974, Astrophys. J., **189**, 469.
Osaki, Y. 1975, Publ. Astron. Soc. Japan, **27**, 237.
Osaki, Y. 1977, Publ. Astron. Soc. Japan, **29**, 235.
Osaki, Y. 1982, in Pulsations in Classical and Cataclysmic Variables, ed. J.P. Cox and C.J. Hansen (Boulder, JILA), p.303.
Osaki, Y. 1986a, Publ. Astron. Soc. Pacific, **98**, 30.
Osaki, Y. 1986b, in Seismology of the Sun and the Distant Stars, ed. D.O. Gough (Dordrecht, Reidel Publ. Company), p.453.
Papaloizou, J., and Pringle, J. E. 1978, Mon. Not. R. astr. Soc., **182**, 423.
Saio, H. 1982, Astrophys. J., **256**, 717.
Shibahashi, H., and Osaki, Y. 1976a, Publ. Astron. Soc. Japan, **28**, 199.
Shibahashi, H., and Osaki, Y. 1976b, Publ. Astron. Soc. Japan, **28**, 533.
Shibahashi, H., and Osaki, Y. 1981, Publ. Astron. Soc. Japan, **33**, 427.
Smith, M. A., and Penrod, G. D. 1985, Proc. 3rd Trieste Conf. on Relationship between Chromospheric/Coronal Heating and Mass Loss, ed. R. Stalio and J. Zirker, Trieste Obs., p.394.
Unno, W., Osaki, Y., Ando, H., and Shibahashi, H. 1979, 'Nonradial Oscillations of Stars' (University of Tokyo Press, Tokyo).
Waelkens, C., and Rufener, F. 1985, Astron. Astrophys., **152**, 6.
Willson, L. A. 1986, Publ. Astron. Soc. Pacific, **98**, 37.

Lively conversation:
Edith Müller
Peter Ulmschneider
Yoji Osaki
Léo Houziaux

DISCUSSION

BURKI: I have three questions concerning the Rossby modes.
1) What is the typical minimum value of the rotation necessary for the existence of a R-mode?
2) Can g-modes and R-modes be simultaneously present in a rotating star?
3) Is there any possible dependence between the existence of the R-modes and/or the g-modes and the class of variable stars (i.e. yellow supergiants, Be-type stars, O-type stars, etc.)?

OSAKI: 1) There is no minimum value of the rotational frequency for the existence of a R-mode. What you have in the case of zero rotation is that the R-modes become simply steady eddy motions.
2) Yes, g-modes and R-modes can exist simultaneously.
3) Yes, I guess so. I suspect that R-modes may possibly be involved in Be-stars while p-modes may be involved in yellow supergiants.

HEARN: For the people who are not experts in this field, could you explain how the Rossby mode is generated?

OSAKI: The Rossby modes are modes of horizontal eddy motion whose restoring force is due to the Coriolis force. It is based on the principle of conservation of the absolute vorticity over the spherical surface. Here the absolute vorticity is the vorticity of fluid in the inertial frame. It is the sum of the relative vorticity of fluid in the rotating frame and the vorticity of rotation itself.

CASTOR: I have two questions about R-modes:
1) Can you do a non-adiabatic stability analysis?
2) Do you consider that the relation of Myron Smith and others that PxM = constant is a good argument for R-mode?
Is the $2/l(l+1)$ term in ω large enough to allow you to reject R-modes?

OSAKI: 1) Yes, I think we can at least in principle. However, I have not done it yet myself.
2) Yes, at least for high l R-modes.
3) In the case of α Vir (Spica) which Myron Smith (1985) discussed, observed modes are those of l=8 and l=16. Then, the term $2/l(l+1)$ is a small correction.

MAEDER: Concerning the violently unstable mode, which you and Shibahashi found in supergiants. Would you please comment more on which one of the excitation mechanisms you considered is responsible for the instability you found.

OSAKI: We suspect that the super-adiabatic temperature gradient in the hydrogen convection zone may be responsible for the violently unstable mode.

HEARN: The κ- mechanism is well known as a radial pulsation driving mechanism. Can you say under what conditions the κ-mechanism will drive predominantly non-radial pulsations rather than radial pulsations?

OSAKI: The classical Cepheids are pulsationally unstable both for radial modes and non-radial p-modes with high degree ($l \geqslant 6$), but apparently radial modes seem dominant in these stars. However, the blue edge of instability against high-l f-modes extends beyond the blue edge for radial modes. Thus there exist some stars which are stable against radial modes but unstable against non-radial modes.

LAMERS: You mentioned that Ando (1986) has shown that the angular momentum transport in non-adiabatic rotating stars can explain the episodic mass loss in Be-stars. Are the calculated timescales in agreement with the observed ones, which seem to be of the order of a decade?

OSAKI: The time scales by this mechanism depend on the amplitude of the oscillations but according to Ando (1986), the timescales estimated based on parameters reasonable for Be-stars are of the order of several years.

HENRICHS: The effects of rotation on NRP published so far has been limited to the case of slow rotation. Has any progress been made in the case of rapid rotation which has been observed in the ζ Oph type stars? If not, what would be the first effects that you expect to happen in a more rapidly rotating star?

OSAKI: As far as I know, there is no investigation for rapidly rotating stars. I would expect that the restoring force due to rotation becomes very important in rapidly rotating stars.

HENRICHS: Your suggestion about Be outbursts transporting angular momentum due to NRP in rotating stars is apparently also restricted to the slow rotation limit. How seriously would your basic conclusion be altered for 'real' Be stars which all seem to rotate rapidly?

OSAKI: My description of angular momentum transport by NRP waves is only qualitative. I do not find any reason why it must be restricted to the slow rotation limit.

NOËLS: Did you find in all your investigations, unstable modes because of the ε-mechanism? If they were, they should have been found in H-shell burning models but was it for trapped modes only or also for low order, low l-modes?

OSAKI: Yes, we found some unstable modes because of the ε-mechanisms. Those found unstable were well-trapped modes.

NOËLS: Would you agree to say that the most probable effect of an unstable trapped mode with a low amplitude at the surface, would be mixing in the region of the star where the amplitude is large?

OSAKI: Yes, I agree.

ZAHN: I wish to recall work done by Papaloizou and Pringle in Cambridge, and by Berthomien, Provost and Rocea in Nice on the g- and r-modes of rapidly rotating stars. The latter find that for given period and spherical number ℓ, the properties of the g-mode and of the r-mode are very similar, in particular the ratio κ between horizontal and vertical amplitude. Thus I am not surprised by your result on the similarity of the line profiles due to both types of mode. Furthermore, concerning their possible excitation mechanism, it seems unlikely that they are driven by the Kappa-mechanism, since the pressure and temperature variations associated with them are very small, for horizontal velocities compatible with the observations.

OSAKI: The line-profiles which I have shown are those calculated based on a pure spheroidal mode and a pure toroidal mode. The former has the vertical velocity component only and the latter has the horizontal velocity component only. Nevertheless both of these two modes give very similar line-profiles. This indicates that the mode identification based on line-profiles may not be unique.

ULMSCHNEIDER: Magnetic fields represent another restoring force. Have you thought about stellar oscillations with this restoring force e.g. for Ap stars or neutronstars?

OSAKI: Yes, magnetic fields will provide another restoring force and they may be important in magnetic Ap stars.

HENRICHS: This remark is concerned with the energy involved with NRP properties of rotating massive stars. As a working hypothesis I proposed some time ago that the variable factor in stellar wind properties of OB stars might be explained by (observed) variability in NRP properties, such as amplitude changes and/or mode switching. (An idea, similar to that of Vogt & Penrod (1983) for the more dramatic Be outbursts.) To investigate the energetics involved a kind of pulsational energy level diagram would be useful where the energy is displayed as a function of ℓ and m, given the pulsation period and amplitude, and rotation velocity. Theoreticians, however, do not commonly publish these energy values, one reason being that the energy is proportional to the a-priori-unknown value of the square of the velocity amplitude of the pulsation. The energy scaling in the next diagram is therefore a free parameter, but is taken to be in the ball park of the observed values. Dr. B. Carroll was kind enough to calculate the desired quantities for a rotating 12 $M_\odot$ ZAMS model, which are displayed in the following 'astroquantummechanical' diagram (see figure). This diagram should not be taken too literally. Its only purpose is to illustrate the amount of energy involved with this type of NRP. The vertical scale runs between 1 and 10 x 10^{44} erg for all ℓ and m values considered. This is the main point I want to make: 10^{44} erg is a lot of energy and even a low efficiency of converting this amount into directed kinetic energy would be able to do the job of ejecting a shell or puff for instance. Such a release of energy

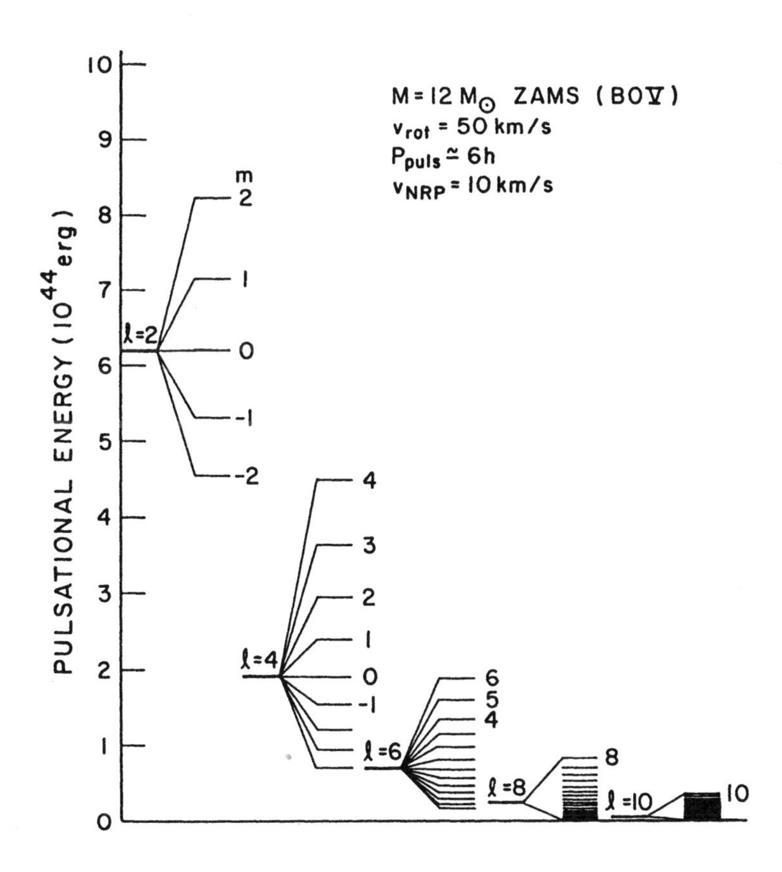

Figure 1. Schematic "Astroquantummechanical" diagram for a slowly rotating Non Radially Pulsating massive star. For each multiplet the pulsational energies are displayed belonging to a pulsation period of about 6 hours. (calculations: B. Carroll)

could be associated with a moderate change in amplitude. As a numerical example, Vogt & Penrod (1983) estimate that (only) about 10^{41} ergs of kinetic energy are necessary to produce a typical outburst of $10^{-7}M_{\odot}$ in ζ Oph. Whether, how and how frequently pulsational energy can be converted (partially) into directed motion are of course very intriguing questions. The point is here that the energy source of the variable mass outflow as observed in UV stellar wind variability might well be found in the NRP energy. Many studies are obviously needed to establish the value of such a hypothesis.

It would be useful to extend this sort of energy calculations to more massive stars and rapid rotators.

BURKI: Concerning the time required for a change of the pulsation mode, we can mention the calculations of J.P. Cox and A.N. Cox on supergiant stars coming in or getting out of the Cepheid instability strip: the stars need roughly 200 to 300 pulsation cycles to achieve the change of pulsating status.

HENRICHS: Thank you. It would be very interesting to know if this also holds for Be stars.

LAMERS: Looking at the numbers for the energies involved in the pulsations, it seems to me that the amount of energy involved in the transition from one mode to another is of the order of 10^{44} ergs. One can estimate the amount of mass which can be ejected when this energy is released: $E = 0.5\ V^2_{esc}\ \Delta M$. Assuming V_{esc} = 600 km/s for a typical OB main sequence star one finds $\Delta M = 3.10^{-5}\ M_\odot$. If part of the energy is also used to give this material an additional velocity, ΔM would be smaller. This estimate shows that mode switching might result in considerable mass ejection.

V.D. HEUVEL: The amount of mass loss which Lamers mentioned, of order $10^{-4} M_\odot$, seems like a nice outburst for a B-emission star outburst. Are you suggesting this mechanism as a cause for the outbursts of these stars?

HENRICHS: This has actually been proposed in the paper by Vogt & Penrod (1983, Ap.J. 275, 661). They give a similar energy argument but derived the amount of energy involved with such an outburst from estimated observed outburst properties, rather than from a purely theoretical calculation, and concluded that only a small fraction of the pulsational energy is needed to be converted into mass loss.

APPENZELLER: Concerning the amount of energy involved in the pulsations: The quoted energy values ($\sim 10^{44}$ ergs) are still small compared to the total heat content of a 12 $M_\odot$ star. If the energy would be evenly distributed over the star, its temperature would increase only very slightly.

HEARN: That is not a completely fair comparison. The energy heating the solar corona is very small compared with the radiative flux of the Sun, but it is not insignificant since it is sufficient to form the solar corona, which is a great spectacle at a total eclipse of the Sun.

CONTI: The mode-switching energy available would seem to be sufficient to be related to Be star 'episodes'. But shouldn't you use a higher rotational velocity? It does seem to me that mode switching will turn out to have some connection with Be star 'episodes'.

HENRICHS: The 50 km/s rotational velocity used in the calculation is about as high as the slow rotation approximation allows. We have no way to scale the energies up to higher rotational velocities, but the general consensus seems to be that the order of magnitude will not change much. That mode switching in Be stars might be directly connected to the Be outbursts was proposed by Vogt & Penrod (1983).

CASTOR: The modal energies in this figure raise the quesion of the mode switching times discussed earlier. This time can be estimated by dividing the pulsation energy by the luminosity of the star. With an energy of 10^{44} erg and $L=10^{38}$, this gives 10^6 sec, which is quite comparable to the observed time scale for mode changes.

HEARN: You have discussed a number of possible exitation mechanisms for NRPs. If you consider B type stars for example, and you were a gambling man, which of the exitation mechanisms would you back as being the most important?

OSAKI: I will bet to the convection-rotation coupling mechanism in the case of non-radially pulsating B stars.

ZAHN: You mentioned recent work by Lee and Saio which, you said, confirmed your theory on the driving of non-radial oscillation through overstable convection. Can you give us more details on this?

OSAKI: What Lee and Saio (1986) have done is that they have calculated complex eigenfrequencies and eigenfunctions of non-radial modes in a uniformly rotating star having a convective core. They have found that convectively unstable g-modes become oscillatory unstable in a rotating star and some of the overstable modes have larger amplitude both in the convective core and near the surface.

PIJPERS: Does the timelag existing between different parts of the star affect the stability of pulsational modes (if this is not usually already taken into account)?

OSAKI: I think that the linear stability analysis we use will automatically take into account the effect you mentioned.

The Swiss: Müller, Burki, Rufener

de Jager explaining turbulent pressure to Appenzeller,
with Tony Hearn thinking about it

THEORY OF VIBRATIONAL INSTABILITIES IN LUMINOUS EARLY TYPE STARS

I. Appenzeller
Landessternwarte
Königstuhl
D-6900 Heidelberg 1
Federal Republic of Germany

ABSTRACT. Because of the high temperature sensitivity of thermonuclear reactions massive blue stars are expected to be vibrationally unstable above a certain critical mass M_{crit} which depends on the evolutionary state. For ZAMS stars the most accurate calculations resulted in M_{crit} values between ~90 $M_\odot$ and 440 $M_\odot$. The origin of this discrepancy is unclear at present. For helium burning WR stars M_{crit} may be as low as 8 $M_\odot$. In the case of main-sequence stars the pulsation amplitudes are limited to relatively small values by nonlinear effects. Although the pulsations may enhance the mass loss from very massive stars, for ZAMS stars radiative driving forces will probably always dominate the mass loss rate.

1. INTRODUCTION

As pointed out already by Eddington (1926) in a pulsating star with a temperature dependent internal heat source heat is converted into mechanical work, if the heat generation rate $\varepsilon(T)$ increases with increasing temperature T. This behaviour is easily understood from a pressure-volume diagram (Figure 1), where the local pressure P is plotted as a function of the specific volume $V = 1/\rho$ (ρ being the local matter density). In this diagram periodic pulsations correspond to closed curves connecting points of maximum compression (i. e. maximum pressure and temperature) to points of maximum volume (or minimum pressure and temperature). For strictly adiabatic pulsations the P(V) relations during the contraction and expansion phases coincide and are given by the adiabatic P(V) relation of the stellar matter (e. g. $P \sim V^{-\gamma}$ for an ideal gas). In the nonadiabatic case the expansion and contraction phases are represented by different lines. A temperature dependent heat source (like nuclear burning), e. g., results in an higher than average heat produc-

H. J. G. L. M. Lamers and C. W. H. de Loore (eds.), Instabilities in Luminous Early Type Stars, 55–72.

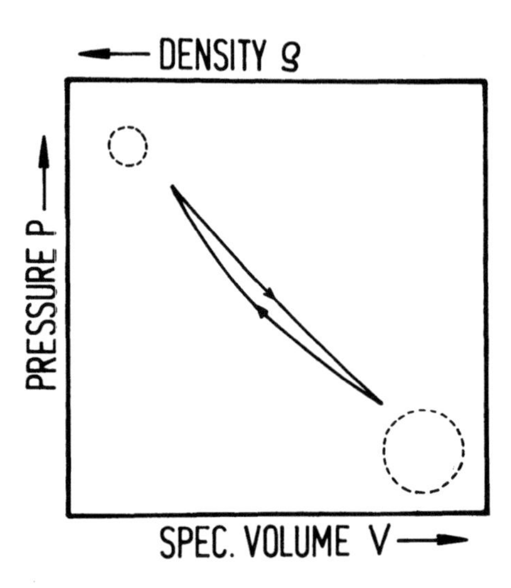

Figure 1. Schematic P(V) diagram

tion during the state of maximum compression and a lower than average heat generation during maximum volume. Thus, the temperature and the pressure are higher during the expansion phase than during the contraction phase. As a result, like in an automobile engine (which works according to the same principle), during each cycle mechanical energy is gained from the nuclear heat production. The amount of mechanical energy produced during one pulsation cycle is given by the work integral

$$W = \iint P \, dV \, dt \qquad (1)$$

where the integration has to be carried out over the full stellar volume and over one pulsation period. In Figure 1 the value of W corresponds to the surface enclosed by the P(V) curve. In the adiabatic case this surface vanishes and we have $W = 0$.

Although (for the reasons outlined above) the nuclear burning zones of pulsating stars usually give positive contributions to the work integral W, in most stars we have $W < 0$ for all eigenmodes. This is due to the fact that during maximum compression the radiative heat flow tends to be higher, resulting in a lowering of the temperature and pressure during the expansion phase. These enhanced heat losses have just the opposite effect of the nuclear burning and cause a negative contribution to the work integral. As the eigenmodes of stellar pulsations generally have smaller amplitudes in the nuclear burning interior than in the outer layers where the radiative losses are most effective, the radiative losses normally dominate the work integral and the loss of mechanical energy (corresponding to $W < 0$) leads to a rapid damping of any stellar pulsations. Only in stars with very high energy generation rates ε and a small increase of the pulsation amplitudes towards the surface Eddington's "ε-mechanism" may be expected to result in a positive work integral. Both these conditions are met in very massive main-sequence stars, which are very luminous (and thus must have high nuclear burning rates) and which (due to the increasing radiation pressure contribution) with increasing mass approach the case of homologeous pulsation (cf. Figure 2).

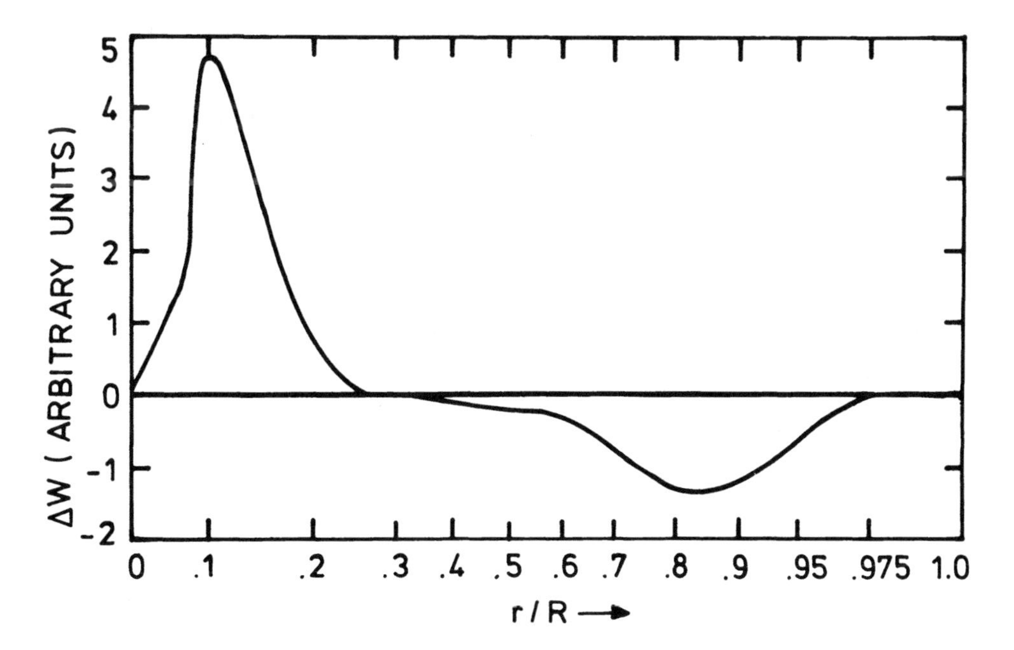

Figure 2. Contribution ΔW of individual mass shells to the work integral as a function of the distance from the stellar center for a 100 $M_\odot$ ZAMS star. (From Ziebarth 1970.) Note the positive contributions of the hydrogen burning interior layers and the negative contribution of the outer layers.

If $W > 0$ for at least one of its (many) pulsational eigenmodes a star is "pulsationally" or "vibrationally" unstable. Any random variation of the pressure and density in its interior (e. g. due to thermodynamic fluctuations) will be amplified and will lead to regular pulsations. Normally the higher harmonic radial modes have high radiative losses and the non-radial modes tend to have small amplitudes in the stellar center. Therefore, in stars with core nuclear burning the fundamental radial eigenmode is most easily energized by the ε-mechanism. However, in stars with shell burning sources, nonradial and higher harmonic modes may become unstable while the fundamental mode is stable (Kirbiyik et al. 1984, Noels and Scuflaire 1986).

Obviously $W < 0$ is a necessary and sufficient condition for vibrational stability. Hence, a vibrational stability analysis is equivalent of determining the sign of W. In principle, this can be achieved by numerically solving the time-dependent dynamical equations of stellar structure for the pulsating stars. W then can be calculated from the computed P(t) and V(t) relations. If we assume spherical symmetry (restricting us to radial pulsation modes) and using the independent variables t (=time) and M_r (=mass inside the radius r) and the dependent variables pressure $P(M_r, t)$, specific volume $V(M_r, t)$, temperature $T(M_r, t)$, energy flow $L_r(M_r, t)$ and radius (i. e. distance to the stellar center) $r(M_r, t)$, the dynamical equations of stellar structure can be written as:

$$-4\pi r^2 \frac{dP}{dM_r} = \frac{d^2r}{dt^2} + \frac{GM_r}{r^2} \quad (2)$$

(equation of motion, G is the gravitational constant),

$$\frac{dr}{dM_r} = \frac{V}{4\pi r^2} \quad (3)$$

(equation of continuity),

$$\frac{dL_r}{dM_r} = \varepsilon_N - \frac{dU}{dt} - P\frac{dV}{dt} \quad (4)$$

(energy balance equation , U(P,T) is the internal energy)

$$L_r = \frac{-64\pi^2 \, ac \, r^4 T^3}{3\kappa} \frac{dT}{dM_r} + L_{rc} \quad (5)$$

(energy transport equation; a is the radiation density constant, κ the opacity, c the velocity of light, L_{rc} the convective energy flow).

Together with the equation of state $P = P(V, T)$ the above equations form a system of nonlinear partial differential equations for which (in the case of realistic stellar models) there exist no analytical solutions. Numerical solutions are possible and have been obtained by several authors. However, as pointed out in more detail in Section 3 such computations are relatively complex and require numerical tricks which limit their accuracy and reliability for a derivation of W. Therefore, most of our present knowledge on the vibrational stability of massive stars is based <u>not</u> on the direct numerical solutions of the basic equations, but on linear approximation procedures. Such linear approximation methods are expected to give reliable results for very small pulsation amplitudes ($\delta r/r \ll 1$), and they allow to derive the work integral W in the infinitesimal amplitude limit. On the other hand, in real stars the sign of W depends on the pulsation amplitude, as the radiative damping increases strongly at large amplitudes (when shock waves start to occur). Thus, while the linear theory allows to determine the onset of a vibrational instability, its consequences can be predicted reliably only from nonlinear model computations.

2. LINEAR STABILITY ANALYSIS

2.1. Linearization and Adiabatic Approximation

The linearized pulsation theory was developed by Eddington (1918, 1919), Rosseland (1931), and Cowling (1936) during the first decades of this century. A detailed modern description of this approximation is given in the excellent review by Ledoux and Walraven (1958). Basically the linear pulsation theory is a classical perturbation method, where the dependent variables are written as sums of the type

$$\begin{aligned} r\ (M_r,t) &= \delta r\ (M_r,t) + r_o\ (M_r) \\ P\ (M_r,t) &= \delta P\ (M_r,t) + P_o\ (M_r) \end{aligned} \tag{6}$$

where r_o, P_o, ... are static equilibrium solutions of the basic equations. By inserting (6) into the basic equations (2) to (5), assuming $\delta r/r_o$, $\delta P/P_o$, ... $\ll 1$, and retaining only the lowest order (linear) terms of the perturbations, we obtain a new set of differential equations, where the time integration can be separated.

If the equilibrium solutions are known, the problem is thus reduced to the solution of a system of ordinary differential equations for the linear approximation eigenfunctions $\delta r\ (M_r)$, $\delta P\ (M_r)$, ... where

$$\delta r\ (M_r,t) = \delta r\ (M_r) \cdot e^{i\sigma t}$$
$$\delta P\ (M_r,t) = \delta P\ (M_r) \cdot e^{i\sigma t}$$

In general σ is a complex eigenfrequency. Only for strictly periodic pulsations σ is real and corresponds to the normal pulsation frequency. In the case of energized or damped oscillations σ contains an imaginary part which has the value $-W/2E\Pi$, where E is the pulsation energy and Π the pulsation period.

The linearized theory becomes particularly simple if we assume as a further approximation that the pulsations are adiabatic. In this case we can use the adiabatic P(V) relation to eliminate P or V and to reduce the number of dependent variables. As a result, with only two unknown functions left, the Equations (2) and (3) can be solved separately. In this way we obtain the "adiabatic" linear eigenfunctions $\delta r_{AD}(M_r)$ and $\delta P_{AD}(M_r)$.

In most layers of pulsating stars the relative change of the entropy during one pulsation period is much smaller than unity. Therefore, for most stellar layers the assumption of adiabatic pulsations is a quite good approximation. A major exception are the optically thin surface layers where radiation flow effects must result in strong deviations from the adiabatic case.

2.2. The Quasiadiabatic Stability Analysis (QAD)

Following the development of the linear pulsation theory by Eddington, Rosseland, and Cowling, it was Ledoux (1941) who first used this theory for a study of the pulsational stability of massive luminous main-sequence stars. Assuming approximate stellar models as static equilibrium solutions and a Kramer opacity law, Ledoux determined fundamental mode radial eigenfunctions in the linear adiabatic approximation. With these eigenfunctions and linearized versions of the energy balance and transport equations he then calculated approximate values of W and thus of Im(σ). This so-called "quasiadiabatic" stability analysis developed by Ledoux has since been applied by many other authors. Compared to Ledoux's pioneering work the more recent calculations use more realistic stellar models of different evolutionary stages, better numerical methods, and improved opacities and equations of state.

That W and Im(σ) can be computed by means of the adiabatic eigenfunctions on first glance appears contradictory, as for adiabatic pulsations we have by definition W = 0. We also note that the assumption of strict adiabasy is inconsistent with the Equs. (4) and (5). However, since (as noted above) in most stellar layers the deviations from the adiabatic situation are very small, the adiabatic eigenfunctions are good approximations in most stellar layers. An exception are the surface layers where (due to the radiation losses) the adiabatic eigenfunctions are very poor approximations. In these layers the use of the adiabatic eigenfunctions results in highly incorrect contributions to the work integral W, and the errors introduced in this way may be orders of magnitude larger than the (true) work integral. In order to avoid such a spurious and disasterously wrong contribution of the surface layers, in Ledoux's QAD method the outermost stellar layers (usually about ≤ 5 % of the radius) are simply ignored when calculating the work integral. Several authors (Rosseland 1931, Stothers and Simon 1970) tried to estimate the errors introduced by this omission. From approximate analytical estimates they concluded that (at least in the case of main-sequence stars) the outer layers have little influence on the work integral and that neglecting these layers does not introduce significant errors. On the other hand, recent numerical computations by Klapp et al. (1986) seem to indicate that the surface layers can well play a major role in determining the value and sign of W.

2.3. Nonadiabatic Linear Stability Analysis

Methods for solving the full set of (nonadiabatic) linear pulsation equations have been developed by Cox (1960, 1963), Baker and Kippenhahn (1962, 1965), and by Castor (1971). As shown by these authors, numerical solutions of the linearized basic equations and the corresponding work integrals can be calculated rather accurately and are good approximations for small amplitude pulsations. As the resulting nonadiabatic eigenfunctions are approximate solutions everywhere in the star, including the surface layers, the effects of the stellar surface is automatically and correctly included in the work integral derived from such solutions. The nonadiabatic linear pulsations theory has been widely used for investigations of the δ Cep variables. For vibrationally unstable massive blue stars there have been only two applications of this theory (Ziebarth 1970, and Klapp et al. 1986). As noted below, there is a serious disagreement between the results of these two studies.

2.4. Results from the Linear Theory

A list of linear stability studies of Population I massive blue stars is given in Table I. In this table the investigations using the quasiadiabatic analysis are labeled "QAD" and the nonadiabatic computations are labeled "NAD". Column 4 gives the critical mass above which stars of the evolutionary state indicated in Column 3 were found to be vibrationally unstable. "evol. sequence" in Column 5 means that the stability analysis has been carried out for several or many different models forming an evolutionary time sequence, usually starting with the state given in Column 3.

As shown by Table I, the more modern (>1970) QAD computations (which all use realistic opacities and static equilibrium models) are in reasonable agreement concerning the critical mass M_{crit} for vibrational instability. For Population I ZAMS stars the result is about $M_{crit}/M_{\odot} \approx 90 \pm 10$. For pure helium stars the most realistic computations give $M_{crit} \approx 16\ M_{\odot}$. But as shown by Noels and Magain (1984) even a very loss-mass hydrogen rich envelope will increase the critical mass of helium stars to much larger values. Maeder (1985) found that realistic WR stellar models may be vibrationally unstable to fundamental mode radial pulsations at

TABLE I. Vibrational Stability of Massive Stars: Results from the Linear Theory

Author(s)	Method	Evolut.State	$\frac{M_{crit}}{M_\odot}$	Remarks
Ledoux 1941	QAD	ZAMS	$\gtrsim$100	Kramer κ
Schwarzschild and Härm 1959	QAD	ZAMS	~60	Thomson κ evol.sequence
Stothers and Simon 1970	QAD	ZAMS	~100	
Ziebarth 1970	NAD	ZAMS	~90	
Papaloizou 1973a	QAD	ZAMS	~85	
Maeder 1985	QAD	ZAMS	~100	evol. sequence
Klapp et al. 1986	NAD	ZAMS	~440	
Boury and Ledoux 1965	QAD	He*	7-8	Thomson κ
Noels-Grötsch 1967	QAD	He*	9.2	Thomson κ
Simon and Stothers 1969, 1970	QAD	He*		evol. sequence
Stothers and Simon 1970	QAD	He*	$\gtrsim$17	
Noels and Gabriel 1981	QAD	WR		evol. sequence
Noels and Masereel 1982	QAD	He*	16	
Noels and Magain 1984	QAD	He*	80	thin ($\Delta M/M = 10^{-4}$) H envelope
Maeder 1985	QAD	WR	$\gtrsim$8	evol. sequence
Noels and Scuflaire 1986	QAD	WR	~40	g-mode instability of H shell source

mass values as low as 8 $M_\odot$. WR stellar models with H shell burning energy sources may become unstable towards certain nonradial eigenmodes (Noels and Scuflaire, 1986). As first shown by Schwarzschild and Härm (1959) vibrationally unstable stars which evolve away from the ZAMS without mass loss soon became stable again during the later stages of the main-sequence phase. This behaviour is easily understood as a consequence of the expanded outer layers of the post-

-ZAMS stars, which results in a larger pulsation amplitude and strongly increased damping in the outer stellar layers. However, if massive stars evolve with sufficiently strong mass loss, the removal of the hydrogen-rich outer layers may keep the stars unstable until (or make them unstable again when) core helium burning becomes the main energy source (Simon and Stothers 1970, Noels and Gabriel 1981, Maeder 1985). As the WR stars are generally assumed to form following such a scenario, it has repeatedly been suggested that at least the more massive WR stars are vibrationally unstable and that the (compared to O stars of similar luminosity) enhanced mass loss of the WR stars is related to this instability (see e. g. Maeder 1986).

The two only NAD studies of massive blue stars (Ziebarth 1970, Klapp et al. 1986) were both restricted to ZAMS models. While the critical mass derived by Ziebarth (1970) is in agreement with the values derived from the QAD approximation, Klapp et al. found a value which is higher by more than a factor of four (cf. Table I). Although M_{crit} is known to depend on the chemical composition and on the opacity (for which somewhat different assumptions were made by Ziebarth and by Knapp et al.) the difference in the critical mass is much too large to be explained by the different model assumptions. A possible source of the discrepancy could perhaps be the considerably finer zoning of the new computations of Klapp et al. (1986). However, as it is often the case for numerically derived results, the published data and graphs do not allow to determine the origin of the large difference in M_{crit}. At the time of this writing the difference must be regarded as unexplained and a careful new and independent NAD derivation of M_{crit} for ZAMS stars is highly desirable to clear up the contradiction. If the result of Klapp et al. were the correct one, the earlier results derived from the QAD approximation could no longer be regarded as trustworthy. This applies in particular to the computations for the helium star and WR models, which so far are based entirely on QAD calculations.

3. NONLINEAR NUMERICAL COMPUTATIONS

Direct numerical integrations of the basic equations (2) to (5) for pulsating massive blue stars have been carried out by Appenzeller (1970, 1971), Ziebarth (1970), Talbot (1971a, b), and Papaloizou (1973b). The objective of these computations was to determine the consequences of nonlinear effects in finite amplitude pulsations and the limiting amplitudes at which the nonlinear damping compensates the nuclear energizing of the pulsations. All these computations were restricted to ZAMS models and in all cases the opacity was approximated by pure Thomson scattering. From the linearized theory it was known that the pulsations of very massive ZAMS stars grow very slowly with relative gains of pulsational energy per pulsation period of the order 10^{-6}. Therefore, in order to make the computations tractable, in all cases the growth time was artificially decreased by suitable numerical procedures and the true growth rates (or work integral) had to be derived by indirect methods.

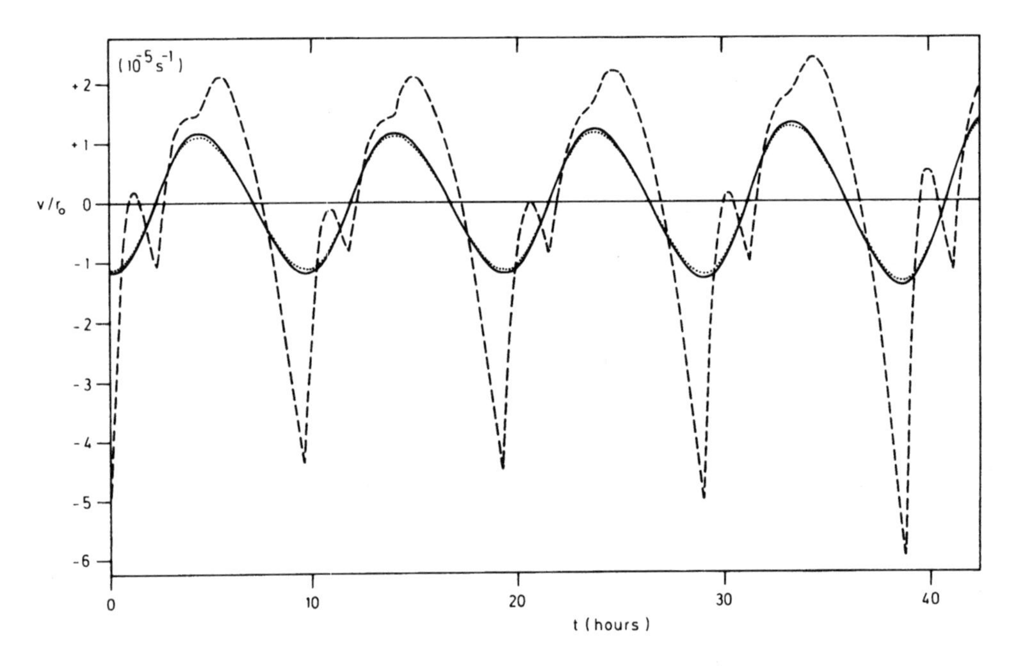

Figure 3. Variation of the normalized velocity of a pulsating 130 $M_\odot$ star at the surface (broken line) and in the interior layers (solid and pointed lines). (From Appenzeller 1970). Note the strong nonlinear effects in the surface layers.

Although the different authors used different numerical procedures, the basic results of the nonlinear model computations are in good agreement. According to all computations the ε-mechanism energized pulsations of massive ZAMS stars are limited by nonlinear surface effects (cf. Figure 3) at relatively small pulsation amplitudes (corresponding to relative radius changes $\lesssim 10$ %). The nonlinear damping becomes important when the velocity amplitude in the surface layers becomes comparable with the local velocity of sound in these layers. In massive ZAMS stars the effective temperature and thus the velocity of sound increase only slowly with mass. On the other hand, the stellar radii increase more rapidly than the pulsation period. Therefore, although the mechanical energy generation of the pulsations increases strongly with mass, the limiting relative pulsation amplitude remains almost independent of the mass.

For a vibrationally unstable massive ZAMS star the numerical solutions predict a considerable variation of the bolometric luminosity (up to about 1 mag). However, as the bolometric correction of a radially pulsating very hot star is expected to vary about as strongly as the bolometric luminosity, the observable visual magnitude (and UBV color) variations are predicted to be very small. Another result of the nonlinear computations is the prediction of a considerable expansion of the outermost layers when the limiting pulsation amplitude is reached. This extended outer layers or "false photospheres" seem to be caused by the momentum input of the shock waves into the atmosphere. In some computations this momentum and heat input resulted in a continuous mass outflow or wind driven by the pulsations. According to the computations this massive wind will hide the pulsating interior of the star, making a detection of the pulsations even more difficult. As expected the pulsationally driven mass loss depends on the mass and increases strongly with increasing mass. A reliable upper limit of the pulsationally driven mass loss can be estimated from the assumption that all the mechanical energy generated is converted into wind energy (Appenzeller 1971). A comparison of this maximum pulsationally driven mass loss with an extrapolation of the empirically determined $\dot{M}(L)$ relation of the radiative mass loss of normal (vibrationally stable) O stars (Garmany 1981) shows that the pulsationally driven mass loss is expected to be of the same order or smaller than the radiatively driven stellar wind of such luminous stars. Hence, the life expectancy of very massive ZAMS stars will not be changed dramatically by the vibrational instability due to the ε-mechanism. In particular, the onset of vibrational instability can hardly explain the rareness or absence of very massive ($M > 200\ M_\odot$) stars in the Milky Way and the nearby galaxies. Very likely such stars either cannot form (because of fragmentation) or they loose most of their mass already during the protostellar phase of star formation (cf. e. g. Appenzeller 1982). Vibrational instabilities may play a more important role during the WR evolutionary phase. However, in the absence of nonlinear pulsation computations for this phase the consequences of finite amplitude pulsations for WR stars at present cannot yet be assessed reliably.

4. CONCLUSIONS

From the observations of luminous stars it seems clear that we know at least some stars with $M > 100\ M_\odot$. On the other hand at present we have no reliable evidence for the existence of stars more massive than about 200 $M_\odot$. Hence, because of the contradictory results on M_{crit} described above, at present it is unclear whether the most massive ZAMS stars in our galaxy and the nearby extragalactic systems are vibrationally unstable or not. However, from the nonlinear numerical results it seems clear that, even if vibrationally unstable massive ZAMS stars exist, the effects of their pulsations will neither be observationally conspicuous nor of much consequence for the evolution of these objects. Although the pulsations may enhance the stellar winds of these stars, their mass loss will probably always be dominated by radiative effects. From the results of the linear adiabatic approximation it appears likely that the vibrational instability can have a more significant effect on the WR evolutionary stage of the massive stars. At least luminous hot WR stars should be pulsating in their interior below the visible optically thick expanding envelopes. That at least part of the WR stars are compact and hot enough to be vibrationally unstable seems to be confirmed by recent derivations of very high effective temperatures for these objects (Cherepashchuk et al. 1984, Pauldrach et al. 1985). On the other hand, the exact T_{eff} values of the WR stars are still somewhat in dispute. If the T_{eff} values are as low as derived recently by Schmutz et al. (1987), the corresponding extension of the outermost layers of the WR stars would probably effectively prevent the onset of vibrational instability. Finally, although during the past 45 years much has been learned on the vibrational stability behaviour of massive blue stars, the present state of our knowledge is still not satisfactory. Obviously it is highly desirable that the contradiction between the results on the ZAMS critical mass derived from the NAD studies is clarified soon. Furthermore, in order to determine the possible effects of vibrational instabilities in WR stars the NAD and nonlinear computations should be extended to these objects.

REFERENCES

Appenzeller, I.: 1970, Astron. Astrophys. 5, 355
Appenzeller, I.: 1971, Astron. Astrophys. 9, 216
Appenzeller, I.: 1982, Fund. Cosmic Phys. 7, 313
Baker, N. H., Kippenhahn, R.: 1962, Zeitschr. f. Ap. 54, 114
Baker, N. H., Kippenhahn, R.: 1965, ApJ 142, 868
Boury, A., Ledoux, P.: 1965, Ann. Astrophys. 28, 353
Castor, J. I.: 1971, ApJ 166, 109
Cherepashchuk, A. M., Eaton, J. A., Khaliullin, K. F.: 1984, ApJ 281, 774
Cowling, T. G.: 1936, Month. Not. Roy. Astr. Soc. 96, 42
Cox, J. P.: 1960, ApJ 132, 594
Cox, J. P.: 1963, ApJ 138, 487
Eddington, A. S.: 1918, Month. Not. Roy. Astr. Soc. 79, 2

Eddington, A. S.: 1919, Month. Not. Roy. Astr. Soc. 79, 177
Eddington, A. S.: 1926, The Internal Constitution of the Stars, Cambridge 1926, p. 200
Garmany, C. D., Olson, G. L., Conti, P. S.: 1981, ApJ 250, 660
Kirbiyik, H., Bertelli, G., Chiosi, C.: 1984, in 25th Liège Colloquium, Eds. A. Noels and M. Gabriel, p. 126
Klapp, J., Langer, N., Fricke, K. J.: 1986, ApJ (in press)
Ledoux, P.: 1941, ApJ 94, 537
Ledoux, P., Walraven, Th.: 1958, Hdb. d. Phys. 51, 353
Maeder, A.: 1985, Astron. Astrophys. 147, 300
Maeder, A.: 1986, Highlights of Astronomy 7, 273
Noels-Grötsch, A.: 1967, Ann. Astrophys. 30, 349
Noels, A., Gabriel, M.: 1981, Astron. Astrophys. 101, 215
Noels, A., Magain, E.: 1984, Astron. Astrophys. 139, 341
Noels, A., Masereel, C.: 1982, Astron. Astrophys. 105, 293
Noels, A., Scuflaire, R.: 1986, Astron. Astrophys. (in press)
Papaloizou, J. C. B.: 1973a: Month. Not. Roy. Astr. Soc. 162, 143
Papaloizou, J. C. B.: 1973b: Month. Not. Roy. Astr. Soc. 162, 169
Pauldrach, A., Puls, J., Hummer, D. G., Kudritzki, R. P.: 1985, Astron. Astrophys. 148, L1
Rosseland, S.: 1931, Oslo Publ. No. 1
Schmutz, W., Hamman, W. R., Wessolowski, U.: 1987, Paper presented at the IAU Symp. 122
Schwarzschild, M., Härm, R.: 1959, ApJ 129, 637
Simon, N. R., Stothers, R.: 1969, ApJ 155, 247
Simon, N. R., Stothers, R.: 1970, Astron. Astrophys. 6, 183
Stothers, R., Simon, N. R.: 1970, ApJ 160, 1019
Talbot, R. J.: 1971a, ApJ 163, 17
Talbot, R. J.: 1971b, ApJ 165, 121
Ziebarth, K.: 1970, ApJ 162, 947

DISCUSSION

NOËLS: Do you have any idea of the behaviour of the adiabatic eigenfunction in the work by Klapp et al.? I would like to mention that when we obtained a critical mass for the stars of 80 $M_\odot$ by adding a small H envelope, it was not just a difference in the evaluation of the work integral but it was a difference in the whole structure of the star which produced a large difference in the adiabatic eigenfunctions, having larger amplitudes at the surface.

APPENZELLER: Klapp et al. do not describe their eigenfunction in detail. They only state that the non-adiabatic eigenfunctions differ considerably from the adiabatic ones.

NOËLS: You mentioned that if WR stars have lower effective temperatures they should be stable. It is perfectly understandable because as long as there is a H burning shell, even if the mass is larger than the critical mass for the stars, it remains stable. WC stars could be unstable if they are formed early enough in the core He burning phase, following Peter Conti's scenario. At that time, they are nearly homogeneous He stars, more massive than the critical mass so they should be unstable. But this depends very strongly on the mass loss rate adopted. Vibrational instability can also come from non-radial oscillations in H burning shell models, as shown in the poster by Noëls and Scuflaire. This could apply to WN stars as well as to early type supergiants at the beginning of the H shell burning phase, if they have an intermediate convective zone.

APPENZELLER: I fully agree.

DAVIDSON: May I ask for additional clarification or reassurance concerning evolved stars? You remarked that extended stars are relatively stable, and that the outer layers provide damping. But the outer layers of very massive stars eventually approach the Eddington limit as their opacities increase. In such cases, do the outer layers still provide adequate damping to preserve stability?

APPENZELLER: If the temperature does not become so low that the κ-mechanism (of the δ Cep stars) sets in, the damping of pulsations will always increase with the extent of the outer layers.

ROUNTREE-LESH: In the case of the Zero-Age Main Sequence stars which have masses only slightly greater than M_{crit}, is there any possibility of observing the pulsations? And what is the period?

APPENZELLER: One would expect to observe small amplitude ($\leqslant$ 0.1 mag) regular light variations with predicted periods of about 8-12 hours. Various observers (including myself) have tried to find such variations in luminous O main sequence stars. But so far no such regular variations could be detected.

V.D. HEUVEL: I would like to follow up on Davidson's remark. The observations show that near the ZAMS even very massive stars are still stable, but towards the right in the HRD they become more and more unstable. Furthermore, the old Eddington book (1923) shows that the upper mass limit for stability against radiation pressure is proportional to $1/\mu^2$, where μ is the mean molecular weight. In a hydrogen star, $\mu = 1/2$, in a helium star, $\mu = 4/3$, such that going from a hydrogen star to a helium star, the upper mass limit for stability goes down by a factor $(8/3)^2 \simeq 7$ (see Van den Heuvel, 1979). This fits excellently with the fact that for hydrogen stars one finds an upper mass limit of 100 $M_\odot$, and Mrs. Noëls finds an upper limit for helium stars of 16 $M_\odot$. So, since stars have increasing μ when moving towards the right in the HRD, I would expect them to become more unstable, just as the observations show. I am therefore very surprised to see that Maeders calculations show that the models become more stable when the stars move towards the right. I do not understand this.

APPENZELLER: Eddington's μ^{-2} law is an approximation based on the assumption that the opacity is due to electron scattering only. With realistic κ values the damping is stronger, but $M_{crit} \sim \mu^{-2}$ is still a useful approximation if one compares stars of similar structure (i.e. homogeneous stars or main-sequence stars). The critical mass of ZAMS stars of given mass e.g. roughly varies like μ^{-2}. However, if a star evolves away from the ZAMS the change of its mass distribution (contraction in the core, expansion in the outer layers) and the corresponding change of the pulsational eigenfunctions usually has a much stronger effect on the stability. It is for this reason that stars evolving away from the ZAMS become more stable again. The observed light variations observed in the supergiant stage therefore must be caused by another mechanism. Only when the star contracts again (after having lost its H-rich layers) in the WR stage, the ε-energized vibrational instability may become important again.

DE JAGER: How to reconcile the statement that stars at the ZAMS are vibrationally unstable, and those further away should be more stable, with the observation that the three near-MS Carina stars (mentioned by Davidson) are stable while nearly all supergiants further away from the main sequence are more and more variable.

APPENZELLER: The most massive (known) O stars in Carina are either below 100 $M_\odot$ on the ZAMS or > 100 $M_\odot$ above the ZAMS. These evolved stars may have been vibrationally unstable, but have now become stable again due to the evolutionary change of their structure (central contraction and outer layer expansion). Therefore, the absence of light variations in these stars is not inconsistent with the theory.
The cooler luminous supergiant stars (like S Dor, R127, R71) are far too extended (and too strongly damped) to be vibrationally unstable due to the ε-mechanism. Their variability must have another reason, like κ-related instabilities in the wind or (as you suggested) turbulence pressure. Both, κ and turbulence related effects will increase rapidly with decreasing effective temperature.

MAEDER: Evolution away from the zero age main sequence during the hydrogen burning phase tends to stabilize the star with respect to the epsilon-mechanism. The reason is that the increasing density contrast makes the ratio of central to surface amplitude lower. Thus, the injection of pulsation energy at the stellar center becomes smaller and smaller as evolution proceeds. However, after the main sequence, in the blue supergiant phase for example other instability mechanisms can intervene, such as the κ-mechanism, and lead to the observed variability.

HENRICHS: Just a brief comment on the O3V star HD 93250. From the CIV UV profiles it is clear that this star exhibits strong wind variability as observed in the so-called discrete high-velocity absorption components. These observations are taken four months apart, but comparison with other O stars indicates that the actual timescale is much shorter. In this respect HD 93250 is not unlike all other O stars.

MAEDER: You recalled that the instability of Wolf-Rayet stars is favoured by a high T_{eff}. But, is it not quite consistent with the high T_{eff} recently found by Cherepashchuk et al. and Pauldrach et al. for WR stars?

APPENZELLER: High effective temperatures would make us more confident that the ε-mechanism is at work in WR stars. The very high values derived by Pauldrach et al. are indeed encouraging in this respect. However very recent computations by the Kiel group (Hamann et al.) resulted in lower T_{eff} values again.

V.D. HUCHT: To add to the confusion on the effective temperature of Wolf-Rayet stars, Joe Cassinelli and I have a poster here addressing in particular the WC9 stars, which show CII emission lines. One wonders how a 90000 K effective temperature could produce CII lines. We favour more something like 20000 K.
Then, with the larger μ values coming from the work of Maeder and De Loore, and with the larger number of nucleons per electron when interpreting the radio data, we conclude to mass loss rates that are a factor 2 to 3 larger than thought previously (e.g. Abbott et al., 1986).
Particulars of our work can be found in Van der Hucht, Cassinelli and Williams (1986).

CONTI: I believe the T_{eff} of WR stars are still uncertain by a factor two. The emergent continua are clearly determined by the wind blanketing processes, which are not yet well modeled. Similarily the observed ionization balance is affected by wind blanketing, hence neither type of data can be yet used to determine T_{eff} until adequate physical models are provided. The eclipsing binary V444 Cyg does provide a relatively model independent estimate of a 'core' for this WN5 star, Cherepashchuk et al (ApJ. 1985) find a radius of some 3 $R_{\odot}$, thus leading to a 'hot' T_{eff} of some 90000 K, this is only one WN5; I wouldn't be surprised if another WN5 might have a different T_{eff}.

DAVIDSON: In IC 1613 there's a Wolf-Rayet star that illuminates an isolated HII region; Kinman and I observed it, several years ago, also Rosa and D'Odorico, if my memory is correct. The emission lines from the nebula allow us to make a Zanstra estimate of the number of ionizing photons. In fact, there are even a lot of HeII-ionizing photons (beyond 54 eV). This object must be hot, using just simple nebular reasoning without even referring to the stellar spectrum itself.

CASTOR: Two comments:
We should remember that the Wolf-Rayet envelope is very optically thick in the photo ionization continua, and that, even though the effective temperature referred to the core radius ($\simeq 2R_\odot$) may be 10^5 degrees, the UV radiation from there will be processed a great deal in the outer part of the envelope. It is a semantic question what we call the photosphere and hence the effective temperature.
The envelope, although opaque still has a density of only 10^{-12} g/cm^3 or so, many orders less than the interior density, so the edge of the core is really guite sharp, and should provide good reflection of the waves within the core. This would justify neglecting the envelope for the pulsation calculation. Some wave leakage will none-the-less occur of course, and could be important for the envelope.

APPENZELLER: I agree that the strong drop in density at the boundary between the 'true surface' and the wind of the WR stars will probably prevent that much energy is lost into the wind region. However, I would expect that the optically thick wind will 'reflect' part of the outward radiation, resulting in a higher temperature and lower density in the outer hydrostatic layers, which again will result in different eigen-functions of the pulsations.

CASTOR: Yes, that is true in principle, but I believe the temperature will be raised much above that without the envelope only in a thin layer at the edge of the core, and the effect on the eigenfunction may not be great. A calculation is needed to see.

LAMERS: How do vibrational instabilities in massive stars appear to an observer: strictly periodic, growing amplitudes?

APPENZELLER: The pulsations will grow until non-linear damping effects will stop the growth. The amplitude of the strictly periodic pulsation will then remain constant. In massive ZAMS stars near M_{crit} the maximum amplitude will be rather small (corresponding to light changes of the order $\leqslant 0.1$ mag). The worst effect on the star will be some moderate mass loss.

MOFFAT: Photometric observations of WR stars indicate that the largest variations tend to occur for WNL stars and as one approaches the He-ZAMS (hotter WN stars, and WC stars) the amplitudes of the variations diminish. If this is due to reprocessed radiation and damping effects by the wind, then one wonders why WN8 stars have the largest amplitudes, since we do not see down to their photospheres either. Admittedly, the time scales are days or several hours at the shortest and need not be related to ε-induced instabilities of ≃ 15 min. Perhaps one should look for instabilities in the UV where we may be seeing down closest to the photosphere (cf. deduced hot T_{eff} of the WN5 star in V444 Cyg by Cherepashchuk et al. 1984, based mainly on the narrowness and great depth of the UV eclipse curve).

Appenzeller and the red dot in the P(V) diagram of a pulsating star

Observed variations in O and Of stars*

Dietrich Baade
Space Telescope-European Coordinating Facility
European Southern Observatory
Karl-Schwarzschild-Str. 2
D-8046 Garching
Fed. Rep. Germany

Summary: There are hardly any features in the electromagnetic spectrum between X-rays and the radio domain which time resolved observations have not detected to be variable in at least some O stars. The as yet most prominent patterns of intrinsic variability are:

(i) variations of flux and flux distribution with a strong increase of the amplitudes towards evolved, luminous stars,

(ii) radial velocity variations,

(iii) in supergiants, quasiperiods of (i) and (ii) which form an extension of the quasiperiod-luminosity-color relation for B- to G-type supergiants and are generally much longer than the period of the radial fundamental mode,

(iv) variable absorption line profiles occurring from ZAMS stars to supergiants and indicating nonradial pulsations (NRP's); the variations are truly periodic, and such periods of supergiants appear generally shorter than the quasiperiods mentioned under (iii),

(v) short-term V/R variations of emission components in a variety of spectral lines, often periodic with the same period as a low-order NRP mode,

(vi) decreased NRP amplitudes during phases of enhanced mass loss rates in Oe/Be stars,

(vii) in relatively narrow-lined supergiants, cyclic variations of the H-alpha emission strength on time scales from 5 to 20 days,

(viii) irregular broad-band polarization variations indicating changing density distributions of the near-stellar wind,

(ix) changes in position and strength of narrow high-velocity absorption components in UV resonance lines.

As indicated, the time scales of the individual phenomena can be different and generally range from few hours (dynamical) to many days (approximately rotational, but a detailed identification of any variation with rotational modulation has not yet been achieved). Intrinsic variability also affects the determination of more static stellar parameters such as macroturbulence and rotational velocity, and occasionally it may provide alternatives to binary models.

*Based in part on observations collected at the European Southern Observatory, La Silla, Chile

H. J. G. L. M. Lamers and C. W. H. de Loore (eds.), Instabilities in Luminous Early Type Stars, 73–80.

From the above enumeration the need for several observational surveys can be deduced, namely (1) completing the list of variabilities and their morphologies, (2) mapping their frequency in the HR diagram, and (3) searching for interdependencies and correlations. For (1) some intuition will be helpful, but (2) and (3) will be expensive since low amplitudes require high S/N while multiply and/or not strictly periodic processes necessitate long continuous data strings. The mapping is ideally done in a multidimensional parameter space since, e.g., points (iv), (v) and (vii) suggest a dependence of the pulsation properties on $v \sin i$, and will evidently have to extend far beyond the boundary to B-type stars because as far as variability is concerned the distinction between O and B stars is rather meaningless. In few individual stars have searches for pairwise correlations between different variabilities already been undertaken. These include (i) and (iv); (iv) and (vii); (iv) and (ix); (vii) and (viii); (viii) and (ix); and others. The results have been mostly negative; but observations were often made when the time scales involved were not yet known so that time resolution and/or coverage were not sufficient.

The required big observational efforts receive much of their justification from (a) the expectation that the study of a star's response to its own variability will provide a much broader basis in parameter space for all derived quantities than does a merely static analysis, and (b) the possibility that the variability has a noticeable effect on the mass loss process and, hence, the stellar evolution. On a first glance, for (a) the discovery of NRP's should prove especially useful. However, the possible multitude of mode types whose effects can become observationally indistinguishable when too few observing channels are used, poses a serious problem so long as the eigenfunctions of rotating stars are not well enough known. As to (b), point (vi) above is clearly encouraging, and (vii) and (ix) add further to the suspicion that the mass loss from early-type stars may rather generally include an episodic component. But for a definitive conclusion, a much better understanding of the propagation of stellar variabilities into the wind and their interaction with the intrinsic instabilities of radiatively driven winds is needed.

An extended version of the talk presented at this workshop will be published elsewhere (Baade 1986).

Reference

Baade, D.: 1986, in *O, Of and Wolf-Rayet Stars*, eds. P.S. Conti and A.B. Underhill, NASA/CNRS *Monograph Series on Nonthermal Phenomena in Stellar Atmospheres*, in press

DISCUSSION

CASSINELLI: The X-ray emission of the main sequence stars ζ Oph O9.5V and τ Sco B0V are interesting. The Solid State Spectrometer observations show two X-ray emission components. There is a dominant component with $T \simeq 6 \times 10^6$ K and a hot component with $T \simeq 2.0 \times 10^7$ K. It is not clear how this X-ray emission is to be explained, but it is interesting that these stars have similar X-ray spectra as one (ζ Oph) is a rapidly rotating star Oe star and the other, τ Sco, is a slowly rotating star ($V \sin i \leqslant 10$ km/sec). Could you comment on these results?

BAADE: I do not suggest that all X-rays are due to some secondary effect of NRP's, but the variable part of it, if any, may be. If ζ Oph and τ Sco show similar X-ray variability remains to be seen. But if they do the hypothesis remains that τ Sco is seen truly pole-on so that sectoral NRP-modes are undetectable. Currently this would appear less unlikely than to assume that a star with spectral type B0 V does not pulsate.

NOËLS: a) What is the order of magnitude of the periods you get for the main sequence stars?
b) Is NRP a general feature in those MS stars, which could be detected if the stars were all carefully studied and could you already suggest a range of mass for which there is NRP on the MS?

BAADE: a) Smith (1978, 1981) found periods of the order of 0.5 to 1 day.
b) In broad-line non-supergiant B stars with spectral type B7 or earlier I think it is difficult to detect a star that does not show line-profile variations. O-type MS stars have hardly been given any attention yet.

MOFFAT: You did not talk about continuum variations from photometry. These variations must typically be very small (cf. work of Van Genderen) $\simeq 0.01^m$ for main sequence O-stars.

BAADE: As I said, the photometric amplitudes increase with luminosity which probably explains the observational bias for supergiants. For O-type main sequence stars there is virtually no published information about variability available.

HENRICHS: Are you able to model the pulsational properties of ζ Pup from the rapidly varying profiles you observe? For instance: mode, velocity amplitude and/or pulsation period?

BAADE: For the mode order, m, and the period, P, I don't see serious problems. For the short period mode $m = 4 \pm 1$, $P = 0.178 \pm 0.005$ day are preliminary estimates. After I have seen that something very similar occurs in the Hα-reversal which is blue shifted by about 100 km/s, I would be more careful and sceptical about all other parameters.

HENRICHS: Is the difference between η CMa and ε Ori regarding Hα/CII variations understandable in terms of a more pole-on view in the case of η CMa?

BAADE: This cannot be ruled out. On the other hand I have selected supergiants with relatives broad lines (v sin i $\geqslant$ 80 - 100 km/s) in order to see NRP's more easily. My sample should therefore at least contain no pole-on stars.

APPENZELLER: Seeing your impressive evidence for non-radial oscillations of O supergiants, we would of course again like to know what energizes these pulsations. As pointed out yesterday, the ε-mechanism will not be able to work in these stars. Could the overstable convection mechanism outlined yesterday by Dr. Osaki explain these pulsations? Or do we know another mechanism to explain the observations?

BAADE: Since NRP's are seen over such a wide range in T_{eff} ($\sim$ 8000 to 42000 K) an envelope mechanism may appear less likely if the driving is the same for all these stars and down to the main sequence. As pulsations are observed in so many stars, it must be easy to excite them. Dr. Osaki's mechanism may well fulfill either criterium.

MAEDER: You have nicely shown that rotation favours the visibility of non radial pulsation as well as their physical existence. Concerning these narrow line supergiants which show evidences of disturbed profiles, would you interpret them as due to appearing non radial oscillations rapidly destroyed by interferences or by something else?

BAADE: I do not usually see absorption line profile variations which call for an interpretation by a single mode. If there are several modes with shorter periods variable amplitudes, their beating might be a reasonable alternative interpretation of the long and irregular time scales seen in these stars.

MAEDER: Do you have new evidences concerning the coupling between non radial modes?

BAADE: No, I have not. But it has been suggested (Osaki, this conference: Penrod 1986, private communication) that one would get the impression of m-commensurate periods if the periods in the corotating frame are long and not commensurate and if at the same time the star is a fast rotator. Then the observed pattern speeds are essentially equal to the rotation speed so that the commensurability is only apparent in the inertial frame.
However if my analysis of μ Cen (Baade, 1984) is right where this effect was seen for the first time but in retrograde modes, the above argument may not always hold.

LAMERS: I want to report same results about fast variations in the UV resonance lines of α Cam (O9.5 Ia) obtained with IUE during a 72 hour observing run in September 1978. After a very careful analysis and recalibration of the data to avoid instrumental effects, we found fast variations in several parts of the profile. The spectra were combined in 5 periods of 5 hours each and separated by about 16 hours.

Figure 1 shows the profiles of Si IV, C IV and N V respectively. Variations can be seen on the blue side of the absorption near -1800 km/s and on the red side near -700 km/s. (This latter absorption appears in the figure at -300 km/s for C IV and +300 km/s for N V because the plots are normalized to the laboratory wavelength of the blue component). The result of a detailed analysis of these variations are summarized:
An absorption component moved from -1670 to -1800 km/s in 26 hrs with an acceleration of $-1.4\ 10^2$ cm/s^2 and remained at that velocity during the next 46 hrs. From the acceleration we can deduce that the blob or shell which produced this component was at a distance of 5 ± 2 R_* when it was first observed and it travelled to a distance of 20 ± 3 R_* in 57 hrs. From the strength of the absorption and the distance, we derived a mass of $2.10^{-12} \leqslant M \leqslant 3.10^{-9}$ $M_\odot$ for the absorbing feature. The lower limit corresponds to the assumption that the absorption is produced by a blob of gas which happens to be in the line of sight to the stellar disc and covers the star exactly. The upper limit corresponds to the assumption that the absorption is produced by a spherical shell. If the absorption is due to a temporarily increase of the quiescent mass loss (5.10^{-6} $M_\odot$/yr) by a factor

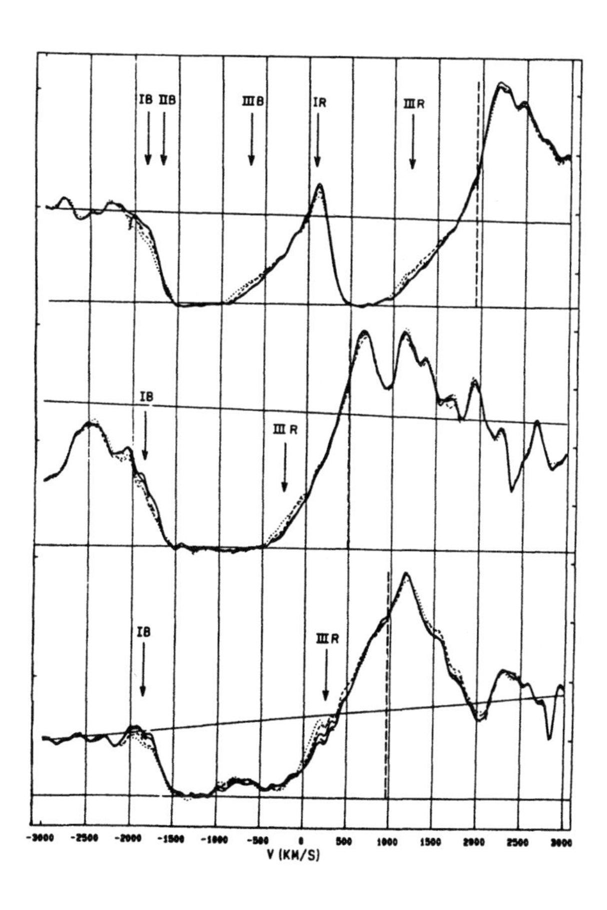

$\alpha \simeq 2$ to 10, the ejection time must have been $5/\alpha$ hrs. This short time scale agrees with the timescale of the variations in Hα, which are produced just above the photosphere.
The absorption component near -700 km/s shifts from -700 to -790 km/s in 17 hrs, which corresponds to an acceleration of -1.5 10^2 cm/s^2. This acceleration is 0.05 times slower than expected for a wind with a standard velocity law. We also found some evidence for changes in the emission part of the profiles, but these are less certain.
These observations, together with the simultaneously observed variation in Hα and in the degree of polarization, show that α Cam, and possibly most other supergiants eject blobs or puffs on timescales of hours. (A full report of this study will be published in Astron. Astrophys. by Lamers, De Jager, Langerwerf and Snow.)

MOFFAT: A potentially important check in your model of radial ejection of blobs in the wind is through linear polarization observations, simultaneously with UV spectroscopic studies of resonance lines. For example radial ejection would lead to constant polarization angle with time, but changing polarization amplitude as the scatterers in the blob separate from the star.

LAMERS: Hayes (1984, Astron. J. 89, 1219) has measured the linear polarization of our program star α Cam in 1978, which includes the period when our UV observation were made. He reported variations in the linear polarization which can be interpreted in terms of variable asymmetric outflow. We still have to combine the study of the simultaneous observations of the UV lines, the Hα profiles and the polarimetry, to trace the evidence of blob ejections in all three types of observations.

HENRICHS: It is interesting to note that in the case of μ Cen B2 IV-Ve the Hα emission showed spectacular changes on very short timescales (Peters, 1985), whereas the UV wind profiles did not show any indication of variability. This is suggestive for a latitude dependence effect: Hα and polarization are preferentially located near the equatorial plane like the NRP; the UV wind inhomogeneities/variability at moderate latitudes and just wind from the polar region. Likewise, τ Sco B 0.2 V might be a pole-on star: we have never seen any wind variability of this very narrow line star and attempts to detect NRP have failed as far as I know.

BAADE: Such a latitude dependence would be expected if NRP's cause puffs which produce narrow components. However, the occurrence of saturated narrow components in Be stars may pose a problem for this picture.

HEARN: You have observed a large number of Be stars with IRAS. Is there not one of these for which the equatorial disc lies between the star and the observer so that the disc could be observed with UV instruments.

LAMERS: Yes, these stars are probably the so-called shell-stars, which show narrow shell like absorption components in their visual spectrum. From a first study of their IR characteristics it seems that they behave similar as the Be-stars in the IRAS bands. We have not yet performed a comparitive analysis between the shell stars and the Be stars.

MAEDER: Remark to Lamers. The difference you and Rens Waters found between polar and equatorial mass loss rates is even more striking if you take into account that the equator should be significantly cooler than the poles. Of course gravity is also lower at equator.

BAADE: a) From Voyager far-UV observations, Polidan (1985, private communication) does not find a need to use a two-temperature model to fit the energy-distribution of Be stars. The effect of rotation should be small.
b) I don't know whether NRP's can explain the IRAS mass fluxes for Be stars. But both spectroscopy (Baade, 1986 and in preparation) and polarimetry (Hayes, 1985) suggest that at least some Be stars experience many small discrete mass-loss events. Their average mass loss rates hence should be larger than those extrapolated from the radiation-pressure dominated case of much more luminous stars.

APPENZELLER: A lower surface temperature at the equator of Be stars needs not to mean a lower mass loss rate from these surface regions. As pointed out several years ago by Kurucz and Schild (1976, IAU Symposium No. 70), at a certain temperature the absorption may drastically increase with decreasing temperature. As radiatively driven mass loss depends on the product of the radiation flux and the opacity, radiatively driven mass loss can under certain conditions be highest from a low temperature region of a stellar surface.

Dietrich Baade giving his review on variations in O-stars

Roberta Humphreys discussing with André Maeder
while husband Kris Davidson keeps on eye on her

Jean-Marie Vreux, Peter Conti and Jean-Paul Zahn

OBSERVED VARIATIONS IN WOLF-RAYET STARS

Jean-Marie Vreux
Institut d'Astrophysique
5, avenue de Cointe
B-4200-Cointe-Ougrée
Belgique

ABSTRACT. Observed variations in WR stars are reviewed with a special emphasis on the ones which could be intrinsic and periodic.

1. INTRODUCTION

Among the twelve brightest (v<8) Wolf Rayet stars nine are presently classified as binaries (SB2, SB1, or WR+C) and a tenth, WR 78, is known to have exhibited periodic variations at a certain epoch. Such a high frequency of binaries - or suspected binaries - among the brightest of these objects most probably explains why WR variability has been reported since the beginning of the century and also why binarity is the first explanation usually put forward to account for periodic variations. However, as we will see, in addition to this extrinsic source of variability there are also indications for the existence of an intrinsic variability.

The study of the intrinsic variability of a single WR can bring useful constraints to the theoreticians who try to model the internal structure of these objects as well as to those studying the wind itself, its origin and its dynamics. As a matter of fact, these last problems are sufficiently complicated, even in single objects, that it is presently better to keep away from the oddities in the variability introduced by the presence of a companion, like tidal effects and streams of gas flowing from one component to the other. This and the fact that the instabilities associated with the flow of matter in close binary systems are certainly not the subject of the present workshop have led me to try to avoid as much as possible the binaries in the present review.

However, we touch here a methodological problem. After the success of Pachinsky's scenario (1967) in explaining the formation of WR stars, a frantic search for binaries started, and, due to the difficulties inherent to the nature of WR

H. J. G. L. M. Lamers and C. W. H. de Loore (eds.), Instabilities in Luminous Early Type Stars, 81–98.

spectra (mainly their extremely broad lines), for some time, nearly any sign of variability has been suggested to be the signature of binarity (de Monteagudo and Sahade, 1970; Moffat and Haupt, 1974; Bisiacchi et al., 1982) Such a broad definition combined with insufficient sampling has led to percentages of binaries which presently looks rather excessive: "maybe much higher than 73%" (Moffat and Haupt, 1974), "up to 100% for the WNE" (Bisiacchi et al., 1982). Presently, with Conti's scenario (1976) we know that single WRs may be produced as the result of the evolution of massive O stars and, in a recent paper, I have questioned the interpretation in terms of binarity of the variability observed in some SB1's with a low semi-amplitude K (Vreux, 1985). This is the reason why such stars will be included in the present review. The other binaries will only be mentioned when they exhibit a variability which apparently is not linked to their binary nature i.e. is not known to be linked to the orbital phase.

I shall also try to concentrate on the variability exhibiting a periodic or quasi-periodic character. It would indeed be rather surprising that the huge amount of material ($\sim 3\ 10^{-5}\ M_{\odot}\ yr^{-1}$) leaving these objects does so like some kind of steady laminar flow; we would rather expect the existence of random fluctuations in the wind. However, due to the present state of the art of wind dynamics this kind of variability is of little use for the modelisations. Periodic or quasi-periodic variations are more likely to be usable in a near future.

2. OBSERVATIONS IN THE VISIBLE PART OF THE SPECTRUM.

2.1. Ultra short time scale periods ($P<1^{min}$)

For three star (WR 11, WR 128 and WR 134) photometric observations have been performed with high time resolution in search of very short time scale periodicities. In every case the result has been negative.

2.2. Short time scale periods ($2^{min}<P<2^{h}$)

In a recent paper, Jeffers et al.(1985) have presented new data concerning the variability of the emission line complex around λ4650 Å in the spectrum of γ^2 Vel, and they suggest that this variability as well as the previously reported periodicities are due to the presence of a neutron star orbiting inside the extended envelope of the WC 8. More recently Moffat et al. (1986) have analysed a collection of 50 spectra and they conclude that there is virtually no evidence for a third low mass body in the system. As a consequence it looks as if we were left with a normal WC 8 which sometimes exhibits a period of the order of 150^s-200^s. To check this, I have carefully searched the literature, consulting more references than the ones given by Jeffers et al. (1985) (for example the important observational effort of Haefner et al., 1977): it appears that the periodicity of 150^s-200^s has only been claimed by one group in two papers (Jeffers et al. 1974; Sanyal et al., 1974). It is possible that we are dealing with a transient phenomenon occuring only occasionally. However a comparison between the two papers claiming the periodicity and the more numerous and later ones failing to find it shows a fundamental difference in the observing procedure. In the papers claiming the periodicity, the temporal coverage (i.e. the length of a run of data points for which time is kept) never exceeds $\sim$10 min for the observations with a sufficient time resolution. On the contrary, the later (and unsuccessful) observing runs using a variety of techniques always include much longer temporal coverages. Some of them report variability, but no periodicity. The point is that it has never been shown that there is a "long" term coherence in the succession of the fluctuations reported in the "discovery" papers i.e. that we are dealing with a truly periodic phenomenon. Taking all this into account, I think it is reasonable to consider that the existence of a period of the order of 150^s-200^s in γ^2 Vel is not proven.

Other stars have been searched for periods ranging from a few minutes to a few hours. This has been done by different authors using different techniques (for more details and references, see Vreux, 1986). The results can be summarized as follows (some papers give a list of stars studied but one does not find them mentioned in the body of the paper nor in the conclusions: they are not taken into account). For the WN stars, nine stars have been searched and no period has been found. In general no variability has been reported either, even if the quality of this statement varies from star to star due to different techniques. For the WC stars, seven stars have been studied and again no period has been found in the time scale presently discussed. However variability on a time scale of minutes is observed in some stars with a tendency of being more frequent in the later subclasses. The sample being limited this conclusion has to be taken with caution.

2.3. Medium time scale periods ($0.3^d < P < 1^d$)

This is a difficult observational problem mainly if P is longer than the length of the night and close to a submultiple of a day. As a consequence the probability to find it by mere luck is small. However it has been quoted as possible for a few stars: WR 40, 71, 103, 136 and WR 138. (see Vreux, 1985, for more details and comments).

2.4. The SB1 stars with a low semi amplitude K.

In a rather speculative paper (Vreux, 1985) I have drawn the attention on some pecularities exhibited by the published periods of the WR+C, notably a strange tendency for clustering around values which can be easily related to each other (aliasing or simple harmonics). It is not easy to understand how such coincidences (generally to the level of a few percents) and such correlations between the periods could exist within a sample of genuine WR+C: we would rather expect a random distribution of periods resulting from the initial distribution of the original periods and from the violent disturbance at the moment of the explosion of the initial primary. I have suggested that for most of these stars the observed variability is due to some internal properties of the objects and is not linked for example to their likely runaway status (assumed to be the signature of a supernova explosion in a binary system). In that context it is interesting to notice that Gies (1985) has recently concluded from the study of 20 OB runaway stars that a cluster ejection model is a better interpretation than the supernova recoil to explain O runaways. I have no space here to come back in detail to the table giving the distribution of periods (see Vreux, 1986). Let me just mention that, after two years, a new star, WR 128, P=3.85 d, has to be added to group C and two stars move from one group to another: WR 123, with a new period of 2.37 d, moves from group A to group B and WR 134 with a new period of 1.81 d moves from group E to group A. No systematic search for periods less than one day has been published so far; my own tentatives have been hampered by poor weather.

Leaving the arithmetical game on the periods I would like now to present to you something new concerning these objects, which again I find easier to accept in the context of non radial pulsations than in the one of binary systems. More details will be given elsewhere (Vreux, 1986) as here I am only able to give a summary. Going through the old and most recent literature I have found that among the so called WR+C it is possible to find evidence for time dependent photometric amplitudes (the variations can be important and happen on a time scale as short as one day), time dependent radial velocity amplitudes (not marginal variations), time dependent periods (direct aliasing changes are excluded but different periods obtained by different techniques are taken into account) and, as previously mentioned, the possibility or multiperiodicity. As for the arithmetics on the periods, some of the published results I have used may be wrong due to the intrinsic difficulties of these objects. I am however confident that all these facts are not due to this cause: there are indisputable situations.

Close binary systems are known to sometimes exhibit

peculiar behaviours; however all this list of pecularities seems to me a little difficult to handle and I cannot refrain from reminding you that for example the Cepheid HR 7308 presents epoch dependent variations of its light amplitude as well as of its radial velocity amplitude.

3. OBSERVATIONS IN VARIOUS SPECTRAL RANGES.

3.1. Infrared observations.

Presently three cases of infrared variability linked to WR stars have been reported. The first one concerns a strong and variable IR source (no periodicity mentioned) the position of which was found to coïncide with a faint stellar image. According to Danks et al. (1983) a near infrared spectrum of that star reveals a heavily reddened WC 9 star. However, comparing the published spectrum with my own collection of near infrared spectra (Vreux et al., 1983 and unpublished data) I consider that spectral identification far from evident and a visible confirmation of the spectral type would be most welcome.

The second case is WR 140= HD 193793, a WC7+abs. This star has been found to go through episodes of fading and brightening between 1970 and 1985, with large time intervals between the episodes. We had to wait until the present meeting to get the clue to that mystery: Williams et al. have now shown that WR 140 is a long period binary (P=7.9 years) with a high eccentricity (e=0.7-0.8) and that the dust formation is linked to periastron passage.

The third case is WR 137=HD 192641, which is also a WC7+abs. The reported IR variations are less numerous than for WR 140. The time scales and some physical properties of the variations also differ from the ones observed in WR 140. Presently this star is not known to be a binary. However, due to the difficulties encountered in the case of the WR 140, I think we cannot consider that matter as settled.

3.2. Radio observations.

Abbott et al. (1986) have recently reviewed the radio emissions from a sample of about 40 WR stars. For many of them, repeated observations are available. The only example of radio "flaring" they have found is WR 140. From this meeting we now know (Williams et al.) that WR 140 is a binary and that most probably the radio variability can be explained in that context.

3.3. X rays observations.

In the literature I have surveyed, I have found reports on

X ray variability for only three Wolf Rayet stars: WR 6, WR 139 and WR 140. The last one is again the binary discussed by Williams et al. From visible observations WR 6 is claimed to be a binary (WR+C) with a period of 3.76 days (Firmani et al., 1979) and WR 139 is the well known eclipsing binary V444 Cyg, a WN5+O6 with a period of 4.21 days. For both stars a correlation with orbital phase is reported (Moffat et al., 1982): minimum X ray flux is observed when the WR component is in front. However, as mentioned by Moffat et al. (1982) this interpretation assumes an otherwise constant wind and is subject to confirmation: it is indeed based on only four observations, one day apart. Concerning the interpretation of the variability of WR 6 as due to the presence of a compact companion let me also point out that its mean X ray luminosity ($\sim 10^{33}$ erg s^{-1}) is comparable to that seen from single hot stars (O and other WR) and is significantly less than the one observed among massive X ray binaries.

3.4. Ultraviolet observations.

3.4.1 Emission line profiles variability. This type of variability has been observed in the spectrum of WR 40 by Smith et al. (1985). It consists in a wavelength shift of the red and blue edges of the emissions. However these shifts are markedly different (the short wavelength one is larger than the long wavelength one) and they cannot be interpreted as a simple radial velocity displacement of the whole line. This kind of variability has been observed in the emissions which are the narrowest in the spectrum and thus which probably arise from highly ionized species, close to the star.

3.4.2 Transient components. A transient component of low velocity ($\sim$ -150 km s^{-1}; time scale $\sim$ 5 hours) has been observed by Gry et al. (1984) in the blue wing of Lyman δ in the spectrum of γ^2 Vel=WR 11, a WC8+O9. However it is not proven that this "puff" is not due to the O9 component of the binary. In some way similar to the puffs are the narrow components observed in many hot stars (Henrichs, 1984). No such component has ever been reported for WR stars most probably because the absorption component of the P Cygni profiles of resonance lines are generally saturated in WR spectra (Henrichs, 1986).

3.4.3 P Cygni profile variability. This kind of variability has been observed in the spectra of WR 6 (Willis et al., 1986), WR 14 (Bromage et al.,1982), WR 40 (Smith et al.1982), WR 48 (Beeckmans et al. 1982) and WR 140 (Williams et al., 1987). WR 140 is the binary already mentioned about IR, radio and X ray variability. WR 48 is probably a triple system and the observed line profiles could even be mainly due to the O star. WR 14 has been claimed to be a WR+C on the basis of the variation observed in a P Cygni line profile but I have not been able to find a later confirmation of that assertion. The two best documented examples are WR 40 and WR 6. In the case of WR 40 the most different profiles of Si IV λ1400 and C IV λ 1550 differ by a large Doppler shift of the violet edge of the absorption component of the P Cygni profile, a small Doppler shift of the red edge of the absorption component and of the blue edge of the emission component and no variation of the red edge of the emission component. No correlation has been found between these variations and any of the suggested photometric periods (visible). In the case of WR 6 the main part of the variation is again observed in the absorption component of a P Cygni profile (N IV λ1718). However for this star the violet edge is not modified: the variation is located closer to the center of the absorption. No correlation has been found with the well known photometric period (visible) of that star; the variability of the P Cygni profile instead appears to occur on a time scale of a few hours.

3.4.4 Numerical simulations of P Cygni line profile variations. Leaving for a while the WR stars, let us remember that in the case of ζ Oph the observed non radial oscillations imply velocities of the order of 21 km s^{-1} with periods in the range of 7 to 15 hours (Vogt and Penrod, 1983). This means that the base of the wind moves radialy with a speed comparable to the local soundspeed. Due to the importance of the location of the sonic point (or its equivalent) in the formation of a supersonic wind it seems reasonable to expect some kind of modulation of the wind in the case of pulsations with that amplitude. Another coïncidence points into that direction: the velocity gradient inside a wave (i.e. between successive layers in the model) can be quite comparable to the one observed at the base of the wind using Abbott's law (Vreux, 1986). This velocity gradient is essentially due to the density gradient in the last layers of the model.

Without waiting for a proper handling this difficult dynamical problem, R. Scuflaire and I have decided to assume that such a situation would in some way lead to a modulation of the wind shaped by the pulsations pattern. Preliminary results are given in this book and will be published elsewhere (Scuflaire and Vreux, 1986)

4. TWO PECULIAR OBJECTS.

In this review on observed WR variability I feel that I cannot avoid mentioning WR 103=HD 164270 which exhibited a gradual fading of over a magnitude during the course of 20 nights in June 1980. Evidence of a previous fading in

September 1909 has been found later on. Different hypothesis have been suggested to explain this behaviour, all of them implying the existence of a companion (Massey et al., 1984) However, how interesting is the solution of these still somewhat mysterious fadings, I feel that the last object I now want to speak about is more important in the frame of the present workshop. It is WR78=HD 151932, a WN7. It has been observed during 8 consecutive nights in July/August 1971 by Seggewiss (1974, 1977). In that star the violet displaced absorption edges of He I lines are clearly visible. Furthermore they are often split into two or even three components, some of them with highly variable velocities. The line He I λ3889 is always split into two components, a short wavelength one of constant velocity ($\sim$ -1250 km s^{-1}) and a longer wavelength variable one. When the radial velocity of the latter is plotted as a function of time, it varies along with the He Iλ4471 and 5876 absorptions in the form of ascending branches, rising from about $\sim$ -600km s^{-1} to nearly the speed of the stationary component i.e. $\sim$-1250 km s^{-1}. The successive branches are separated by about 3.5 days. This behaviour is epoch dependent (Seggewiss and Moffat, 1979) and WR 78 has never been claimed to be a binary. Arguments against a binary interpretation of this phenomenon (in addition to the fact that it is epoch dependent) are also given in Seggewiss (1977). These observations are interpreted as pseudo periodically starting clumps of higher density causing the beginning of a new ascending branch. It may be (again) a pure coïncidence: the pseudo period of about 3.5 days is precisely equal to two times the period of group A in the table of WR +C (Vreux, 1985). In some way these ascending branches are reminiscent of the pulsating β Cepheï star σ Sco. In that star the radial velocity of the C IV line never becomes positive which means that the C IV absorbing layers are not seen falling down, they flow out constantly as a successive series of separate ejection (Burger et al., 1982).

5. LINK WITH THE THEORETICAL MODELS.

In different parts of this review I have raised the question of the possible existence of non radial pulsations as an explanation of some of the observed variabilities: about the peculiar behaviours of the observed periodicities in the binaries with a low K, about the numerical simulation of P Cygni profile variations and about the cause of the starting of the ascending branches observed in WR 78. The main problem with that suggestion has always been that, according to the theory, non radial pulsations are not very probable in WR stars, at least in He burning stars (Maeder, 1985). Quite recently some relief came from Noels and Scuflaire (1986) who have shown that unstable non radial modes can develop in some models of WR stars i.e. the ones which still have a H burning shell. In that context it is worth looking again to the candidates. If we forget everything about the SB1 with a low K except the spectral types we face the fact that the sample includes mainly late type WN (WN 6-7-8). This is also true for WR 78, the non WR+C candidate. If we now turn to the work of Conti et al. (1983) we find that the highest percentage of detection of H in WR spectra is among the late WN. And, interestingly, the presently two best candidates, WR 40 (2 periods simultaneously) and WR 78 (ascending branches) are among the WN exhibiting the highest H/He ratio. The theoretical necessity of a H burning shell for the existence of unstable non radial modes and the observational fact that the suggested candidates belong the subclasses exhibiting the highest percentage of H may be another mere coïncidence. Nevertheless I have the feeling that there are too many coïncidences in that story and it seems to me that this hypothesis deserves further investigations.

Among the unsettled questions concerning that hypothesis, one concerns the fact that the observed periods are always longer than the predicted ones. There is presently a hope of a way out in that the new theoretical predictions of Noels and Scuflaire (1986) are rather close of the shortest reported periods which are of the order of a fraction of a day. Beating is another way out: it has been observed to occur between different modes in non radial pulsators. The remaining question is then: why do we observe only the beating periods and not the fundamental ones. A solution may lie in the fact that in most (all?) the WR star we do not see the "surface" of the star. We are observing the wind blowing out of these objects, and, depending to the wavelength, we are looking at different depth in the wind. This is true for the lines (and the different parts of the line), this is also true for the continuum. We can imagine that the wind acts as a low frequency filter. There are two ways to do it. One could be called a dynamical filter. It is possible that the wind is not able to transmit everywhere any kind of frequency: from the spectrum of frequencies available at the "surface" of the star the wind may be only able to carry the ones lying in a certain domain, with the possibility that the passband is function of the distance of the star. The other filtering process is less speculative: it is what I could call an observational filter. This filter is operating for the line profile variations for example. It is due to the fact that each point of a line profile is the result of an integration on a line of constant projected velocity. To be seen, a perturbation has to occupy a sizeable fraction of such an integration line, at least of the order of R_*. Due to the particular shape of the curve of constant projected velocity this implies a perturbation which has a duration at least of the order of R_* / v_∞ Perturbations with higher frequencies have a signature which is probably too small to be extracted from the noise of the observed profiles.

6. CONCLUSION

I think that there is some hope to link some of the variabilities observed in WR stars with pulsations. This is a potentially powerful tool. However it looks as if we were facing a very difficult challenge. From the presently available data on WR variability (and also from what we have heard about the Of stars) it seems that the observed periodicities may not only be epoch dependent but also depth dependent. Consequently, numerous and carefully planned observations are the only hope for progress. This most probably implies large scale collaborations. The most promising objects are the late WNs with the highest H/He ratios. Logically the most appealing are the WN+abs because these stars are generally thought to be single, and to have a very thin atmosphere i.e. to be the only WR in which we have a chance to see the star itself. Let's hope that their very thin wind does not mean that they are particularly quiet...

Acknowledgements: I would like to express my thanks to all those who have kindly answered my call for preprints and have allowed me to present you with an up to date review of the subject. I am specially grateful to Tony Moffat and his collaborators.

Tony Moffat and André Maeder

REFERENCES.

Abbott, D., 1986, to appear in NASA/CERN Monograph Series on *Nonthermal Phenomena in Stellar Atmospheres* volume *O, Of and WR Stars*, eds.: P. Conti and A. Underhill.
Abbott, D., Bieging, J., Churchwell, E., and Torres, A., 1986, Astrophys. J., 303, 239.
Bisiacchi, G., Firmani, C., and de Lara, E., 1982, in IAU Symp. 99, *WR Stars: Observations, Physics, Evolution*, eds.: C. de Loore and A. Willis, Reidel, p. 583.
Beeckmans, F., Grady, C.A., Machetto, F., and van der Hucht, K.A., 1982, in IAU Symp. 99, *WR Stars: Observations, Physics, Evolution*, eds.: C. de Loore and A. Willis, Reidel, p. 311.
Bromage, G.,Burton, W., van der Hucht, K., Macchetto, F., and Wu, C., 1982, in *Proc. 3th. Europ. IUE Conf.*, ESA-SP-176, p. 269.
Burger, M., de Jager, C., and van den Oord, G., 1982, Astron. Astrophys., 109, 289.
Conti, P., 1976, Mem. Soc. Roy. Sci. Liège, Serie 6, 9, 192.
Conti, P., Leep, M., and Perry, D., 1983, Astrophys. J., 268, 228.
Danks, A., Dennefeld, M., Wamsteker, W., and Shaver, P., 1983, Astron. Astrophys., 118, 301.
de Monteagudo, V., and Sahade, J., 1970, Observatory, 90, 198.
Firmani, C., Koenigsberger, G., Bisiacchi, G., Moffat, A., and Issertedt, J., 1980, Astrophys. J., 239, 607.
Gies, 1985, Ph. D. Thesis, University of Toronto.
Gry, C., Lamers, H., and Vidal-Madjar, A., 1984, Astron. Astrophys., 137, 29.
Haefner, R., Metz, K., Schoembs, R., 1977, Astron. Astrophys., 55, 5.
Henrichs, H.F., 1984, in *Proc. 4th. Europ. IUE Conf.*, Rome, ESA-SP 218, p. 43.
Henrichs, H.F., 1986, private communication.
Jeffers, S., Stiff, T., and Weller, W.G., 1985, Astron. J., 90, 1852.
Maeder, A., 1985, Astron. Astrophys., 147, 300.
Massey, P., Lundström, I., Stenholm, B., 1984, Publ. Astron. Soc. Pac., 96, 618.
Moffat, A., and Haupt, W., 1974, Astron. Astrophys., 32, 435.
Moffat, A., Vogt, N., Paquin, G., Lamontagne, R., and Barrera, L., 1986, preprint.
Noels, A., and Scuflaire, R., 1986, Astron. Astrophys., in press.
Paczynski, B., 1967, Acta Astron., 20, 47.

Sanyal, A., Jeffers, S., and Weller, W., 1974, Astrophys. J. Lett., 187, L 31.
Scuflaire, R., and Vreux, J.-M., 1986, submitted Astron. Astrophys.
Seggewiss, W., 1974, Publ. Astron. Soc. Pac., 86, 670.
Seggewiss, W., 1977, in IAU Coll. 42, *The Interaction of Variable Stars with their Environment*, eds.: R. Kippenhahn, J. Rahe and W. Strohmeier, p. 633.
Seggewiss, W., and Moffat, A., 1979, Astron. Astrophys., 72, 332.
Smith, L., Lloyd, C., and Walker, E., 1985, Astron. Astrophys. 146, 307.
Vogt, S., and Penrod, G., 1983, Astrophys. J., 275, 661.
Vreux, J.-M., 1985, Publ. Astron. Soc. Pac., 97, 274.
Vreux, J.-M., 1986, in preparation.
Vreux, J.-M., Dennefeld, M., and Andrillat, Y., 1983, Astron. Astrophys. Suppl. Ser., 54, 437.
Williams, P., Longmore, A., van der Hucht, K., Talevera, A., Wamsteker, W., Abbott, D., and Telesco, C., 1985, Mont. Not. Roy. Astr. Soc., 215, 23.
Williams, P., van der Hucht, K., Wamsteker, W., Geballe, T., Garmany, C., Pollock, A., 1987, in *Instabilities in Luminous Early Type Stars*, eds.: C. de Loore and H. Lamers, Reidel.
Willis, A., Howarth, I., Conti, P., and Garmany, C., 1986, in IAU Symp. 116, *Luminous Stars and Association in Galaxies*, eds.: de Loore, Willis and Laskarides, Reidel.

Ed van den Heuvel discussing Wolf-Rayet binary stars

DISCUSSION

V.D. HEUVEL: One way to distinguish a (WR + compact) system is: by looking whether it is a runaway object. Which fraction of the systems that show the periods of order of days are indeed runaways?

VREUX: Except for one, WR78, all the stars I have discussed are runaways. The point is that they have been searched for small semi-amplitude variations because they were runaways! It has not been done at random, presently.

V.D. HEUVEL: To settle the matter whether or not they are binaries: Should not one do the same search for periods among the late type WN stars that are not runaways, i.e.: that are very close to the galactic plane?

VREUX: I perfectly agree that the sample should be increased and that as a priority all the late type WNs should be searched for the kind of variability I have described. This is under way for photometry. However we already know that the amplitude of the variations is epoch dependent. So it will take some time before we get the answer to your question and this for photometry only, which I am convinced is a less powerfull tool than spectroscopy for that kind of problem.

MOFFAT: Lamontagne, Moffat and Seggewiss did look for variations in a sample of WNL stars selected without bias. Stars of largest separation + radial velocity tend to be the runaways.

LAMERS: The work of Seggewiss (1971) shows a series of moving absorption components in a velocity versus time diagram of WR78. Exactly the same behaviour was found in P Cygni. The difference is in velocity (-200 km/s for P Cygni and -1300 km/s for WR78) and in periodicity ($P \simeq 200$ days for P Cyg and $P \simeq 2$ days for WR78). It may be that we are observing the same mechanism in P Cygni, which is supposed to be an immediate progenitor of a WR star, and in the WR stars themselves.

CASSATELLA: On the basis of the available observations, you have suggested the existence of allowed pulsation periods in WR stars. I have two questions in this respect: a) do you think the sample of stars used is statistically significant, and b) is there any evidence for the existence of allowed pulsation periods in other luminous hot stars?

VREUX: a) I perfectly agree that the sample is limited. The point is that I have used all the data available in the litterature, I mean all the observations which have been handled in such a way that semi amplitudes of the order of 20 km s^{-1} could be found. During the next few years we will try to increase that sample. It has already begun but it takes a lot of time. For example since my initial paper (PASP) only one new WR + C has been claimed and it fits the pattern. Personally I am presently more engaged in the checking of some of the data through new observations. b) I don't think so.

STICKLAND: Puffs are not just a modern phenomenon. Hiltner found an excellent example in CQ Cephei in the 1940s where the HeI line was seen to accelerate away and to a reach a constant velocity of recession the next day.

VREUX: I do not know about CQ Cephei but narrow absorption components moving blueward have been reported in an even older paper by Perrine around 1920. During many years they were considered as the signature of episodic ejections of material. Later on it has been shown that this phenomenon was connected to the phase and is now interpreted as the signature of a stream of gas flowing between two components. The variation is due to geometrical effects. The star I have shown to you is a narrow line WR, it means radial velocity measurements more accurate than usual, and it has never been claimed to be a binary.

DE JAGER: When dealing with 'puffs' or other stellar ejecta one normally assumes the wind velocity law on which a density disturbance is superimposed. But are we allowed to assume the same velocity for the disturbance as for the wind? Are there observational indications for or against that assumption.

VREUX: If the wind is optically thin in the driving lines, it can handle the perturbations and we keep the same velocity law. If it is not, which is the best guess in most of the cases, then we most probably get a different velocity law, may be like the one described by Lucy. We are aware of it but for the moment it is a necessary simplifying assumption to be able to handle the computations in a situation where spherical symetry has been removed.

HENRICHS: ξ Per is the only case among the O stars where we were able to measure the acceleration of a high-density layer expanding in the stellar wind (Prinja et al in preparation). The result is that it can be described by a velocity law $\beta = 2$ rather than $\beta = 1/2$, i.e. a much slower acceleration than the ambient wind. This does not seem to be unreasonable.
The graph you showed containing the recurrent radial velocity variation in WR78 does seem to indicate streams, rather than shells. Is this comparable to the case of P Cygni?

LAMERS: Observationally the variations seen in WR78 and P Cygni are very similar in nature: both have more or less periodical narrow components, whose velocity increases with time, and a stable component possibly due to a distant shell. The differences are in the velocities (larger for WR78) and the timescales (larger for P Cygni).

MOFFAT: Lamontagne (1983, Ph.D. thesis Univ. de Montreal) has recently (1980-82) monitored WR136 over some 2.5 years for radial velocity changes. On the bases of mean velocities of the 5 HeII Pickering lines (4100, 4201, 4338, 4542, 4860Å) on 59 photographic spectra no evidence is found for <u>any</u> periodic variation over short (hours) or long (years) time scales above $K \simeq 10$ km s^{-1}. Thus, compared with the Vreux et al. $K = 32$ km s^{-1} for this star from subsequent observations we conclude that either the star shows epoch-dependent variations or someone's data is of questionable quality.

<u>VREUX</u>: Independently of the fact that these stars seem to exhibit a variability which is epoch dependent, it is also evident from the published literature that lines formed at different depths can exhibit a different variability (phase lag effects or absence of periodicity). So the influence of taking the mean of many lines should be carefully checked. As some of them are blended the measuring technique is also important. As far as the quality of the data is concerned we have used a fully automatic method taking into account the whole real profiles in order to remove as much as possible the human influence. We have also used a line (NIV) that you have yourself advocated te be a good indicator for other stars. We do not see what we can do to do it better. When we found that period we were reluctant about it: we did not like such a short period. We have searched a reason for it and were not able to find one, so we concluded that its most probable origin is the star itself. There are other sources of error than the quality of the data when one looks for variations in periodicities. Careful and homogeneous data handling is also important. May I remind you that using your own published numbers, L. Smith (1985) was not able to find any of your claimed periods but has shown that another period was present. In that context it would be useful to have somebody handling with the same technique all the published numbers.

<u>MOFFAT</u>: It is not obvious to me what the physical mechanism is to explain why the period for one star should be related (in my view through numerical games) to that in another star. To take the example of the 19 known ZZ Ceti stars, white dwarfs in a narrow instability strip, known to be non-radial pulsators, there exist no such correlations from one star to another although their spectra are much more alike than the WR stars. White dwarfs are also 'boring' in that they lack fusion processes and have negligible winds compared to WR stars.

<u>VREUX</u>: This is again the question of statistics. I again agree that the sample is dangerously limited. The point is (again) that I have used all the data available in the literature, I mean all the observations which have been handled in such a way that semi amplitudes of the order of 20 km s^{-1} could be found. In that sample my point is that it looks like that there are too many coïncidences. If it means something, it seems to me that there is a high degree of homogeneity of some of the properties of the observed objects. What is precisely that property, what is the physical mechanism involved, I do not know. I have guessed that it could be the pulsational properties of most of these objects. It was just a guess. During my talk I have given what I consider as new arguments for it, independent of that numerical game.

<u>DE GROOT</u>: Two remarks. First, we should really think in terms of shells,

not of streams in a binary, in the viewgraph you showed about HD 151932. In P Cygni we see very much the same thing on a different time-scale. Of course, we do not yet know how to scale this according to the differences in mass, temperature and other relevant parameters.
Second, I don't claim to answer De Jager's question, but we should remember that in the case of the Sun we have one obviously important ingredient - magnetic fiels. Until we receive more than theoretical ideas about magnetic fields in hot stars we should not equate the solar situation with the stellar.

OWOCKI: 1) Solar mass ejections observed with white light coronagraphs do indeed apparently propagate relative to ambient wind speed expected in the low corona; they are not directly observed beyond a few solar radii, however. By the time they reach interplanetary space, they are observed in situ to move near the ambient solar wind speed, so some coalesence must have occured in between.
2) These solar mass ejections are now believed to be generally driven by magnetic fields reconnection. I agree with the previous comment that we have little idea whether the solar case is applicable to those early type stars, since much of the physics is so different (radiative driving, etc.).

NOËLS: Am I right in thinking that you discard WC stars for variability and that your best candidates are late WN stars?
This has a strong implication on the kind of modes responsible for the pulsation. In a WC star or a WN star without any H, there can be radial pulsations amplified by the ε-mechanism, with periods of about 1/4 to 1/2 hour. In late-type WN stars, there can only be non radial low l, low order g^+ modes amplified by the same mechaninism, with periods in agreement with the few hours you find, and high l g^+ modes with shorter periods, trapped in the H shell burning region.

VREUX: The fact is that most of the data concern late WN. As I think that, at this point, we can only look for a general pattern, it means that the late WN are the best candidates. We cannot exclude mere coincidence or an error due to the handling of noisy data. For example we have only one WC in the sample: so I do not like to speak about WC.
Of course I feel it extremely encouraging that the most probable candidates i.e. the late WN are precisely the ones for which theory predicts periods which are of the same order of magnitude (hours) as the ones we have some hope to be able to observe.

MOFFAT: A summary of very recent photometric and polarimetric monitoring of significant samples of WR stars, updating the review by Vreux, is presented. Details are given elsewhere in the poster sessions.

CONTI: 1) HD 197406 is certainly a binary but I am not so sure it is at 800 pc, if one adopts a fainter M_V within the WN7 range, its distance would be smaller.
2) I suppose the system velocity is well determined and this is a runaway on this basis.

MOFFAT: 1) We answer this question in Drissen et al. (1986, Ap.J. in press). When M_V is assumed to be say -5 instead of -6.5, the star will come in 2x closer ($Z \simeq 400$ pc). This is certainly less extreme. However, the peculiar radial velocity increases as one brings the star in so that it is a runaway no matter what value one takes for M_V, at least as a pop. I object.
2) The systemic velocity is based on the relatively narrow weak emission line of NIV 4058 Å. This line usually gives a good estimate of the γ-velocity, all the more so when the line is narrow as is the case for HD 197406.

WILLIAMS: In addition to the three long term infrared variables mentioned in the review, we may be able to add WR70. Pik Sin Thé, v.d. Hucht and I have observed this to have faded since first observed about 10 years ago. Like WR 140 and WR 137 it has a WC + absorption spectral classification. We have continued to observe Danks' star WR 48a which is still fading in the infrared.
Only in the case of WR 140 have two certain (and one probable) infrared outbursts been observed. Calling the variation periodic might be premature but, if we phase the absorption line radial velocities against the infrared period, we see there is the same period variation in velocity in 60 years of data. The modulation of the X-ray and Radio emission by the binary motion is shown in a poster.

MAEDER: Regarding the origin of the high observed mass loss rates of WR stars, there is a large debate in litterature and in this meeting as to whether this high $\dot{M}$ is due to pulsations or to radiative effects. What I want to emphasize is that both pulsations and radiative effects are highly releated to the very large L/M ratio of WR stars. Let us remember that both theoretically and observationally it was found that the WR stars obey a mass-luminosity characterized by a very large L/M ratio. Thus, whether the high mass loss is driven by pulsations or radiation, I think the basic physical reason for the sustained high mass loss rates of WR stars is the very high L/M ratio.

NOËLS: I would like to call your attention on the importance of the mass loss rate adopted to compute the WR models, in the problem of the vibrational instability of WR stars towards radial oscillations. Such an instability can only occur if the star is massive enough, more massive than the critical mass for He burning stars, and if it is nearly homogeneous, which can only happen if mass loss has ejected enough hydrogen rich material.So, late type WN stars are very bad candidates for a vibrational instability towards radial oscillations. The situation is much more favourable in massive early type WN stars.WC stars can be unstable if they are still massive enough and if there is not a too steep gradient in the mean molecular weight. We see that some WR stars are likely to be stable, some are likely to be unstable. The effect of such an instability on the mass loss rate is not known. It certainly plays an important role, maybe enhancing it during the unstable phase but it cannot be thought of as the first cause of the mass loss itself, in the WR stars.

Ready for the bicycle excursion to the Kröller-Müller museum with the van Gogh paintings: André Maeder and Peter Conti

Cees de Jager and Françoise Praderie

VARIATIONS IN LUMINOUS BLUE VARIABLES

Henny J.G.L.M. Lamers
SRON Laboratory for Space Research
Beneluxlaan 21
3527 HS Utrecht, The Netherlands
and
Sonnenborgh Observatory
Zonnenburg 2
3512 NL Utrecht, The Netherlands

ABSTRACT. The photometric and spectroscopic variability of Luminous Blue Variables is discussed and compared with predictions. The photometric variations occur on all timescales from weeks to centuries. There are three kinds of photometric variations: large variations with $\Delta V \gtrsim 3^m$, due to large eruptions; moderate variations of $\Delta V \sim 1^m$, due to shell ejections; and microvariations with $\Delta V \sim 0.1^m$, probably due to pulsation. <u>If</u> the large eruptions are recurrent, the time of recurrence is of the order of centuries. The moderate variations occur on timescales of months to decades. During shell ejections the bolometric luminosity remains constant but the temperature of the pseudo-photosphere decreases and the radius increases. The observed variations of V, B-V and T_{eff} of S Dor and AG Car agree reasonably well with the predicted effects of variable mass loss. The semiperiods of the microvariations are about twice as large as for normal supergiants of the same L and T_{eff} and 4 to 20 times larger than the fundamental mode for radial pulsation.

There are three kinds of spectral variations: changes of the spectral type; large variations of the shapes of the line profiles; small variations of absorption components superimposed on the overall line profiles. The first two variations are due to large variations in the mass loss, and are accompanied by moderate photometric variations of $\Delta V \sim 1^m$. After an increased mass loss, the spectrum often shows forbidden emission lines formed in the ejected shell. The small variations of the absorption components of P Cygni indicate recurrent ejections of shells or blobs on a timescale of about 2 months with a shell mass of about 10^{-5} $M_\odot$.

I. INTRODUCTION

The study of Luminous Blue Variables (LBV's) has gained a considerable momentum in the last decade when it was realized that they play a critical role in the evolution of the most massive stars. Their high mass loss rate may explain why stars with initial masses of $M_i \geqslant 50$ $M_\odot$ do not evolve into red supergiants, but turn into Wolf-Rayet stars after having

H. J. G. L. M. Lamers and C. W. H. de Loore (eds.), Instabilities in Luminous Early Type Stars, 99–126.

been blue supergiants and LBV's for some 10^4 to 10^5 years.

Apart from their significance for stellar evolution, the study of LBV's is of interest because their atmospheres seem to show all the strange and fascinating effects of normal early type stars, but on a much larger or more violent scale: mass loss, shell ejections, photometric variations, line profile variations, pulsations and variable polarization. It is clear that the LBV's are less stable than normal stars so that any internal instability or disturbance has a much more severe effect on their atmosphere and the observable characteristics. LBV's are very extrovert stars! Therefore the physical processes can be studied more clearly than in normal stars.

There is one characteristic which the LBV's do not share with their more introvert early type relatives: they can change their spectral type drastically over a period of years or decades. When they get more excited and lose more mass, they disguise themselves by means of a thick pseudo-photosphere as stars of later spectral type and become visually brighter. Spectral type changes from early-B to A and F or from O to A have been observed.

In this review I will describe the general characteristics of the LBV's (Section II), their photometric variations from nova-like brightening by large eruptions to microvariations (Sections III to VI), and their spectroscopic variations (Section VII).

II. LUMINOUS BLUE VARIABLES

The name Luminous Blue Variables was suggested by Conti (1984) to indicate a group of stars which was known by different names, but shows very similar characteristics in terms of spectral type, luminosity and variability:

absolute visual magnitude: $M_V \lesssim -7$
spectral type : O, B, A
photometric variations : $\Delta V \gtrsim 0.2$.

Within this definition we combine groups of stars such as the Hubble Sandage variables in M31 and M33, the S Doradus variables in the LMC and the P Cygni variables in our galaxy. The most famous stars in this group are η Car, and P Cygni. These two stars have shown very drastic photometric variations due to outbursts in 1837, and 1600 respectively. Not all the LBV's have shown such dramatic outbursts in the recorded history: most of them are irregular variables on a smaller scale.

The photometric variability of the LBV's is often accompanied by spectroscopic variability: the spectral types may change from O to late-A or even F and reverse. Therefore the restriction on spectral class of the LBV's is not a strict one.

The known LBV's are listed in Table 1, together with some of their characteristics. In this table I also indicate which stars belong to the class of P Cygni Type (PCT) stars, as defined by Lamers (1986): "O B A stars with $M_V \lesssim -7$ which in their visual show or have shown P Cygni profiles of not only the strongest Balmer and HeI lines (Hα, Hβ, Hγ, HeI 6678, HeI 5875)." It is clear that most of the LBV's are also P Cygni Type stars. It is even possible that all the LBV's are P Cygni type

TABLE 1 LUMINOUS BLUE VARIABLES

System	HD	Name	Type or T_{BB}	M_{bol}	Class	Ref.
GAL	193237	P Cyg	B1 Ia$^+$	-9.9	PCT	1
	94910	AG Car	B0 I - A1 I	-7.8	PCT	2
	90177	HR Car	B2eq	--	PCT	3
	--	η Car	O	-11.9		4
LMC	268835	R66	B7 - Aeq	-8.9	PCT	5
	269006	R71	B2.5 Ieq - A1 Ieq	-10.6	S Dor, PCT	6
	269128	R81	B2.5 eq	-10.0	PCT	7
	35343	R88 = S Dor	B8 Ieq - A5eq	-10.6	S Dor, PCT	8
	269859	R127	OIa fpe - B5	-10.6	S Dor	9
M31	--	AF And = Var 19	25000	-10.9	HS, PCT	10,11
	--	AE And	15000	-8.3	HS, PCT	10,11
	--	Var A-1	25000 - 30000	-11.0	HS, PCT	10,11
	--	Var 15	8300	-8.4	HS	11
M33	--	Var A	22900	--	HS	11
	--	Var B	27400	-10.3	HS, PCT	10,11
	--	Var C	--	--	HS, PCT	10,11
	--	Var 2	25000	-10.8	HS	11
	--	Var 83	37000	-11.7	HS	11

References: 1 = de Groot (1969); 2 = Wolf and Stahl (1982); 3 = Bond and Landolt (1970); 4 = Viotti (1969); 5 = Stahl et al. (1983b); 6 = Wolf et al. (1981b); 7 = Wolf et al. (1981a); 8 = Leitherer et al. (1986); 9 = Stahl et al. (1983a); 10 = Kenyon and Gallagher (1985); 11 = Humphreys et al. (1984).

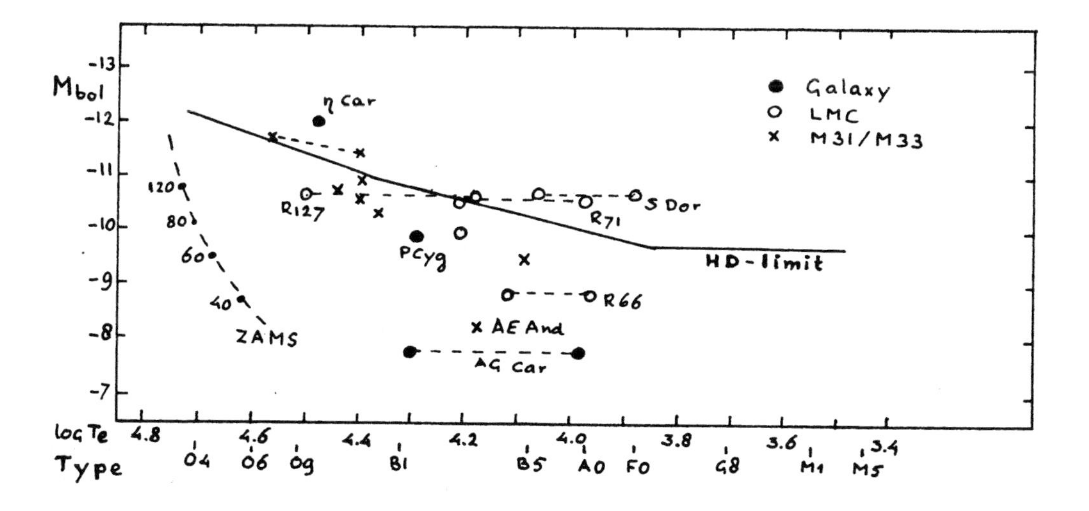

Figure 1. The location of the LBV's in the HR Diagram. The location of main sequence and the Humphreys-Davidson limit is shown.

stars and that the two classes are identical. This is not sure at present, since there are some LBV's which have not (yet?) shown P Cygni profiles in their visual spectrum, e.g. η Car.

The location of the LBV's in the HR diagram is shown in Figure 1. Notice that the LBV's are all close to the Humphreys-Davidson luminosity upper limit. This upper limit reflects the location at which the almost horizontal evolutionary tracks return to the left after the post-main sequence expansion, due to a high mass loss rate (Maeder, 1983). Therefore it is very likely that the LBV's represent a phase in the evolution of the massive stars, $M_i \geqslant 50\ M_\odot$ at which the stars become unstable and lose so much matter that the expansion is stopped and the tracks eventually reverse when the Helium core contains 70 per cent of the stellar mass (see Maeder, 1983; Lamers, 1986).

III. PHOTOMETRIC VARIATIONS IN LBV'S

a. Brief review

Luminous Blue Variables are photometrically variable on a very wide range of timescales; from weeks to centuries, and with a large range of amplitudes: from $\Delta V \simeq 0.1$ to $\Delta V \gtrsim 3^m$. The larger the variations, the longer is the timescale at which these variations occur. These types of variations will be discussed in more detail below. Here I present a brief review.

i. <u>The large variations with $\Delta V \gtrsim 3^m$</u> are associated with "eruptions", i.e. sudden ejection of a large amount of mass. For example, the outbursts of η Car in 1837-1860 have produced the presently observed nebulosity around this star. Such large eruptions have not been observed in all LBV's. This can be due to two effects: either not all the LBV's suffer large outbursts, or the timescale between large outbursts is so long (e.g. centuries) that they have not been observed <u>yet</u> in all LBV's. The discovery of ring nebulae around a number of LBV's for which no large eruptions have been observed in the recorded history shows that these stars must have suffered outbursts in the past, thus supporting the second possibility. Examples of such nebulae are AG Car (Thackeray, 1956) and R 127 (Appenzeller et al.; these proceedings). The lack of nebulae around many other LBV's cannot be considered as evidence that these stars did not have large outbursts, because there are counter examples. For instance, the large outburst of P Cygni in AD 1600 did not leave a visible remnant nebula. So I conclude that all LBV's <u>may</u> have suffered large outbursts.

ii. <u>The moderate variations of $\Delta V \simeq 1^m$</u> have been observed in almost all LBV's. In fact many of the LBV's are classified as such because of these variations. The timescale for such variations seems to be on the order of decades. For instance the stars S Dor and AG Car have gone through such active phases during the last 50 years. These will be discussed in more detail below. The LBV's do not show such variations continuously; some of them go through quiet phases which can last as much as a century. For example, the star P Cygni is stable within $\Delta V \simeq 0.2$ since 1870. The moderate variations are also associated with mass ejection. This is clearly demonstrated by S Dor and AG Car which reached their visual maxima in 1983 and 1981 respectively (Leitherer et al. 1986; Whitelock et al. 1983). Both stars showed evidence for increased mass loss in their IR energy distribution and UV line strengths at the time of their visual maximum.

iii. The small scale photometric variations with $\Delta V \lesssim 0.1$, which are called "micro-variations" by van Genderen, occur on timescales of weeks to months. They have been found in all LBV's which have been observed sufficiently accurately. In fact, they have been found in many luminous normal supergiants as well (van Genderen, 1985; van Genderen, 1986a). This kind of variation is probably due to some pulsational instability of the star. It may be that these variations also produce mass loss, but the evidence for this is not convincing. For instance, the recurrent ejection of shells by P Cygni occurs on about the same timescale (~ 2 months) as the photometric variations, but the existence of a correlation between maximum brightness and shell ejection has not been proven yet (van Gent and Lamers, 1986).

b. The nature of the photometric variations

A major breakthrough in our understanding of the photometric variability of LBV's was achieved by the investigations by the Heidelberg group which showed that the bolometric magnitude of the LBV's remained constant during periods of drastic changes in the visual magnitude. For instance from a study of the visual and UV energy distribution of the LMC star R71 during minimum and maximum visual brightness Wolf et al. (1981) and Appenzeller and Wolf (1982) showed that M_{bol} remained almost constant while V varied by a few magnitudes. A similar conclusion was reached for AG Car by Viotti et al. (1984) who also found that M_{bol} is constant during strong visual variations.

These results show that the variations in m_V are due to changes in the temperature of the photosphere which must be due to changes in the radius of the photosphere in such a way that the product $R^2 T_{eff}^4$ remains constant. Another way of expressing the same conclusion is: the variations in m_V are due to variations in the bolometric correction BC in such a way that

$$V + BC = \text{constant} \quad \text{and} \quad \Delta V = -\Delta BC. \tag{1}$$

Although this relation is only found for the medium size variations of $\Delta m_V \simeq 1$ to 2^m, of a few LBV's only, it is generally accepted to be valid also for the large variations with $\Delta m_V \simeq 3^m$ of LBV's. However, Davidson (these proceedings) has argued that during the giant eruptions of η Car the luminosity of the star was increased by more than one magnitude. (See Viotti's poster paper for counter arguments). It is not clear at present that the small variations with $\Delta m_V \simeq 0.1$ also occur at constant M_{bol}. The anticorrelation in the variations of V and U-B, found by van Genderen (1985), shows that the temperature decreases when the visual brightness increases. Whether this implies a constant M_{bol} remains to be proven.

Let us consider the quantitative consequences of the medium size variations in M_V at constant M_{bol}. They turn out to be very drastic! Leitherer et al. (1985) has discussed the example of S Dor and compared the characteristics of the photosphere during minimum and maximum visual brightness. The photosphere is called "pseudo-photosphere" during maximum, because it is the layer where the continuum is formed in an optically thick shell or wind during periods of high mass loss rates.

Phase	year	m_V	M_V	M_{bol}	BC	T_{eff}	$R_*/R_\odot$	dM/dt
Min	1965	11.3	-7.3	-9.3	2.0	22000	44	?
Max	1983	9.3	-9.3	-9.3	0	8000	330	$5\ 10^{-5}$

(values slightly adapted). We see that T_{eff} dropped from 22000 K to 8000 K because the pseudo-photosphere increased the radius from 44 to 330 $R_\odot$! This very large increase in R_* is due to a mass loss rate of $\simeq 5.10^{-5}$ $M_\odot$/yr during maximum.

c. Predictions of photometric variations due to mass loss variations

The very large changes in radius during periods of increased mass loss rates are quantitatively in agreement with the predictions by Davidson (1987), who studied the dependence of the temperature of the pseudo-photosphere on the mass loss rate. He adopted an analytic expression for the distribution of the opacity and density in the photosphere of $\kappa(r)\rho(r) \sim r^{-n}$, with n = 2, 3 or 10, and calculated the radiative transfer in the diffusion approximation. This resulted in a dependence of the characteristic temperature, T_o, in the photosphere and the mass loss rate. The temperature T_o is defined by the mean energy of the emergent photons, $E = 2.7\ kT_o$, and should be quite close to the value of T_{eff} derived from the observations. From these predictions we can derive the expected dependence of the stellar parameters on the luminosity and the mass loss rate. The results are shown in Figure 2 for n = 2 and 3. I adopted a velocity at the base of the pseudo-photosphere of $v_o = v_{sound}$. The "characteristic temperature" T_o of the pseudo-photosphere is not exactly T_{eff}, but I will assume $T_o \simeq T_{eff}$. For the conversion of M_b to M_V the BC (T_{eff}) dependence for supergiants of Schmidt-Kaler (1982) was adopted.

For small values of $\log \dot{M} + 0.3\ M_b \lesssim -7.2$ the temperature is strongly dependent on this parameter. So if a star with $M_b = -10^m$ changes $\dot{M}$ from 10^{-5} to 10^{-4} $M_\odot$/yr, T_o will drop from 25000 K to 6000 K (for n = 2). However, a further increase of $\dot{M}$ will hardly change T_o: 6000 K is the minimum temperature which can be reached by an increase of $\dot{M}$. The lower part of the figure shows the variation of $\log \dot{M} + 0.3\ M_V$ on $\log \dot{M} + 0.3\ M_b$. At large mass loss rates, $\log \dot{M} + 0.3\ M_b \gtrsim -7.2$, this relation is linear with slope + 1, as expected, since T_o and BC are almost constant. For smaller mass loss rates, however, the value of $\log \dot{M} + 0.3\ M_V$ is approximately constant. This is due to the coincidence that the variation in $\log \dot{M}$ is compensated by a variation in 0.3 BC. The fact that $\log \dot{M} + 0.3\ M_V$ is about constant for small mass loss rates has a very interesting consequence, as it implies that variations in m_V can be translated into $\dot{M}$-variations in such a way that

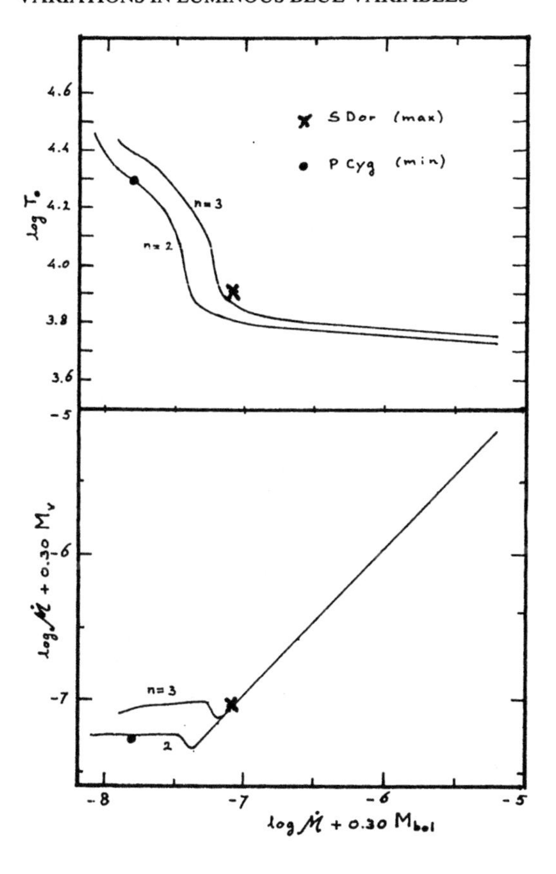

Figure 2. The predicted relations between the photospheric temperature T_0 and $\log \dot{M} + 0.3\ M_V$ as a function of $\log \dot{M} + 0.3\ M_b$ for two values of n (from Davidson, 1987) is compared with the observations for two LBV's.

$$\Delta \log \dot{M} \simeq -0.3\ \Delta m_V \tag{2}$$

The reliability of these predictions depends on the accuracy of the models, in particular on the $\kappa\rho \sim r^{-n}$ dependence; the absorption coefficients, and the relation between T_0 and T_{eff}.

In order to test these models, I have plotted in Figure 2 the location of two LBV's with accurately determined parameters: S Dor during maximum and P Cyg in quiescence (Leitherer et al., 1985 and Lamers et al., 1983). The observations agree with the predictions if $n \simeq 3$ for S Dor at maximum and $n \simeq 2$ for P Cyg. It would be very interesting to plot the path of LBV's throughout their variations in these diagrams to test and improve the theory.

IV. LARGE ERUPTIONS WITH $\Delta m_V \gtrsim 3^m$

Two galactic LBV's (P Cygni and η Carinae) and one extragalactic LBV (V12 in NGC 2403) have suffered large eruptions with $\Delta m_V \gtrsim 3^m$ during the recorded history. The eruption of η Car between 1837 and 1860 is discussed in detail by Kris Davidson in the next review. I will briefly describe the eruption of P Cygni as an example of large photometric variations.

The star P Cygni suddenly brightened in 1600 AD as a "nova" to the third magnitude, whereas it was invisible to the naked eye before that date, so $\Delta m_V \gtrsim 3^m$. After irregular fading and brightening, the star had a second outburst in 1654 when it again reached $m_V \simeq 3^m$. Both of these bright phases lasted 5 to 6 years. In the second half of the seventeenth century the star had irregular brightness variations, sometimes even being invisible, and was described as a "reddish" star. Between 1700 and 1900 P Cygni gradually brightened from $m_V \simeq 6$ to 4.9^m, which is its present brightness (see Zinner, 1952, de Groot, 1969, van Genderen and Thé, 1985, Lamers, 1986). It is interesting to estimate the luminosity of P Cygni during these outbursts. Assuming a distance of 1.8 kpc, an interstellar extinction of $E(B-V) = 0.60$ and $A_V = 1.8^m$, we find $M_V \simeq -10$ when the star was at $m_V = 3^m$. If the bolometric correction at the time of maximum brightness was 0^m, then $M_{bol} \simeq -10^m$ which is close to the present value of $M_{bol} = -9.9$ (Lamers et al., 1983).

The rapid fading of the star after 1606 and 1659, i.e. after maximum, may have been due to dust formation. A decrease by $\Delta m_V \simeq 3^m$ would correspond to a change in colour of $\Delta E(B-V) \simeq 1^m$ which implies that the colour of the star would change from $B-V \simeq 0.6$ at maximum (if $(B-V)_o = 0$) to $B-V \simeq +1.6^m$ when the star was at $m_V \simeq 6^m$. This would agree with the "reddish" colour observed in the 17th century after the outbursts.

The eruption did not leave a visual remnant. The radio arc near P Cygni (Wendker, 1982) is probably due to an eruption of 1500-6000 years ago if the mean expansion velocity was 200 - 50 km/s.

V. MODERATE PHOTOMETRIC VARIATIONS: $\Delta m_V \simeq 1$-2^m

Moderate variations of $\Delta m_V \simeq 1^m - 2^m$ are the most commonly observed photometric variations of LBV's. They have been described in the following papers and in the references to older observations mentioned therein:

AG Car : Mayall (1969; Whitelock et al. (1983)
S Dor : van Genderen (1979); Leitherer et al. (1985)
R71 : Wolf (1975); van Genderen et al. (1985)
R81 : Appenzeller (1972)
R127 : Stahl et al. (1983)
LBV's in M31 and M33: Hubble and Sandage (1953); Rosino and Bianchini (1973); Humphreys et al. (1984).

I will discuss this kind of variations for two LBV's, S Dor and AG Car which both have well observed drastic photometric variations.

a. Photometric variations of S Dor

The visual lightcurve of S Dor in the period 1954-1985 is shown in Figure 3. The data are from compilations by van Genderen (1979), Stahl and Wolf (1982) and Leitherer et al. (1985) and van Genderen (1986b). Older observations are listed by Gaposchkin (1943). If successive observations on a short timescale are available, I have connected them with a dotted line to guide the eye.

The figure shows that the star varied between 9.2 and 11.6 magn, and that the variations occur on different timescales. There is a vague indication for a slow large amplitude variation with maxima around 1955 and 1980 and a minimum around 1964. If this trend persists, the star may reach a minimum again in the next decade. The amplitude of this variation is about two magnitudes. In addition to this, there are variations on timescales of 3 to 5 years with an amplitude of about $0^m.5$. This is most clearly noticeable in the observations between 1974 and 1986, which showed three minima and three maxima. There are also variations on a shorter timescale on the order of a few months with amplitudes of about 0.2 magn. These are visible as small narrow dips or peaks in the lightcurve near 1974.5, 1983.0 and 1984.6 when the star was well observed. The largest fast variations occurred in 1964 when the star changed by $0^m.5$ several times within one year.

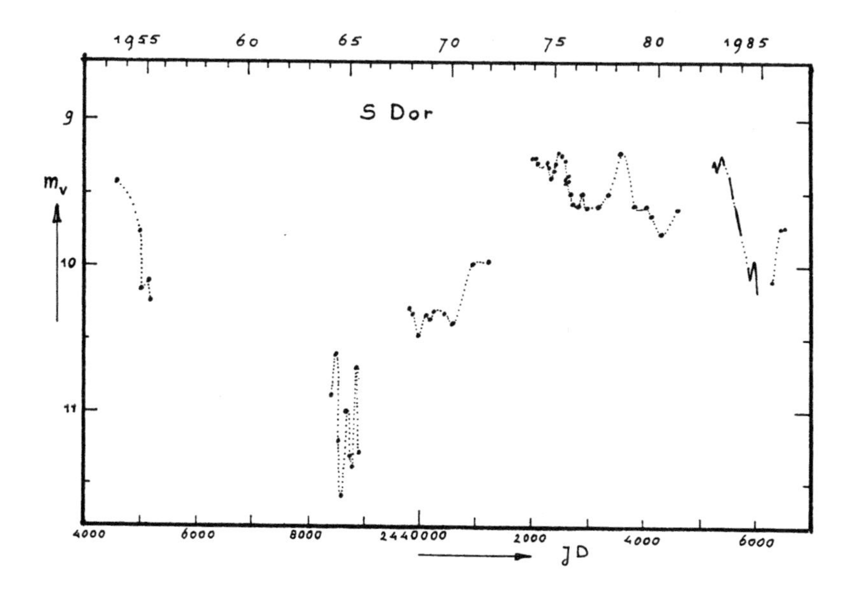

Figure 3. Photometric variations of S Dor between 1954 and 1985.

The variations of S Dor are summarized:

Amplitude	Timescale
1 - 2 magn	20 - 30 yrs
0.5 magn	3 - 5 yrs
0.2 magn	0.1 - 0.3 yrs

b. Photometric variations of AG Car

The star AG Car has been photometrically active during the last decade, with variations of $\Delta m_V \simeq 2^m$ in the period of 1979 to 1983. The variations have been monitored in the near-IR by Whitelock and in the K-band

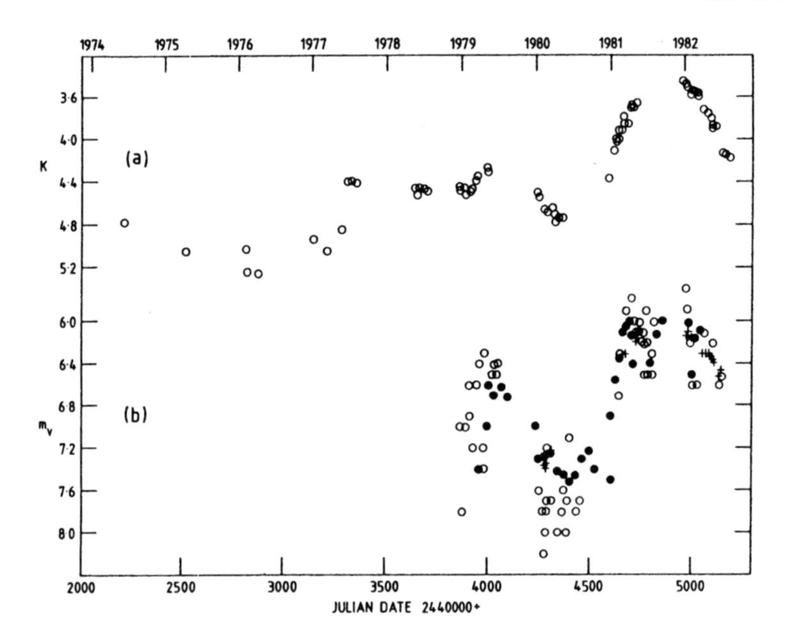

Figure 4. The variations of AG Car in K and m_v between 1974 and 1983 (Whitelock et al. 1983).

(2.2 μm) as shown in Figure 4, from Whitelock et al. (1983).

There is a good correlation between the lightcurves in the visual and in the near-IR. The timescale of the variations between 1979 and 1983 is about 2.5 yrs (max to max) and the "width" of the maxima is about 1.3 yrs. This is of the same order, but slightly less than the timescale of the variations of $\Delta m_V \simeq 0.5$ found in S Dor, which was 3 to 5 yrs.

Despite the large variations in m_V and K, the colours in the near-IR changed only marginally. They are given for two dates close to minimum and maximum:

JD-2440000	V	B-V	V-J	V-H	V-K	V-L
4295 ± 10	7.3	0.54	1.8	2.2	2.6	3.3
5000 ± 10	6.1	0.66	1.8	2.4	2.8	3.2

The energy distribution on the two dates is shown in Figure 5 (from Whitelock et al.) and compared to those from a normal B9 I star when AG Car was faint and an A5 I star when AG Car was bright. The interstellar reddening was assumed to be E(B-V) = 0.55, (from Bensammer et al., 1981). The figure also shows the difference between the observed near-IR energy distribution, and the one expected from the pseudo-photospheres at the two phases.

In both phases the energy distribution is considerably flatter than expected from the pseudo-photosphere. This is obviously due to the free-free emission from the stellar wind. In fact, the energy distributions are similar to those of P Cygni (Waters and Wesselius, 1986) and S Dor (Leitherer et al., 1985), which have been explained by a pseudo-

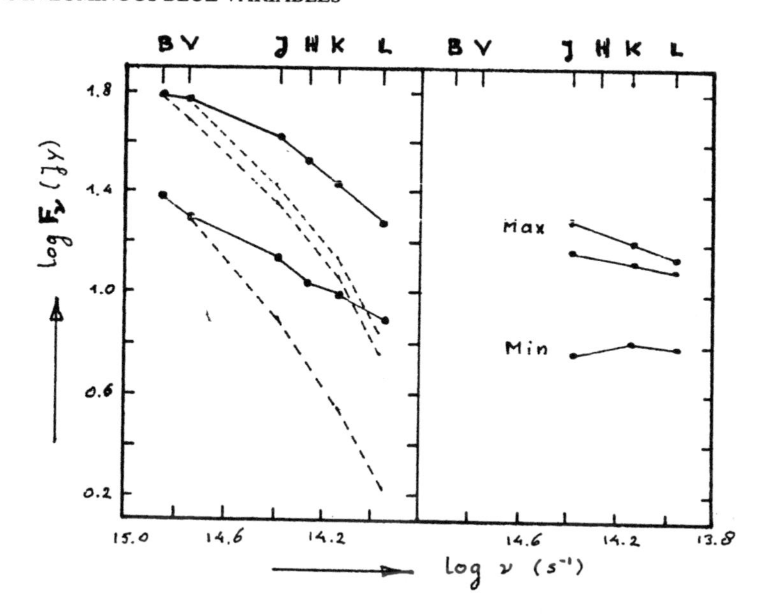

Figure 5. The energy distribution of AG Car at min (March 1980) and max (Febr 1982), compared with normal supergiants (dashed lines). At maximum two possible comparisons are shown, normalized at B or V. The right hand figure shows the energy distribution of the IR excess.

photosphere surrounded by a slowly accelerating wind. When comparing the energy-distributions: of AG Car in Figure 5 during min and max, we see that two changes have occurred simultaneously: the pseudo-photosphere became brighter (B and V fluxes increased by a factor 3.0) and the IR excess flux increased by about a factor 2. Both changes indicate an increase in the mass loss rate. We can now apply the results of Section IIIc to determine the required changes in $\dot{M}$. The spectral type of the photosphere, B9 at min and A5 at max, suggests that AG Car is in the region of the diagram shown in Fig. 2, where Eq. 2 is valid, so $\Delta \log \dot{M} \simeq -0.3\ \Delta V \simeq 0.36$. On the other hand, the flat distribution of the free-free radiation (Fig. 5) suggests that this excess comes from a region of the wind which is optically thin at $\lambda \lesssim 3.7\ \mu m$. For such a region, the free-free radiation is proportional to $F_{ff} \sim EM \sim \dot{M}^2/R_*$ if the velocity distribution remains the same. The increase of the free-free radiation by $\Delta \log F_{ff} \simeq 0.4$ between min and max indicates an increase of $\log \dot{M} \simeq 0.25$ if $\Delta \log R_* \simeq 0.1$.

From this simple estimate we can conclude that the change in V and in the IR radiation are both due to an increase in the mass loss rate, between min and max by about a factor two.

VI. MICROVARIABILITY $\Delta m_V \lesssim 0.1$

Some LBV's are found to show irregular small scale photometric variations on a timescale on the order of a month or smaller. This kind of

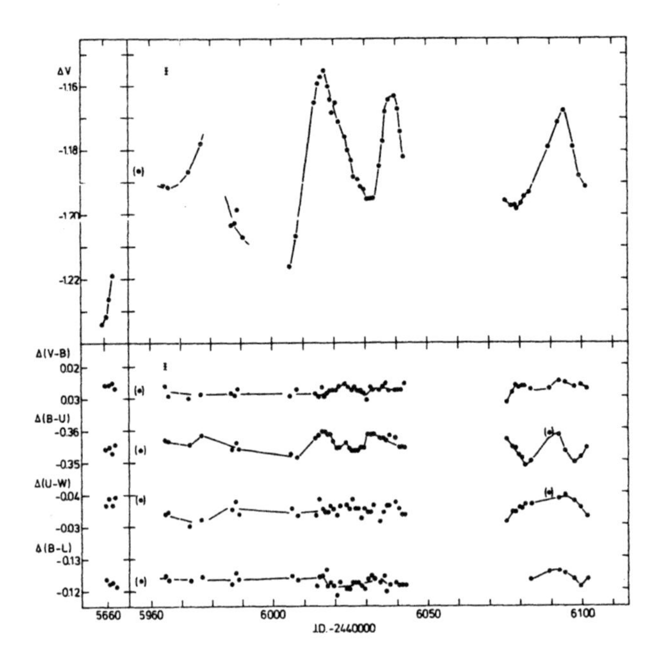

Figure 6. The microvariations of R71 in Walraven photometry (from van Genderen et al. 1985).

variability may occur in all LBV's, but the observations required for its detection are available for one LBV only (R71).

This microvariability is of the same type as the semi-irregular photometric variations in normal supergiants (see e.g. Maeder and Rufener, 1972; Burki, 1978; review by De Jager, 1980; Chapter 8.2; van Genderen, 1986 and references therein). Maeder and Rufener (1972) and Maeder (1980) found that the amplitude of the variations increases with increasing luminosity. So it is likely that all LBV's will show microvariations.

The microvariations observed in R71 are shown in Figure 6 (from van Genderen et al., 1985). The amplitude of the variations is about 0.06^m in Walraven's V and 0.01^m in colour, which corresponds to $0\overset{m}{.}15$ and $0\overset{m}{.}025$ respectively when expressed in normal magnitudes. The variations are not strictly periodic but they show a "characteristic time" or "semi-period" τ which is 24 days for R71.

A study of the small amplitude variations of P Cygni by van Gent and Lamers (1985) based on observations collected from the literature showed a possible range of timescales with $12^d \leqslant \tau \leqslant 60^d$, with a mean value of about 35^d.

The observed semi-periods of R71 ($\tau = 24^d$) and P Cygni ($\tau \simeq 12 - 60^d$) can be compared with those observed in normal supergiants, and those predicted for the fundamental radial pulsations. This is shown in Figure 7 (from Lovey, 1984), which shows the upper part of the HR diagram with lines indicating the expected periods of the fundamental radial modes and the observed semi-periods of normal supergiants. From the position of R71 and P Cygni in this diagram we see that:

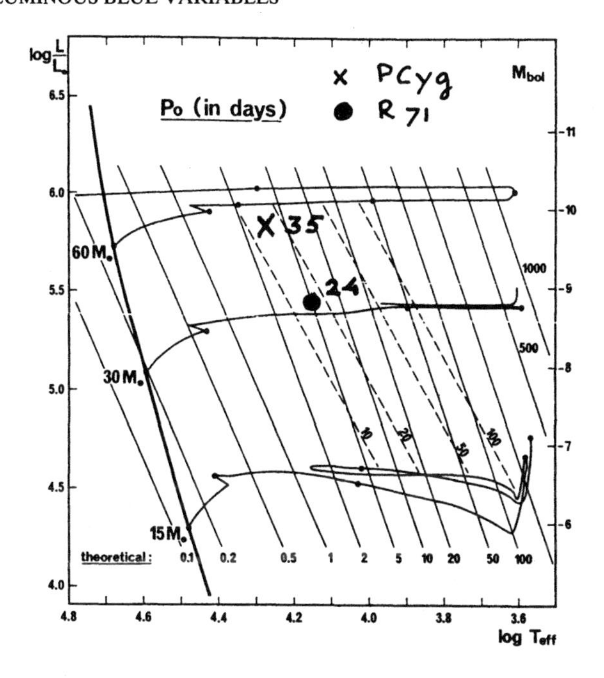

Figure 7. The semiperiods of P Cygni and R71 are compared with the iso-period lines in the HR diagram. The solid lines indicate the theoretical relations and the dashed lines indicate the observed relations for normal supergiants (after Lovey et al. 1984).

a. The semi-periods of the two LBV's are about 2 or 3 times as large as the one for normal supergiants of the same L and T_{eff}.
b. The semi-periods of the two LBV's are about 4 to 20 times as large as the predicted fundamental modes.

These observations show that the microvariations of the LBV's, and of the normal supergiants as well, are not due to fundamental radial pulsations. They also show that the microvariations of the LBV's are different from those in normal supergiants. At present it is not clear whether this implies a different pulsation mode, or the same mode with different periods due to different internal structures.

VII. SPECTRAL VARIATIONS

The LBV's show spectral variations of three different types: i Variations in the spectral type of the star, which can change from early-B to A or from O to B. ii Large variations in the shapes of lineprofiles which can change from normal absorption to strong P Cygni profiles. iii Small variations in lineprofiles which often show multiple absorption components in visual and UV lines due to outward moving shells or blobs.

The first two variations are due to large changes in the mass loss rate which can produce a cooler pseudo-photosphere at maximum $\dot{M}$ (as described in Section III, b) and strong P Cygni profiles. After such outbursts the spectrum sometimes shows the emissions of forbidden lines which are formed in the ejected shells. The third type of variations indicates that the mass loss of the LBV's is not a stationary process but occurs in an irregular way, even when the stars are in quiescence. I will describe a few examples of each of these variations.

a. Spectroscopic variations of S Dor

The visual spectrum of S Dor during the minimum (V ≃ 10.2) in 1945-46 was classified as B8 eq by Martini (1969). The visual spectrum is dominated by emission lines of [Fe II] and Fe II. The strongest lines of Fe II show P Cygni profiles. The lower Balmer lines up to H11 also have P Cygni profiles, but the higher Balmer lines, the Mg II lines and the Si III lines are in absorption. These observations indicate that the star is surrounded by an extended, relatively dense envelope in which the emissions of H and Fe II are formed, and a less dense outer region where the [Fe II] originate. The P Cygni profiles indicate outflow with a velocity difference between the centers of the absorption and emission of 87 km/s.

During the subsequent maximum of 1952 (V = 9.4) the visual spectrum changed to A5 eq. At that epoch most of the lines of H, Fe II, Ti II and Ca II have P Cygni profiles with a velocity difference between absorption and emission of 130 km/s. The Mg II 4481 lines are weakly in absorption, and the [Fe II] lines are absent (Martini, 1969). These characteristics at maximum can be explained by assuming the ejection of a thick shell. This accounts for the development of the P Cygni profiles and the change to later spectral type of a visually brighter star. The disappearance of [Fe II] lines is due to the higher density in the shell.

An example of similar changes in the visual spectrum observed by Leitherer et al. (1985) is shown in Figure 8.

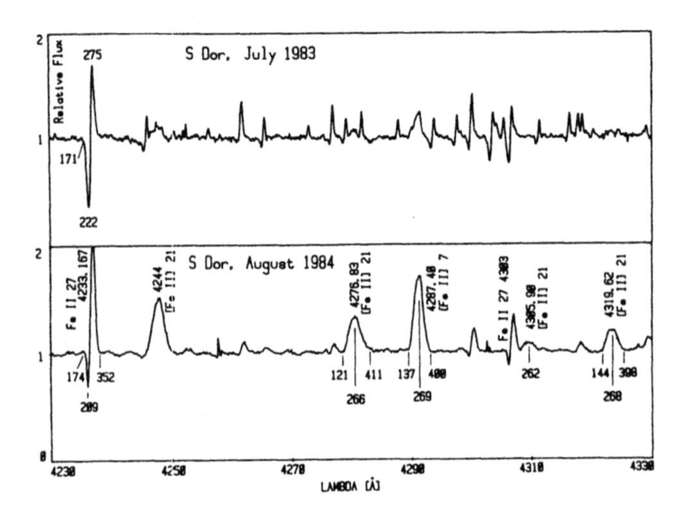

Figure 8. Spectral changes of S Dor between maximum (July 1983) and minimum (Aug 1984). Notice the strong [Fe II] lines at minimum which are formed in a shell ejected at the previous maximum.

The [Fe II] lines are strongest during minimum phases. They are formed at densities $N_e \lesssim 10^{11}$ cm^{-3} and originate probably in shells which were ejected previously. On August 1984 (min) the [Fe II] lines have a rounded profile with a FWZI of 280 km/s, indicating that they are formed in a shell with an outflow velocity of 140 km/s. This agrees perfectly with the largest outflow velocity reached during the previous maximum of July 1983, as measured by the edge velocity of the low ionization UV lines.

During the minimum of August 1984 the spectrum of S Dor also showed the [N II] 6548, 6583 lines in emission. They are formed at $N_e \lesssim 10^5$ cm^{-3}, i.e. at much lower density than the [Fe II] lines. Their width indicates an outflow of the shell of about 60 km/s. These lines are formed in a shell which must have been ejected at least 10 years ago, probably at a lower velocity than the 1983 outburst.

b. Spectral changes in R127

The LMC star R127 is the hottest LBV apart from Var 83 in M33. It was classified as O I afpe extr. in 1977 by Walborn (1977), on the basis of the emission lines of He II λ4686, N III λ4640, N II and Si III. Most He I lines have P Cygni profiles but the Balmer lines are purely in emission. The Si IV λ4089 and N III λ4097 absorption lines were exceedingly narrow and of "interstellar" appearance.

The spectrum has changed drastically in 1982 when it was studied by Stahl et al. (1983). At that time the star was 0.75 magn brighter than before. The typical Of emission features of He II and N III have disappeared. Instead the spectrum is now characterized by P Cygni type lines of H I, He I, N II, Si III and Fe III. This indicates that the visual spectrum is formed in a very thick envelope exhibiting the characteristics of a supergiant of type about B5. The expansion velocity derived from the P Cygni profiles is about 110 km/s.

The most remarkable features in the spectrum of R127 during the 1982 maximum are the broad and shallow emission wings of the Balmer and He I lines, with a full width up to 3000 km/s. Such wings have been observed in the strong Balmer lines of other LBV's where they are attributed to electron scattering of the emission line photons. However, in the case of R127 these wings are also present in the profiles of the weaker lines such as He I 4471, He I 5015 and H8 which do not have strong emission peaks. Therefore they cannot be due to electron scattering, but must be formed in a region where velocities of at least 1500 km/s occur. These extended wings were not present in the spectrum studied by Walborn in 1977 when the star was fainter. There is no obvious explanation for the simultaneous presence of such high velocities and the much lower ones of the P Cygni profiles, unless one adopts a large deviation from spherical symmetry in which the high and low velocities occur at different latitudes, e.g. as in Be stars. Stahl et al. (1983) conclude from the large width of the emission lines of ~ 3000 km/s and the N-rich nebula surrounding the star that R127 is a LBV of spectral type Of during minimum, which has developed towards the WN stars. This confirms the suggestion that the LBV's are the transition between Of and WR stars.

c. Variable absorption components: evidence of recurrent shell ejections by P Cygni

The star P Cygni, which is in a quiescent phase since at least a century, shows regular variations of the profiles in the visual and in the UV which indicate recurrent shell ejections.

The Balmer lines have typical P Cygni profiles, with several variable absorption dips in the absorption part of the profile. At any moment two to four absorption components can be detected in each Balmer line. These were studied by e.g. de Groot (1969), Markova and Luud (1983), Markova and Kolka (1984) and Markova (1986). The absorption dips vary in strength and velocity. The velocities always increase to more negative values. The variations are due to components which start at small velocity ($v \simeq -100$ km/s) and are accelerated up to about -220 km/s in about half a year. Figure 9 shows the radial velocities of the absorption components in the Balmer lines as a function time. The connection of the velocities of the various components has some ambiguity (e.g. from 1981.5 to 1981.6) but the general trend is obvious. The components are accelerated with a mean value of 0.6 cm/s^2 until they reach $v \simeq -220$ km/s. The ejection rate is about 6 shells per year.

The UV absorption lines of once and twice ionized metals (Fe II, III, Cr II, III, etc.) also show variable absorption components. They were studied by Lamers et al. (1985) who found that the UV absorption components have approximately the same characteristics as those in the Balmer lines with two important exceptions: the acceleration of the UV components is 0.17 cm/s^2, which is four times as small as that of the Balmer components, and the recurrence of the shells which produce the UV components is about 1 shell per year. The UV shells reach a final velocity of 206 km/s which is similar to the velocity of 220 km/s reached by the Balmer components.

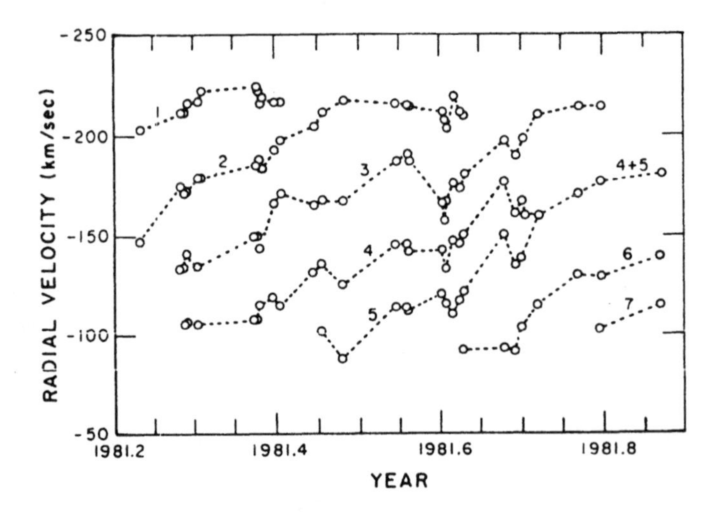

Figure 9. The velocities of the shell components observed in the Balmer lines of P Cygni by Markova and Kolka (1984) (from van Gent and Lamers, 1986).

The difference in behaviour (acceleration and recurrence) of the shells observed in the UV and in the Balmer lines can be explained by differences in the mass of the shells. The components in the UV metal lines are detectable only if the shells are rather massive (columndensity $N_H \gtrsim 10^{20}$ cm^{-2}; Cassatella et al. 1979) whereas the absorptions in the Balmer components can also be measured for smaller column densities ($N_H \gtrsim 10^{18}$ cm^{-2}) because of the higher abundance of H compared to Fe. Obviously, the ejection of massive shells occurs less frequent than the ejection of less massive ones and the first ones are accelerated more slowly than the later ones.

The total mass loss due to shell ejections is ~ 3.10^{-5} $M_\odot$/yr. This is about the same as the quiescent mass loss of P Cygni, which is estimated to be 1.5 10^{-5} $M_\odot$/yr (White and Becker, 1982).

VIII. CONCLUSIONS AND FUTURE PROJECTS

Photometric and spectroscopic variations of LBV's occur on all timescales from centuries to a month. The larger photometric variations with $\Delta V \gtrsim 3^m$ occur on a timescale of centuries; the moderate ones of $\Delta V \simeq 1 - 2^m$ on timescales of decades; the small variations $\Delta V \simeq 0.5$ on timescales of years and the micro variations $\Delta V \simeq 0.2^m$ on timescales of months or weeks. The variations of $2^m \gtrsim \Delta V \gtrsim 0.5$ seem to occur at constant luminosity. They are due to temporary enhancements of the mass loss which produces an extended pseudo-photosphere which is cooler, redder and has a smaller BC so that the stars become visually brighter.

The micro variations are very similar to those observed in normal supergiants, although the periods of the LBV's are about 2 or 3 times as large as in normal supergiants of the same L and T_{eff}. These variations are probably due to some kind of pulsation which has a period about 4 to 20 times longer than the period of the fundamental mode.

The spectroscopic variations of the LBV's cover a range from changes of spectral type, changes of types of lineprofiles from absorption to emission and P Cygni profiles, and changes in strength and velocity of small absorption components which are superimposed on the spectral lines. The first two types of variations are due to variations in the mass loss rates and the resulting changes in the temperature and density of the pseudo-photosphere. At maximum brightness the spectral type is later and more lines show P Cygni profiles than at minimum. The forbidden lines are strongest during the minimum after a strong maximum. These lines are formed in the shells which are ejected during maximum. A most puzzling observation is presented by the extended emission wings of some of the weaker lines in R127 during maximum. They show evidence for outflow at a velocity of about 1500 km/s which is much larger than the velocity of 110 km/s derived from the P Cygni profiles at the same time. This may indicate a very severe deviation from spherical symmetry of the envelope, with regions of very high and low velocities.

The variable multiple absorption components in the visual and UV lines indicates multiple ejection of shells, even when the LBV's are in quiescence. Although they have been studied most extensively in P Cygni, such components have been observed occasionally in several LBV's, e.g.

R81, R127 and S Dor. The study of P Cygni has shown that the star ejects about 6 shells per year with a "thick" shell about once per year. The masses are on the order of 10^{-5} $M_{\odot}$ for the thick shells and possibly a factor 10 smaller for the other ones. The thick shells are more slowly accelerated than the thinner ones, but they all reach velocities of about 210 km/s. Although the timescale of the shell ejections, ~ 2 month, is on the same order as the photometric micro variations of P Cygni, it is presently not clear that the shell ejections are produced by the micro variations which are probably the due to pulsation.

The first quantitative predictions of temperature and magnitude changes of LBV's due to variable mass loss (III, c) are reasonably in agreement with the observations (V, b), but obviously more quantitative interpretation of the observations during phases of variability are needed. Even if we understand how the stellar photosphere changes when the mass loss varies, we still do not understand why and how the mass loss changes. It is clear that the LBV's are at the limit of stability due to radiation pressure or turbulent pressure and that any internal disturbance may have a large effect on the outer layers, but the nature of these internal disturbances on all timescales from months to centuries is not clear at all.

Let me finish with my (subjective) list of requests for future work on LBV's:

- a quantitative study of the photometric and spectroscopic variations throughout a full cycle through maximum and minimum, in terms of T(t), $\rho(t)$, v(t) and $\dot{M}(t)$, in order to understand the way in which the photosphere reacts to changes of $\dot{M}$.
- a theoretical study of the possible pulsation modes in LBV's and an attempt to identify the observed photospheric variations with the predicted internal instabilities.
- a determination of the mass loss in the various phases of the LBV's, in order to determine the consequences for the evolution of the LBV's.
- a statistical study of the characteristics and numbers of LBV's in several galaxies, in order to estimate the duration of the LBV phase and the masses of their progenitor stars.

REFERENCES

Appenzeller, I.: 1972, Publ. Astron. Soc. Japan 24, 483.
Appenzeller, I.: 1987, These proceedings.
Appenzeller, I., Wolf, B.: 1982,
Bensammer, S., Gaudenzi, S., Rossi, C., Johnson, H.M., Thé, P.S., Zuiderwijk, E.J., Viotti, R.: 1981 in Proc. IAU Coll. 59, eds. C. Chiosi and R. Stalio, p. 67.
Bond, H.E., Landolt, A.U.: 1970, Publ. Astron. Soc. Pacific 82, 313.
Burki, G.: 1978, Astron. Astrophys. 65, 357.
Conti, P.S.: 1984, in Observational Tests of the Stellar Evolution Theory, eds. A. Maeder and A. Renzini, (Dordrecht: Reidel) p. 233.
Davidson, K.: 1987, These proceedings.
Davidson, K.: 1987, Astrophys. J. (in press).
de Groot, M.: 1969, Bull. Astron. Inst. Netherlands 20, 235.

de Jager, C.: 1980, The Brightest Stars, (Dordrecht: Reidel) p. 296.
Duemmler, R., Markova, N.: 1987 (preprint).
Feast, M.W., Thackeray, A.D., Wesselink, A.J.: 1960, Monthly Notices Roy. Astron. Soc. 121, 354.
Gaposchkin, S.: 1943, Astrophys J. 97, 166.
Hayes, D.P.: 1985, Astrophys. J. 289, 726.
Hubble, E., Sandage, A.: 1953, Astrophys. J. 118, 353.
Humphreys, R.M., Blaha, C., D'Odorico, S., Gull, T.R., Benvenuti, P.: 1984, Astrophys. J. 278, 124.
Humphreys, R.M., Davidson, K.: 1979, Astrophys. J. 232, 409.
Kenyon, S.J., Gallagher, J.S.: 1985, Astrophys. J. 290, 542.
Lamers, H.J.G.L.M.: 1986, in Luminous Stars in Associations and Galaxies, eds. C. de Loore et al. (Dordrecht: Reidel) p. 157.
Lamers, H.J.G.L.M., de Groot, M., Cassatella, A.: 1983, Astron. Astrophys. 128, 299.
Lamers, H.J.G.L.M., Korevaar, P., Cassatella, A.: 1985, Astron. Astrophys. 149, 29.
Leitherer, C., Appenzeller, I., Klare, G., Lamers, H.J.G.L.M., Stahl, O., Waters, L.B.F.M., Wolf, B.: 1985, Astron. Astrophys. 153, 168.
Lovey, D., Maeder, A., Noëls, A., Gabriël, M.: 1984, Astron. Astrophys. 133, 307.
Maeder, A.: 1980, Astron. Astrophys. 90, 311.
Maeder, A.: 1983, Astron. Astrophys. 120, 113.
Maeder, A., Rufener, F.: 1972, Astron. Astrophys. 162, L3.
Markova, N.: 1986, Astron. Astrophys.
Markova, N., Kolka, I.: 1984, Astrofiz. 20, 250.
Markova, N., Luud, L.: 1983, Publ. Tartu Astrophys. Obs. 32, 55.
Martini, A.: 1969, Astron. Astrophys. 3, 443.
Mayall, M.W.: 1969, J. Roy. Astron. Soc. Canada 63, 221.
Rosino, L., Bianchini, A.: 1973, Astron. Astrophys. 22, 453.
Schmidt-Kaler, T.: 1982, in Landolt-Börnstein, New Series, Group VI, Vol. 2b; eds. K. Schaifers and H.H. Voigt (Berlin: Springer) p. 15.
Stahl, O., Wolf, B.: 1982, Astron. Astrophys. 110, 272.
Stahl, O., Wolf, B., Klare, G., Cassatella, A., Krautter, J., Persi, P., Ferrari-Toniolo, M.: 1983, Astron. Astrophys. 127, 49.
Stahl, O., Wolf, B., Zickgraf, F.J., Bastian, U., de Groot, M.J.H., Leitherer, C.: 1983, Astron. Astrophys. 120, 287.
Thackeray, A.D.: 1956, Vistas in Astron. 2, 1380.
van Genderen, A.M.: 1979, Astron. Astrophys. Suppl. 38, 381.
van Genderen, A.M.: 1985, Astron. Astrophys. 151, 349.
van Genderen, A.M.: 1986a, Astron. Astrophys. 157, 163.
van Genderen, A.M.: 1986b, Private Communications.
van Genderen, A.M., Steemers, W.J.G., Feldbrugge, P.T.M., Groot, M., Damen, E., van den Boogaart, A.K.: 1985, Astron. Astrophys. 153, 163.
van Genderen, A.M., Thé, P.S.: 1985, Space Sci. Rev. 39, 317.
van Gent, R.H., Lamers, H.J.G.L.M.: 1986, Astron. Astrophys. 158, 335.
Viotti, R.: 1969, Astrophys. Space Sci. 5, 323.
Viotti, R., Altamora, A., Barylak, M., Cassatella, A., Gilmozzi, R., Rossi, C.: 1984 in Future of UV Astronomy based on Six Years of IUE Research, NASA Conf. Publ. 2349, p. 231.

Walborn, N.R.: 1977, Astrophys. J. 215, 53.
Waters, L.B.F.M., Wesselius, P.R.: 1986, Astron. Astrophys. 155, 104.
Wendker, H.J.: 1982, Astron. Astrophys. 116, L5.
White, R.L., Becker, R.H.: 1982, Astrophys. J. 262, 657.
Whitelock, P.A., Carter, B.S., Roberts, G., Whittet, D.C.B., Baines, D.W.T.: 1983, Mon. Not. Roy. Astron. Soc. 205, 577.
Wolf, B.: 1975, Astron. Astrophys. 41, 471.
Wolf, B., Stahl, O.: 1982, Astron. Astrophys. 112, 111.
Wolf, B., Stahl, O., de Groot, M.J.H., Sterken, C.: 1981a, Astron. Astrophys. 99, 351.
Wolf, B., Appenzeller, I., Stahl, O.: 1986b, Astron. Astrophys. 103, 94.
Zinner, E.: 1952, Kleine Veröff. Remeis-Sternwarte Bamberg 1, nr. 7.

Henny Lamers giving his review on Luminous Blue Variables

DISCUSSION

HEARN: In one of your slides you showed that the LBV's are sitting in the same region of the HR diagram as more stable, normal stars. Do you have any evidence for the direction in which the LBV's are evolving. Is it possible that the LBV's are evolving to the left in the HR diagram and the more normal stars are evolving still to the right.

LAMERS: Yes, this is possible although it is difficult to find observational evidence for it. I suspect the longer a star has stayed in this region of the HRD, the more mass it has lost, the more unstable the star may be. If this is correct, the more quiet or normal supergiants in that region of the HRD represent the 'new-comers' and the LBV's are the 'veterans'. It is interesting to note that some stars e.g. ζ^1Sco are spectroscopically similar to the LBV's but have not shown large outbursts yet. This indicates that there may be a gradual transition from normal supergiants to LBV's, which would argue against a very different evolutionary stage. Of course, it is also possible that such stars as ζ^1Sco are LBV's which are in a quiet phase in between outbursts.

HUMPHREYS: The LBV's and normal OB supergiants overlap in temperature and luminosity on the HR diagram. I suggest that they are in a different evolutionary state. The LBV's are the massive stars that have already encountered the limit to their stability while the normal supergiants are still evolving to the right and have not yet encountered the stability limit.

BURKI: There is roughly 50% of binary stars among the main sequence OB stars and 30% among the yellow and red supergiants. By comparing the distribution of the orbital periods in these two samples, it is predicted that one third of the early main sequence stars do not become normal cool supergiants because of the mass exchange in binary systems during stellar evolution. Are there any binary stars among the LBV stars and can we expect that some part of the observed vaiability in these objects could be due to binary star evolution?

LAMERS: As far as I know, none of the LBV's are known close binaries. For some LBV's binarity has been proposed in the past to explain radial velocity variations (e.g. in P Cygni), but a more careful analysis indicates that these variations are not really periodic and probably due to ejections (Van Gent and Lamers, 1986, Astron. Ap. 158, 335). Obviously, we cannot exclude binarity, and it would be very important if it can be detected.

STAHL: We have been monitoring R81, which is the counterpart of P Cyg in the LMC, for more than three years within the 'Long-term photometry of variables' program. We found that it shows periodic brightness dips which are most likely due to eclipses with a period of 74.6 days. For more details see the poster paper by Stahl, Wolf and Zickgraf.

V. GENDEREN: I like to mention that according to the Gen. Cat. of Variable Stars of Kholopov (4 ed. 1985) the Luminous Blue Variables are called α Cygni variables. I would like to propose to use this name instead of LBV's. The P Cygni stars and the S Dor type stars are the subclasses of the α Cygni variables. Strange enough α Cygni is poorly investigated photometrically so far, but it is a famous spectrum variable since nearly a century.
Then I like to comment of the viewgraph of the HR diagram. I think that the luminosity of η Car should be somewhat higher viz. log L = 6.8 instead of 6.4.

LAMERS: I do not agree with your suggestion to call the LBV's α Cygni variables. The star α Cygni is a normal supergiant and although it is photometrically and spectroscopically variable, its variations are of much smaller amplitude than in the LBV's. Moreover α Cygni does not show outbursts which are characteristic for LBV's.

V. GENDEREN: I like to emphasize that the excursions of the S Dor variables in the HR diagram are caused by their shells and that thus the position of S Dor now is caused by its present shell. Therefore it lies so far to the red part of the Humphreys-Davidson limit.

HENRICHS: Are the uncertainties in M_{bol} for the LBV's also ± 0.5 mag? If so their positioning in the HR diagram with respect to the upper luminosity line is rather uncertain.

HUMPHREYS: The uncertainty in M_{bol} is typically 3 to 4 mag.

DAVIDSON: Regarding η Car its M_{bol} is -12 mag ± .2 mag and is very well determined from its infrared radiation. So log L = 6.7 ± .1.

HOUZIAUX: A short comment about the association of observed shells in the Balmer lines with photometric variations in the visible. Photometric variations are due to continuous spectrum variations, dominated by hydrogen opacity. In the visual range, the Paschen continuum is formed in relatively high layers of the atmosphere and is likely to be affected by shells or winds and this would explain the association of Balmer shell lines with photometric variations. In the ultraviolet however, if the lines are formed in high layers and thus very sensitive to superficial phenomena, the continuum is formed much deeper and one looks at photospheric radiation, which should be less affected by external shells which are transparent in the ultraviolet continuum, and might remain constant.

JERZYKIEWICZ: You mentioned an anticorrelation of the V and B-V variation in the case of LBV's. Then you said that this anticorrelation is due to a radius effect. I would like to point out that in the case of the red semi-regular variables a similar anticorrelation is observed. In that case, however, it is caused by the temperature sensitivity of the TiO bands. The question is: could the anticorrelation in the case of LBVs be caused by the temperature sensitivity of some spectral features?

V. GENDEREN: I do not think so. The band widths of the filters are much too broad. So I am sure that it is mainly the continuum.

JERZYKIEWICZ: Is there a definite proof for a radius variation in the LBV's, for instance observed radial velocity variations?

LAMERS: The LBV's can make large excursions in the HRD by changing their spectral types drastically. The observations with IUE and ground-based instruments have shown that the luminosity remains about constant. This implies that when the star has a later spectral type (lower T_{eff}) the radius must be larger.

DE GROOT: You mentioned that no photometric variations of P Cygni have shown a 200-day time scale. In view of the scarcity of good photometric observations of P Cygni, how can such a statement be made?
In fact, this problem of the lack of photometry for this bright star has prompted me to encourage more observations of P Cygni, which can well be done by experienced amateurs with photo-electric equipment. Also, recently I have initiated regular observations of P Cygni with the Automated Photo-electric Telescope Service in Arizona.

LAMERS: I agree with you that we cannot exclude the presence of a "period" or "characteristic time" of about 200 days from the photometric data. What struck me was that the timescale of about 200 days is rather prominent in the study of the radial velocities of the UV lines, but that it has not yet shown up in the, admittedly scarce, photometric data.

DE GROOT: You mentioned the observed variety in photometric amplitudes and time-scales for the LBV's. Can you say anything more about the distribution of these parameters over the HR diagram i.e. whether small ΔV occurs in one part of the HRD and larger ΔV in some other part, e.g.?

V. GENDEREN: The light amplitude of the α Cygni variables are smallest for O type stars ($\sim 0^m.01$) and largest for the B type stars up to $\sim 0^m.2$. This trend was already shown statistically by Rufener and Maeder many years ago. Thus for P Cygni I expect considering its place in the HR diagram an amplitude of $\sim 0^m.1$ as is actually observed. I am not amazed that so far the expulsions of thick shells by P Cygni are not recorded photometrically. I am sure they can be found, but then long term base line observations should be made. I expect that in view of similar events in η Car, the light amplitudes should be smaller than $0^m.1$.

DE GROOT: A comment on Tony Hearn's question about the direction in the HRD of the evolution. We must not forget that Maeder's evolutionary calculations show that on their way back from the cool part of the HRD to the left, they describe a blue loop right there where we find the LBV's in the HRD. Thus, a star can both move left or right as a temporary interruption of its more general long-term evolution to the left in the HR diagram. Finally, I do not like the term α Cyg variables. The stars we are talking about today have too different characteristics to be included under α Cyg variables. I still prefer LBV.

CASSINELLI: The LBV's lie in a rather narrow band in the HR diagram. It appears that the stars maintain a roughly constant degree of inhomogeneity, or have 'stalled' in their evolution to the right or left on the HR diagram. I wonder if this is not related to the mechanism that Maeder has proposed for the mass loss of the Wolf Rayet stars. In that stars lose mass at just the right rate to maintain the needed degree of inhomogeneity for pulsations to occur.

MAEDER: The numerical stability analyses clearly show that the epsilon mechanism probably at work in WR stars is prevented from operating in LBV variables by the high density contrast existing in stars lying to the right of the main sequence. Thus, we have to look for something else in the LBV stars. A scheme we may conceive for the apparently recurrent instability of LBV stars is that due to internal evolution the LBV star evolves to the right in the HR diagram, until it reaches the De Jager limit where the resulting surface gravity is close to zero. Then, it loses some amount of mass, which restores the stability and brings the star to the blue side of the instability limit. Due to internal evolution, the star again evolves redwards. Many open questions remain concerning such a possible scheme. The main one is: what makes the process a discontinuous one? Also, is there any relation between the amount of material ejected and the time interval between two consecutive ejections? Let me conclude by saying that the various proposed schemes have to be quantitavely modelized and compared to the observations in future years. What is clear now is that it is when, due to peeling off, the helium core mass fraction is more than some critical value of about 0.7, that the star will definitely start to evolve to the WR stage.

OWOCKI: When you showed that the gravities for LBV's were less than for 'normal' stars with same L and T, was this corrected for the effect of radiation, i.e. were these 'effective' gravities or 'geometric' gravities?
What is the effective gravity of stars near the Humpreys-Davidson luminosity limit? Is it near zero?

LAMERS: The difference in effective gravity (i.e. corrected for radiation pressure) between the LBV's and the normal supergiants which I derived can be misleading and has to be considered with care. Obviously, when two stars (one LBV and one normal supergiant) are in the same place of the HRD, they have the same value of T_{eff}, R and L. They may have the same mass, but that we don't know for sure.
My estimate of the mass and gravity of the LBV's, (see my review in: Luminous Stars in Associations and Galaxies, eds. De Loore et al. 1986) is based on the assumption that the mass must be at most as large as those of stars just leaving the main-sequence and at least as large as those of the Wolf-Rayet stars in which they evolve. (I adopted the evolutionary tracks of Doom, 1982, Astron. Ap. 116, 303). All we can say is that the mass must be within that range. We don't know the real masses

nor the gravities.

DE JAGER: The effective gravity accelaration near the Humphreys-Davidson limit can only be determined observationally by spectroscopic investigations. Such determinations are rare, and their results not in all cases fully reliable, but as a working hypothesis I would advocate to take $g_{eff} \simeq 1/2$ to 1/3 of the geometrical gravity.

OWOCKI: The characteristic speed of a stellar wind tends to scale with the effective surface escape speed, almost independently from the detailed theory for the wind driving. The fact that LBV's seem to have both lower wind speeds and lower gravities then normal stars is thus not too surprising.
How does $\dot{M}v_\infty$ compare with L/c for these stars? If $\dot{M}v_\infty/Lc \geqslant 10$, then a radiative driving model seems unlikely.

LAMERS: For the best studied star, P Cygni, we find the following parameters: log L = 5.9 $\pm$ 0.2, log $\dot{M}$ = -4.70 $\pm$ 0.10 (from radio data), and v_∞ = 250 ± 50 km/s. This gives $\dot{M}v_\infty \simeq 0.4$ L/c, which is well within the possibility of the radiation driven wind theory. For all LBV's for which we know the parameters L, $\dot{M}$ and v_∞ with reasonable accuracy, we find $\dot{M}v_\infty \simeq L/c$ (in: Luminous Stars and Association in Galaxies). This refers to the quiescent mass loss rates. During the large outbursts $\dot{M}v_\infty$ is much larger than L/c.

DE JAGER: The value of v_∞ for LBV's is smaller than those for normal supergiants, as was shown in a diagram presented by Henny Lamers. It is however important to notice that this applies to all stars at or near to the Humphreys-Davidson limit - so that it is a property that the LBV's have in common with non - (or only slightly) variable stars near the HD limit.

WOLF: For the stars R71 and R127 which have been mentioned several times during Humphreys' talk we have presented two poster papers (see Wolf and Zickgraf; Appenzeller, Wolf and Stahl; this volume).
1) R71 could be identified as an IRAS point source. This S Dor variable is surrounded by a very cool (T = 140K) and extended ($R_{shell} \simeq 8000\ R_*$) dust shell.
2) The spectacular S Dor variable R127 which was previously classified as an Of star by Walborn has again increased in brightness and is now the second brightest star of LMC in the visual. Interestingly its spectral appearance is now practically identical with the one of S Dor when it is of similar brightness. Another interesting finding of our recent long-slit spectroscopy is a resolved nebula around this star (found in [NII] 6548, 6583 [SIII] 6717, 6751 and also in Hα) of an extension of 1 pc. We would derive a kinematic age of 20000 years, which is an interesting number since it gives us some idea how long this star can stay in the active S Dor type stage.

VIOTTI: The optical spectrum of R127 is presently characterized by a large number of FeII emissions. The UV spectrum of this and other

P Cyg/LBV stars is dominated by a great deal of FeII absorption. This means that singly ionized metals play an important role in the wind structure, as also suggested by Henny Lamers. But this object (as well as other ones such as AG Car) has changed from Of to Beq in a short time scale. I believe that the simplest explanation of this behaviour, is that these stars have atmospheres at a critical point, so that minor structural changes deep in the atmospheres result in a sudden change of opacity at constant bolometric luminosity, and probably without a large change of mass loss rate.

WOLF: At least during active stages S Dor variables are characterized by an enhanced mass loss rate. Typical values have been given by Henny Lamers in his talk.

HENRICHS: Is the typical lifetime of 20000 yrs you found consistent with the known number of these kind of objects?

WOLF: The question is difficult to answer. The number of S Dor variables is far from being known with completeness. E.g. in the LMC Otmar Stahl has found two S Dor variables within the last few years. During quiescence S Dor variables are sometimes not too different from 'normal' supergiants as the spectrum of R71 shows.

LAMERS: For the determination of the upper luminosity limit due to radiation pressure it is important to know the flux-mean absorption coefficient κ_f as a function of T and ρ. The classical Eddington limit is usually calculated for electron-scattering only, but this is a bad assumption if T $\leqslant$25000 K. I have derived the values of κ_f as a function of T_{eff} and ρ from the modelatmosphere calculations by Kurucz, at $\tau_R = 1$. In this calculations the radiation pressure due to many absorption lines (about 2.10^6) was taken into account. The result is shown in the figure. The full lines show κ_f in cm^2/g for three values of the density. The dashed line shows σ_e, the electron scattering coefficient. Notice that the opacity increases with decreasing temperature and has a maximum at $T_{eff} \simeq 10000$ K. This suggests that the Eddington limit, which is defined as $L = 4\pi GMc/(\kappa_f + \sigma_e)$ will decrease with decreasing temperature from $T_{eff} \simeq 40000$ K to $T_{eff} \simeq 10000$ K. This agrees qualitatively with the observed Humphreys-Davidson upper limit, which also decreased to $T_{eff} \simeq 10000$ K. At lower temperatures the Eddington limit increases again, but this is not important, since stars with a horizontal evolution in the HR diagram will not be able to pass the Eddington limit when they come from the left (high-T) side. (The exact location of the limit in terms of L depends on the density structure in the teneous photospheres of stars close to this limit and is now studied.) A detailed report of this study will be published in Ap. J.
I have suggested on the basis of these calculations that the Humphreys-Davidson limit is due to radiation pressure by numerous metal lines in the Balmer continuum for stars with $T_{eff} \leqslant 25000$ K (Lamers, 1986, in Luminous Stars and Associations in Galaxies, eds. C. de Loore et al.). This is supported by the study of the acceleration of the wind of P Cygni, close to the Humphreys-Davidson limit, which also indicates

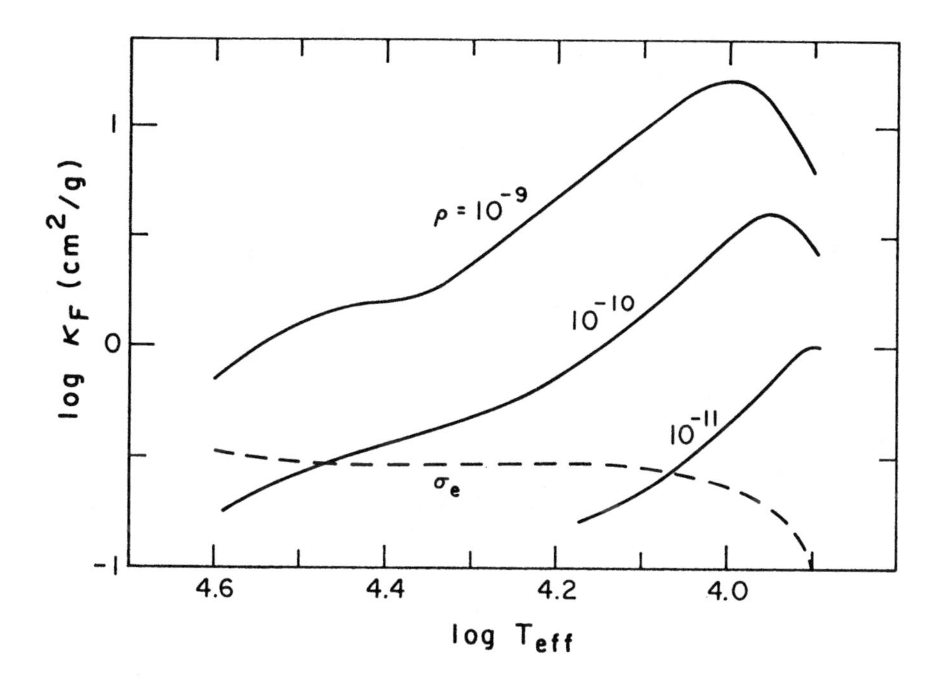

a strong radiation pressure by metal lines (Lamers, 1986, Astron. Ap. 159, 90). The large radiation pressure by metal lines can also explain the S Dor type of variations, as proposed by Appenzeller (1986 in Luminous Stars and Associations in Galaxies) and reported in the discussion of Session 8.

The audience during one of the sessions

The students who did an excellent job for the local organization:
Pieter Mulder, Jenne Wollaert, Jacqueline Coté, Ankie Piters

GIANT OUTBURSTS OF THE ETA CARINAE - P CYGNI TYPE

Kris Davidson
Department of Astronomy
University of Minnesota
116 Church St. S.E.
Minneapolis, MN 55455
U.S.A.

1. GENERAL REMARKS

It is time for us to contemplate more dramatic instabilities! Certain very massive stars, including η Car and P Cyg, give observational hints that their most crucial mass loss occurs in giant eruptions. During such an event, the mass loss rate can be of the order of 0.1 $M_\odot$ yr^{-1} for several years. We do not yet understand this phenomenon, partly because most relevant theoretical work has been devoted instead to more familiar, less dramatic stellar winds.

Before we consider the observed events, let me sketch a naive, intuitive scenario -- a simple framework for discussion. The luminosity/mass (L/M) ratio of a very massive star is safely below the Eddington limit when the star is young. As the star evolves, its luminosity increases or remains roughly constant while its mass slowly decreases (see, e.g., Maeder 1983) so L/M tends to increase. More important, perhaps, is the fact that the Eddington limit involves opacity. When the effective temperature falls below 30000 K, there is a tendency for atoms to recombine and for the opacity to increase. (Let us naively ignore the density-dependence of opacity; our ultimate justification is observational, not theoretical, as will become clear later.) Thus, intuitively, the surface of a very massive star seems endangered by the Eddington limit as its temperature decreases below 30000 K. Prof. de Jager's turbulent-pressure mechanism may enhance the danger.

Therefore we should not be surprised if a sloping instability limit is found at the top of the HR diagram. This may be somewhat below the rough empirical limit emphasized by Humphreys; Lamers (1986) has proposed that the theoretical limit nearly coincides with the positions in the HR diagram of certain relatively inactive P Cyg stars. Anyway, we may expect stars at or beyond this limit to develop very extended, unstable envelopes with high mass loss rates. The point that I want to make here is that something else also happens: sporadic giant eruptions occur.

H. J. G. L. M. Lamers and C. W. H. de Loore (eds.), Instabilities in Luminous Early Type Stars, 127–142.

Sometimes it is amusing to invent a machine that behaves like a specific astrophysical phenomenon. At this meeting, for such an exercise we should employ components appropriate to the Netherlands: ships, windmills, pumps, canals ... Figure 1 represents my own attempt. I hope that the analogy will be clear after the following explanation!

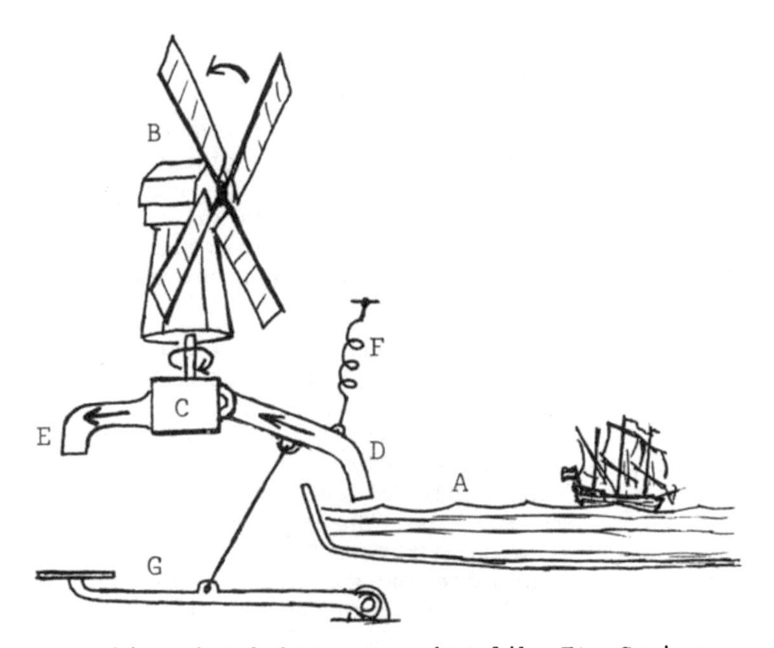

Figure 1. A machine that behaves somewhat like Eta Carinae.

...Water level A slowly rises; this symbolizes the evolution of a very massive star as seen from outside. A fairly constant breeze, vaguely representing the star's internal energy flux, turns the windmill B, which drives pump C. When the water level reaches the pump intake D, then water will be pumped out through outlet E, symbolizing an unusually strong stellar wind. This tends to limit the level A. Let us imagine that the flow rate is proportional to the height of water level A above the bottom end of intake D.

Note that pipe D can move up or down and is suspended by a spring F; I won't specify what this represents but it gives the theorist some equations. Surface waves in A, vertical oscillations of D, and friction throughout the machine all produce fluctuations, maybe even instabilities. So far, this machinery resembles the discussions that we've heard at this meeting.

But we have not yet heard about a strange and sinister lever G, at the bottom of the apparatus. When water begins to gush down from outlet E, it may push G downward, which would pull D far below the water level A. This would cause the flow rate to increase dramatically until the water level has been reduced by a substantial amount. Then, when level A thus falls to the new, low position of D, the flow decreases and G suddenly relaxes upward, causing a gap to develop between A and D so that the flow stops. The water level then resumes its slow secular rise. Evidently, if lever G is significant, our machine produces sporadic outbursts of mass loss.

What physical process in the outer layers of a very massive star is symbolized by lever G? -- I certainly do not know; but the existence of such a process is suggested by the historical behavior of η Car, P Cyg, and maybe some other luminous stars.

(A digression: Fig. 1 may also be relevant to the evolution of a star that has a close neutron star or black hole companion. In such a case, device G might conceivably represent X-ray heating of the stellar surface; Ostriker and I speculated about such a phenomenon back in the early 1970's. But there is no obvious photometric or radial-velocity evidence that η Car and P Cyg are close binaries of this type.)

2. ETA CARINAE

This star -- known in the 19th Century as η Argus, η Navis, and even η Roboris -- is to the southern hemisphere what the Crab Nebula is to the north; i.e., even though it is very famous and is widely believed to be interesting and important, nevertheless we have only a few modern spectrophotometric observations and very few high-resolution calibrated images. Modern observers seem perversely reluctant to study such a bright famous object. Before considering the great 19th Century eruption of η Car, we should review its present appearance.

Most of the relevant data have been reviewed or presented by de Jager (1980), van Genderen and Thé (1985), and Davidson, Dufour, Walborn, and Gull (1986). The total luminosity can be estimated from infrared photometry, because the shell of 19th Century ejecta contains dust which thermally recycles most of the ultraviolet and visual-wavelength luminosity into IR radiation mainly between 5 μm and 50 μm. If the distance of η Car is 2.5 kpc, then its total observed luminosity is about 1.7×10^{40} erg $s^{-1} = 10^{6.65}$ $L_{\odot}$, corresponding to bolometric magnitude $M_{bol} \simeq -11.9$. Various continuum and emission-line data in the wavelength range 0.1 μm to 3 μm are consistent with the idea that the intrinsic continuum crudely resembles a 30000 K blackbody (i.e., before it is absorbed by the dust). Since this temperature is significantly cooler than a main sequence star of comparable luminosity, naturally we suspect that η Car is near the hypothetical stability limit in the HR diagram.

There is no doubt that the star is evolved and internally mixed, because nebular emission lines in the ejecta show abundance ratios N/C >> 1 and N/O >> 1, consistent with the CNO cycle. Moreover, there is evidence that the helium abundance in the ejecta, presumably representative of the star's outer layers, is roughly $Y \simeq 0.40$. All of these data seem consistent with an initial mass around 170 $M_\odot$ and a present age somewhat less than 3 million years -- cf. models by Maeder (1983). The present mass loss rate cannot be much more than 10^{-3} $M_\odot$ yr^{-1} and may be much less (Davidson et al. 1986; van Genderen and Thé 1985).

Weigelt and Ebersberger (1986) have recently found evidence that η Car may be a multiple system. If we interpret their speckle-interferometric data in this way and also invoke other data, then we find that the main component has roughly $M_{bol} \simeq -11$ and a surface temperature around 25000 K. Its initial mass would have been 100 $M_\odot$ or so. The other components would be hotter, less luminous, and more stable. These hypothetical stars are too far apart to affect each others' evolution, so the nature of the main component is qualitatively almost unchanged in the revised picture. At present we do not yet have enough information to decide whether this interpretation of Weigelt and Ebersberger's data is justified.

Now, having reviewed the present state of η Car, we come to the topic that I want to emphasize here: Its photometric history and in particular its major outburst seen in the 19th Century. Apparent visual magnitudes and comparisons with other stars have been reported or quoted by Herschel (1847), Innes (1903), Gratton (1963), de Jager (1980), van Genderen and Thé (1985), and others; Fig. 2 summarizes these.

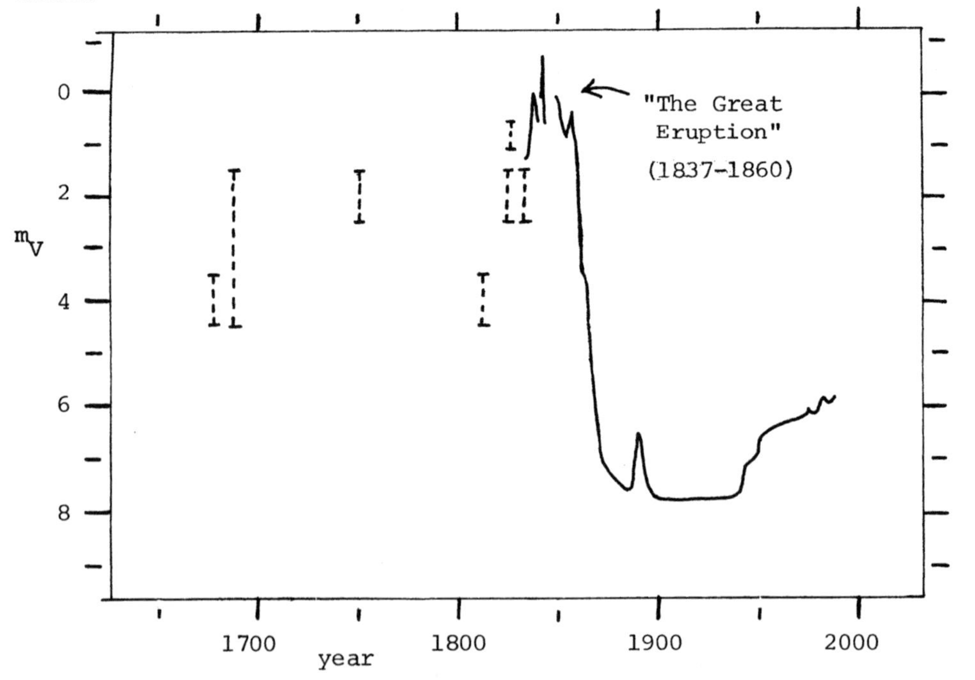

Figure 2. Apparent visual magnitude of η Car during the past 3 centuries. Since 1860 the curve has been strongly affected by circumstellar extinction in the ejecta.

In order to see the quantitative significance of Fig. 2, it is useful to remember some particular numbers. Suppose that the dusty shell of 19th Century ejecta were not present today. Then we would have approximately $m_V - M_V \simeq 13.5$ magnitudes (distance ~ 2.5 kpc, interstellar extinction $A_V \sim 1.5$). The apparent visual magnitude would now be roughly $m_V \sim 4$ or 4.5. If, in addition, the bolometric correction were zero, then we would see $m_V \simeq 1.6$; this is practically the maximum visual brightness allowed by the present luminosity.

As Fig. 2 indicates, before its major eruption η Car sometimes had $m_V \sim 4$. On those occasions the star may have been in about the same condition as it is today, but without as much circumstellar dust. Sometimes, however, it was brighter, $m_V \sim 2$. Probably those were occasions when the mass loss rate was large enough to make the stellar wind opaque, giving "cool" emergent radiation ($T \sim 10000$ K ?) and therefore a reduced bolometric correction. A mass loss rate of the order of 0.01 $M_\odot$ yr^{-1} would have done this. No 2-magnitude changes of this type have been seen in this century; therefore we suspect that η Car was basically less stable before the Great Eruption. It is a plausible guess that the mass loss in that outburst has temporarily relieved the worst of the surface instabilities.

Now consider the Great Eruption itself, observed in the years around 1840. Since the modern infrared discovery (Westphal and Neugebauer 1969), we have grown fond of saying that "the luminosity did not change much during the major outburst -- it was mainly the bolometric correction that changed." But Fig. 2 shows that this common remark is not exactly true. In 1843 the apparent magnitude reached about -1, implying $M_V \sim -14.5$ and probably $M_{bol} \lesssim -14.5$. The average brightness between 1830 and 1860 corresponded to $m_V \sim +0.6$ or $M_V \sim -12.9$ or (presumably) $M_{bol} \lesssim -12.9$, at least a magnitude more luminous than we observe today. The extra energy radiated in that 30-year interval was at least $10^{49.2}$ ergs. Where did this energy come from? I suppose that it must have been the thermal and radiative energy that had been trapped in the layers that were ejected from the star. This idea can be used to estimate the ejected mass. Nobody has yet done this calculation properly, so far as I know; but very crudely, I deduce that more than 1 $M_\odot$ was ejected, probably 2 or 3 $M_\odot$, in order to liberate enough radiated energy. This would be enough to make the expanding "homunculus" that we see today, with its dust content inferred from the circumstellar extinction. At the presently observed mass loss rate (see Davidson et al. 1986), η Car would take more than 700 years (probably more than 2000 years) to lose 2 $M_\odot$. Therefore we suspect that the mass loss, which is crucial in this stage of evolution, occurs mainly in giant outbursts.

...Which brings us to some tantalizing questions: What is the recurrence rate? Are there any ancient records of bright outbursts of η Car? When precession is taken into account, we note that 2000 to 5000 years ago this star was far enough north to be seen by astronomers in Egypt, Mesopotamia, India, China, etc. Indeed, Jensen (1890) speculated that certain Sumerian/Babylonian records referred to η Car!

Unhappily, I must report that R. van Gent (private comm.) has explained in detail that Jensen's conjecture is now known to have been too naive and was almost baseless. (I have heard in less detail from 3 other persons who also understand the old records and who generally agree with van Gent.) It appears that we have no real information, either positive or negative, about η Car in ancient times. This is very frustrating. Note, by the way, that some outlying blobs of ejecta have proper motions consistent with ejection dates 250 to 900 years ago if they have not been decelerated very much (Walborn, Blanco, and Thackeray 1978).

Finally, regarding η Car, since about 1860 circumstellar dust has strongly affected the light curve shown in Fig. 2. It is customary to say that the major decline in brightness between 1858 and 1870 was mainly due to dust formation in the ejecta. In fact, two things must have occurred after 1855: the eruption subsided and dust began to form. I suspect that the eruption subsided first, and the consequent decrease in luminosity triggered the condensation of dust grains in the ejecta. Later, an outburst in 1889 showed that the surface was still unstable. The envelope soon after that event, not surprisingly, had a spectrum like an early F supergiant (see Walborn and Liller 1977). Since that time no similar events have been seen. Aside from minor fluctuations (van Genderen and Thé 1985; Zanella, Wolf, and Stahl 1984), the secular increase in brightness since 1940 may be largely due to a decrease in the circumstellar extinction as the dusty homunculus expands. Eta Carinae may again be a 4th magnitude star forty years from now.

3. OTHER STARS

De Jager (1980), Lamers, de Groot, and Cassatella (1983), and Lamers (1986b) have described the history and present condition of P Cyg. It has, very roughly, $M_{bol} \sim -10$, $T_{eff} \sim 19000$ K, a mass loss rate around $10^{-4.8}$ $M_{\odot}$ yr^{-1}, and a present (not initial) mass of perhaps 30 $M_{\odot}$. The apparent and absolute visual magnitudes are $m_V \simeq 4.9$ and $M_V \simeq -8.3$. Thus the star is less grandiose than η Car but nevertheless is near the hypothetical stability limit in the HR diagram.

In 1600 W.J. Blaeu noticed P Cyg because it had brightened to $m_V \sim 3$. Six years later it began to fade, until between 1630 and 1650 it was invisible to the unaided eye. Then, between 1654 and 1659 P Cyg was bright again, $m_V \sim 3$. It faded thereafter, and for the following 30 years it fluctuated with maxima usually around $m_V \sim 5$. Very likely the outbursts of 1600 and 1654 were η Car-like eruptions, followed by dust formation in the ejecta. There is some evidence for the present existence of a remnant about 0.6 pc across. (As in the case of η Car, the ejection speeds in major outbursts were probably around 800 km s^{-1}.) P Cyg gradually approached its modern appearance during the years from 1700 to 1900. Probably $M_V \sim -10$ during each major outburst, so there may have been extra luminosity as in the case of η Car's major eruption.

Today the intrinsic spectrum of P Cyg is more readily observable than that of η Car, because the latter is complicated by dense circumstellar gas. However, in giving clues to the outburst phenomenon P Cyg has several practical disadvantages relative to η Car: (1) We cannot measure the luminosity of P Cyg from IR-emitting circumstellar dust, as we can for η Car. (2) The 17th-Century ejecta from P Cyg are not as observable as the 19th-Century ejecta from η Car. (3) The 17th-Century eruptions of P Cyg were not recorded in as much detail as the later η Car events, and were not as conspicuously over-luminous.

Now we suspect that some Luminous Blue Variables in other galaxies are susceptible to outbursts of the same type, but so far we have not been lucky enough to observe such an event with modern instruments. For several years during the 1950's, an outburst of the variable star V12 in the galaxy NGC 2403 was seen. That star rose from $M_V \sim -8$ to $M_V < -11$. Tammann and Sandage (1968) noted its possible resemblance to η Car; but today we do not yet know much more about V12.

An outburst in the galaxy NGC 1058, seen in 1961, has sometimes been included in the same class. Zwicky (1964) thought that this event had $M_V \sim -13$ and called both it and η Car "type V supernovae." However, according to a recent discussion (Fesen 1985), the event in NGC 1058 was far more luminous than Zwicky thought and therefore, presumably, was indeed a supernova -- not like η Car at all.

4. SOME QUESTIONS FOR THEORISTS

Let us assess the situation specifically in the case of η Car. I must acknowledge that my description of the nature of this star might be wrong. Conceivably it is instead a unique close binary system including a massive star and a black hole. But this seems unlikely; it is not a very luminous X-ray source. Probably η Car (or its primary component, if there is a multiple system) is indeed a very massive star whose surface has reached the hypothetical instability limit. Probably, in this stage of evolution, unsteady mass moss is crucial and will eventually result in a Wolf-Rayet star. If this is true, then we should be very curious about the mechanisms of instability (recall Fig. 1). Here are a few obvious questions.

(1) Is the stellar atmosphere unstable when $T_{eff} \lesssim 30000$ K, as suggested near the beginning of this review? Lamers (1986a) has shown that radiation pressure in many weak absorption lines can accelerate P Cyg winds when $T_{eff} \lesssim 30000$ K, but this is not enough to predict sporadic eruptions.

(2) Was η Car switching between two well-defined modes or states with different mass loss rates, in the years before 1830 (see Fig. 2)? If so, was the temperature dependence of opacity involved, or is there some other explanation?

(3) Why did the Great Eruption begin (see Fig. 2)? Why did it continue? The last layers to be ejected were initially deep within the star.

(4) Why did the Great Eruption stop? [Here is a specific though possibly naive calculation worth considering: Suppose that a star like

η Car, in its pre-1837 state, loses mass Δm in time Δt, where (dynamical timescale) << Δt < (thermal timescale) and Δm is of the order of ∿3 percent of the star's mass. Does the stellar surface become less stable, perhaps, for small Δm and then more stable when Δm becomes larger?]

(5) Did the outer layers of η Car become more stable as a result of the 1830--1890 outbursts? If so, when will the star's evolution cause serious instability to recur? How many Great Eruptions occur during the evolution of a star like η Car, and does the star really become a Wolf-Rayet star afterward?

Aside from technical details, there is not very much more to be said until theorists have given us their answers to these questions. My own feeling is that the correct answers may allow us to understand the top of the HR diagram.

REFERENCES

Davidson, K., Dufour, R.J., Walborn, N.R., & Gull, T.R. (1986) Astrophys. J. **305**, in press

de Jager, C. (1980) *The Brightest Stars* (D. Reidel, Dordrecht)

Fesen, R.A. (1985) Astrophys. J. Letters **297**, L29

Gratton, L. (1963) *Star Evolution/Evoluzione delle stelle* (Academic Press, New York), p. 297

Herschel, J.F.W. (1847) *Results of Astron. Observations...at the Cape of Good Hope* (Smith, Elder & Co., London), pp. 32-37

Innes, R.T.A. (1903) Cape Annals **9**, 75B

Jensen, P.C.A. (1890) *Die Kosmologie der Babylonier* (K.J. Trübner Verlag, Strassburg), pp. 24-28

Lamers, H.J.G.L.M. (1986a) Astron. Astrophys., in press

Lamers, H.J.G.L.M. (1986b) in *Luminous Stars and Galaxies, IAU Symp. 116* (ed. C. de Loore & A. Willis; D. Reidel, Dordrecht).

Lamers, H.J.G.L.M., de Groot, M.J.H., & Cassatella, A. (1983) Astron. Astrophys. **128**, 299

Maeder, A. (1983) Astron. Astrophys. **120**, 113

Tammann, G.A., & Sandage, A. (1968) Astrophys. J. **151**, 840

van Genderen, A.M., & The, P.S. (1985) Space Sci. Rev. **39**, 317

Walborn, N.R., Blanco, B.M., & Thackeray, A.D. (1978) Astrophys. J. **219**, 498

Walborn, N.R., & Liller, M.H. (1977) Astrophys. J. **211**, 181

Weigelt, G., & Ebersberger, J. (1986) Astron. Astrophys., in press

Westphal, J.A., & Neugebauer, G. (1969) Astrophys. J. Letters 156, L45

Zanella, R., Wolf, B., & Stahl, O. (1984) Astron. Astrophys. 137, 79

Zwicky, F. (1964) in Stars & Stellar Systems, Vol. 8, Stellar Structure (ed. L.H. Aller & D.B. McLaughlin; University of Chicago Press, Chicago), pp. 410-412

Kris Davidson presenting his ideas about Eta Carinae

DISCUSSION

MAEDER: I think it is fair with respect to Kris Davidson to mention the following. Fifteen years ago, there was a great confusion concerning the real status of Hubble Sandage variables. For example some authors said Eta Carinae is a star in formation, or a special class of supernovae or even a neutron star with a surrounding nebulosity. Now, it is the observations by Davidson, Walborn and Gull of CNO ratios and helium content which gave us the answer, confirming the original Burbidge's suggestion that HS variables are post-main-sequence supergiant stars.

DAVIDSON: The history of this topic is interesting and, as usual, non-linear. Before 1975, various authors remarked that Eta Carinae itself has almost no oxygen lines. But the gas was dense and complicated, so we could not be sure that this was really an abundance effect. In the 1970's Nolan Walborn and the late A.D. Thackeray noticed very strong nitrogen features in a lower density outer condensation. Walborn suspected that this proved a high N/O abundance ratio, but, since he was nervous about nebular details, he asked me about his spectra. One look at the spectrogram convinced me! Because nebular spectra of this type are simpler to analyze than photospheres, stellar winds, etc. Nebular blobs are very nice this way. Anyway, I got appropriate IUE data as soon as possible (in a slightly illicit manner, with Roberta's and Ted Gull's help) and these proved the case regarding N/C as well as N/O. High N abundances had been proposed, on other grounds, for P Cyg-related stars in previous years. But it was pleasant to prove a point about a major star in such a simple, indirect, nebular manner. Today we know something about He/H at Eta Car's surface by similar means.

DE LOORE: Concerning the chemical abundances observed in Eta Car, do similar observations exist for P Cygni? This is important since the nitrogen and helium abundance could tell us something about the evolutionary stage of P Cyg.

DE GROOT: The only abundance analysis I know of was done by Luud around 1970 (Sov. Astr., 11). He found N somewhat overabundant, C and O somewhat underabundant. In fact, from inspection of the visual spectrum one would classify P Cyg as a mild OBN star. This seems to fit with the situation for Eta Car. There much matter was lost and C, N, O very much exposed. In P Cyg evolution has not proceeded so far, less mass has been lost, and the abundance effects are smaller.
I do not know about helium in P Cyg, but seem to remember it is rather normal.

CASSINELLI: Would you state again your explanation of the phases when $|M_v| > |M_{bol}|$ i.e. when the outflow of energy exceeded the Eddington Limit luminosity?

DAVIDSON: I didn't mean to suggest that $|M_v| > |M_{bol}|$ to any significant extent; presumably the bolometric luminosity did increase during the major eruption. Naively, of course, we should not be very surprised if the Eddington limit is exceeded in a non-static situation, e.g. in material being ejected! But where did the 'extra' luminosity come from? I think that the energy was merely normal radiation within the gas layers that were ejected.If the gas had not been ejected, the radiation within those layers would have taken 30 to 100 years longer to escape; but instead the ejection of the gas allowed the radiation within it to escape earlier, temporarily enhancing the luminosity. Admittedly this hypothesis may have some quantitative difficulties - for instance, the radiative acceleration of ejecta must have been rather efficient. Somebody should attempt to model the hypothetical event in a computer.

NOËLS: The estimation of the whole duration of the LBV phase is generally given at 20000 yr which is of the order of the time necessary to form a He core with a mass fraction greater than the critical mass to burn He on the blue side for an initial mass of about 100 $M_\odot$ with a mass loss rate of about 10^{-3} $M_\odot$/yr. Are there statistical or other arguments which could support this idea of recurrence of ejections during such a period of time?

DAVIDSON: Suppose (merely picking plausible numbers from a hat) that Eta Carinae loses 2 solar masses in an eruption, with 1000 years between eruptions. Then 50 solar masses would be lost in 25 eruptions, in a time of 25000 years. Maeder can comment on whether this is OK in view of the interior evolution; my impression is that it may be theoretically OK. Observationally, though, only about 1 very massive star in 100 should be in an Eta-Carinae-like phase, if the assumed timescales are correct. Maybe the time between massive eruptions is really longer than 1000 years. In that case the 19th century astronomers were unusually lucky in seeing the big Eta Carinae event.

CASTOR: It is important to keep in mind that steady stellar models can be constructed with L just up to the electron-scattering Eddington limit. The atmosphere becomes just extended enough to lower ρ so that the opacity contributions of bound-free and bound-bound are small. The amount of extension may be so large that a strong wind results, but this itself is not an instability or outburst. This steady structure may not be stable, of course. Can you see how the steady structure can be unstable, and the observed kind of relaxation oscillation can be produced?

DAVIDSON: Quite true; back in the days before stellar winds became fashionable, most astronomers would have expected an evolved very massive star merely to develop a huge low density atmosphere. Apparently this structure is unstable, even though we do not clearly understand why.
Appenzeller remarks that the opacity decreases again, at sufficiently lower temperatures of $T \lesssim 10^4$ K; so we can imagine switching between two states.

APPENZELLER: A mechanism which may possibly explain the 'outbursts' of LBV's can be inferred from the Eddington Luminosity vs. T_{eff} relation mentioned by Dr. Davidson. Qualitatively this relation is indicated in the schematic diagram (below) by the solid line. The minimum is due to an increased line opacity (mainly FeII) in the corresponding temperature range. A massive star evolving (approximatively) horizontally from 1 towards the right will encounter an instability at 2 when it approaches $L = L_{Edd}$. The instability is due to the fact that at 2 L_{Edd} decreases with decreasing T_{eff}. If the star comes close enough to L_{Edd}, it will (due to the lower g_{eff}) expand and increase its $\dot{M}$. This further lowers T_{eff}, increasing again $\dot{M}$, etc. The resulting runaway on a dynamical timescale will be stopped only when the star reaches (at 3) a region where L_{Edd} increases again with decreasing T_{eff}. At 3, the star is in a stable dynamic equilibrium again. But, due to the rapid expansion of the outer layers, it will be out of thermal equilibrium and the layers just below the photosphere will now adjust on a thermal time scale. The main effect probably will be a heating of the layers at the base of the photosphere, which may bring the star (in the diagram) back into the region between 2 and 1. Then a new cycle of this relaxation oscillation can start. A (slightly) more quantitative discussion of the above suggestion has been given in my contribution to the proceedings of the IAU Symposium No. 116.

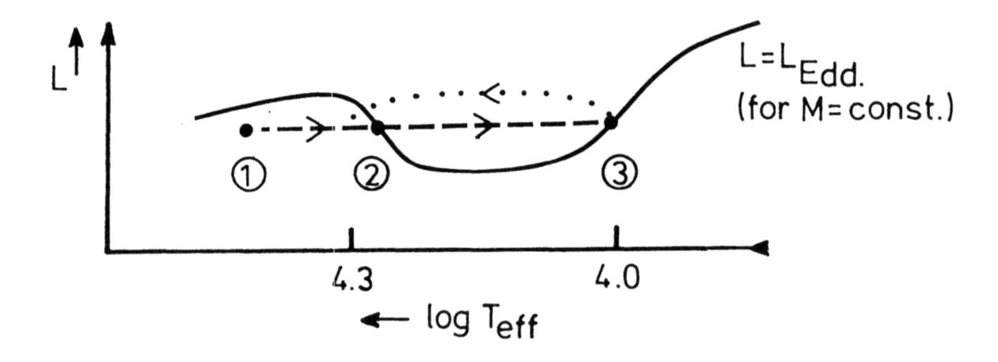

DE JAGER: For model photospheres of cool extreme supergiants the Eddington limit moves way up with decreasing temperature. This was my main reason for looking for an other mechanism to make these atmospheres unstable. I agree that for hotter stars, near 10^4 K effective temperature, the effect of lines, particularly those in the winds, will decrease L_{Edd}, thus supporting the unstability valve mechanism.

DAVIDSON: Here is a puzzle, or a coïncidence, where Prof. De Jager's views are of interest: A few years ago we suspected that the brightest red supergiants correspond to initial masses around 50 or 60 solar masses, - the highest masses that can avoid the Eddington crisis. This idea is roughly consistent with the maximum observed brightnesses of red supergiants. Today, however, we have an alternative idea, namely, De Jager's turbulence-gradient limit. This idea gives about the same red supergiant limit - I think that Maeder has invoked De Jager's mechanism this way. The Eddington and De Jager limitations on red supergiants seem to coincide, even though there is no obvious physical reason. Should we view this coincidence as good or bad? Fortuitous coincidences usually make me nervous. (This question is relevant to the use of red supergiants for galaxy distances: chemical composition affects the Eddington and De Jager limitations differently.)

DE JAGER: I do not think that the Eddington limit applies to red supergiants because the continuous absorption coëfficient is small in such stars, as well as T_{eff}, making

$$-g_{rad} = \frac{4\pi}{c}\int \kappa_\nu \pi F_\nu d\nu \simeq \frac{\kappa_{Ross}}{c}\sigma T_{eff}^4$$

a small value. Thus one calculates that $\log(L_{Edd}/L_\odot)$ for Alpha Sco is about 9.5, while the observed value $\log(L/L_\odot) = 4.7$, making a factor of about 10^5! Adding the contribution of the line absorption helps, but not sufficiently.
On the other hand the micro-turbulence in such stars is large and the dissipation of turbulent energy causes outward acceleration g_{turb}, comparable and opposite to g_{grav}. Therefore I think that the criterion $g_{grav} + g_{rad} + g_{turb} = 0$ does apply to the cool supergiant atmospheres, with $g_{turb} \gg g_{rad}$.

ROUNTREE-LESH: Can you say anything about the other stars that are present in the Eta Car system according to the recent speckle results? In other words, can you put limits on their mass and luminosity, and how does this affect the scenario you have sketched for the Eta Car outbursts?

DAVIDSON: Weigelt and Ebersberger state that Eta Carinae has one major component, plus 3 minor components each about 1.5 or 2 magnitudes fainter than their bigger sister. The separation is several hundred a.u. To a limited extent this is plausible. Assume 4 things: (1) The total observed luminosity, (2) the observed brightness ratios, (3) Suppose that all 4 components are stars of the same age, and (4) use Maeder's evolution models. Then we find a 'unique' solution: Age = 3 x 10^6 years, primary initial mass = 100 solar masses (present mass is less), initial mass of each companion = 65 solar masses.
However, this seems unsatisfying; because the tremendous brightness seen in the 1840's is more difficult to explain if the erupting star was less massive than 100 solar masses.
Incidentally, there are energy-supply objections to the idea that the companion objects are ejected gas blobs rather than stars. Generally

speaking, the speckle results are puzzling. And today we cannot yet discuss suggestions that Eta Car is a close binary producing 'bipolar' ejecta!

THÉ: Have you considered the possibility that the faint components of Eta Car, found by speckle-interferometric method, are actually background stars?

DAVIDSON: Very crudely, each small component has an apparent magnitude around 9 or 10, according to Weigelt and Ebersberger's description. There are three of them, all within a region less than 0".5 across. Statistically, this would be a very unlikely coincidence.

V. GENDEREN: I just show you the light curve of the integrated magnitude of Eta Car during the last 12 years obtained in collaboration with Thé, indicating a new maximum which is appearing just now. I suggest spectrocopists to do spectroscopy to check whether a new-shell is expelled like has been confirmed by Zonella et al. in 1981/1982.

VIOTTI: I was formed in the school of Physics of the Roma University, and what I have expecially learned is 'to stand with our feet on the ground' (= English translation of a well known Italian expression). Then, I moved to astrophysics, and started to work on Eta Carinae - probably the worst subject to understand the physics of what is going on! What we have tried to do is to put the results in a clear scheme which can be used by those involved in models, and to avoid conclusions not supported enough by the observations.
For instance in the case of the mass loss rate, only two papers have treated the argument in an objective way = Andriesse et al. (1978) and Hyland et al. (1979), while in other works it has been derived from empirical $\dot{M}$ vs L laws, which cannot be applied to this case: it would be in any case a circular argument. In addition we are unable to explain the fading of Eta Car with one single event of mass ejection = dust should have continued to form from 1856 until now. Our studies of Eta Car and AG Car suggest that in these cases - as well as in other objects including the Hubble Sandage variables - those variations occur at constant bolometric magnitudes, and are due to structural changes of their winds and/or by dust formation, and subsequent destruction.

DAVIDSON: It seems to me that Eta Car must have increased its luminosity during the 1840's outburst. Its apparent distance modulus (distance combined with ordinary interstellar extiction) is between 13 and 14 visual magnitudes. During the 1840's Eta Car often appeared about as bright as Canopus. Therefore, at that time, $M_V < -13$, most likely $M_V \lesssim -14$; and presumably $M_{bol} \lesssim M_V$. In saying that the luminosity has not varied significantly, do you mean to imply that $M_{bol} < -13$ during the entire time since the great outburst?

VIOTTI: The answer is yes. One should remember however, as discussed by Andriesse et al. in 1978, that according to our mass loss estimates the mechanical power associated to the mass outflow is comparable to the

present radiative power.
This brings the $M_{bol} = -12$ derived from IR (the difference with Van Genderen & Thé of -0.3 is due to the different adopted distance: 2500pc instead of 2800pc), to $M_{bol} \simeq -13$ in agreement with the luminosity at the bright phase assuming BC = zero. Our conclusion is that there are evidences for M_{bol} = constant $\simeq -13$ from beginning 1800 to present time.

VREUX: This morning we have heard about timescales which are longer than the lifetime of a healthy astronomer and even longer than astronomy itself. In that context I would like to remind you of the existence of RCW58 (central star WR40, WN8). According to the classification by Chu, this is the only example of a bubble containing ejecta from the stars (the other example given by Chu presently has some problems: population I or II for the WR). Rays of materials are seen linking the star to the bubble. These rays are reported to be clumpy. It could be interesting to see if it is not possible to extract a time scale between large scale eruptions from an evaluation of the distances between the clumps and from the velocities.

MOFFAT: The star at the centre of the ring nebula RCW58 discussed by J-M. Vreux is HD96548 = WR40, a WN8 star which has been the subject of much discussion at this workshop.Indeed WN8 stars may be very relevant as the most likely candidates as immediate descendants of high luminosity variable LBV stars. Properties of WN8 stars that distinguish them from other WR stars and make them most similar to the LBV stars:

1) WN8 stars have strong, narrow P Cygni profiles in the visible
2) WN8 stars generally have the slowest of all WR winds
3) WN8 stars tend to be the most variable of WR stars

Thus if LBV's evolve into WR stars, they might be expected to be WN8.

Appenzeller, Davidson and Wolf discussing large scale variations. At the other end of the table Baade and Osaki are worried about the small scale non-radial pulsations.

The Belgians: Léo Houziaux and Bert de Loore

INSTABILITIES DUE TO CONVECTION AND ROTATION

J.-P. Zahn
Observatoires du Pic-du-Midi et de Toulouse
14, avenue E. Belin
31400 Toulouse
France

ABSTRACT. In early-type stars, efficient thermal convection arises only in the central region, but even a convective core may be the cause of various instabilities. Some of the instabilities which are associated with convection are discussed, with the possible consequences on the behavior of the surface layers.

The instabilities due to rotation are briefly reviewed. Some of them are likely to generate a mild turbulence, whose nature and efficiency are still a matter of conjecture. In early-type stars, this turbulence may mix the radiative envelope to some extent, and it may also deliver mechanical energy to their outer layers.

1. INSTABILITIES DUE TO CONVECTION

The purpose of this review is to examine whether some of the instabilities observed in the early type luminous stars can be caused by thermal convection or else due to the rotation of those stars.

At first glance, convection seems a very good candidate. In the Sun, the convection zone emits waves which carry mechanical energy to the outermost layers, and this energy serves to create the chromosphere, the corona and the wind. The global acoustic modes, whose study has grown in recent years to a major subject, heliosismology, are also, in all likelihood, excited by the convective motions. And the convection zone is believed to play an important role in the solar activity, although the precise mechanism of the generation of the magnetic field and its exact location are still not well understood.

However, in contrast with the Sun, the early type stars have a very shallow outer convection zone, if any, due to the second ionization of helium, and it can hardly be made responsible for the instabilities displayed by those stars. The prospect looks better with the inner convective core, which is much more efficient in converting thermal energy into kinetic energy, but it remains to be seen whether it can efficiently couple with the surface layers.

To answer this question, one has in principle to solve the following problem. Let us assume that the amplitude of the waves emitted by the considered convection zone is small enough so that their coupling

H. J. G. L. M. Lamers and C. W. H. de Loore (eds.), Instabilities in Luminous Early Type Stars, 143–158.

with the convective motions obeys an equation of the form

$$d^2\xi/dt^2 + L(\xi) = F(\underline{u},\underline{u}) \qquad (1)$$

The type of the waves need not to be specified at this point ; ξ is the displacement vector, and L a linear operator in ξ. The source term on the right hand side, $F(\underline{u},\underline{u})$, is in general a quadratic form in $\underline{u}$, the convective velocity. It is the small amplitude assumption which allows to linearize in ξ, and also to neglect the feedback of the waves on the convective motions (this is the so-called weak-emission approximation.

The explicit expressions for $L(\xi)$ and $F(\underline{u},\underline{u})$ may be found for example in Goldreich and Keeley (1977), who have applied this equation to calculate the damping of the global acoustic modes in the solar convection zone.

The various treatments of the coupling between waves and convection all deal with an equation of this kind, but they differ by the way they model the convective motions in the forcing term $F(\underline{u},\underline{u})$.

1.1. No forcing.

A preliminary step is to examine just the effect of an unstable temperature stratification, without the forcing term $F(\underline{u},\underline{u})$, as was done by Souffrin and Spiegel (1968). They consider two superposed regions : an unstable one, whose entropy gradient is ∇S_1, and a stable one, with gradient ∇S_2. Such a stratification hosts two sets of gravity modes, which in the adiabatic limit are the unstable convective modes (g^-), and the stable gravity modes (g^+). For simplicity, thermal dissipation is taken into account through Newton's law :

$$d\theta/dt + q_i\theta = 0 \qquad (i = 1,2) \qquad (2)$$

θ being the temperature fluctuation and q_i the thermal relaxation rate characterizing each zone.

Souffrin and Spiegel found that the g^+ modes become unstable, due to the presence of the superadiabatic zone, and they proposed this mechanism to interpret the variability of the Beta Cephei stars. The condition for this instability to occur can be written.

$$q_1 \, |\nabla S_1| \; > K \, q_2 \, |\nabla S_2| \qquad (3)$$

where K is of order unity when the two zones are of comparable size. This condition seems difficult to achieve in a real star : in general, $q_1$and q_2 are of same order, but the superadiabatic gradient ∇S_1 is much smaller than ∇S_2, the subadiabatic gradient in the stable zone (in absolute value).

However, this criterion has been derived for an ideal case, with two superposed homogeneous layers, and using the Boussinesq approximation. In a real star, the density stratification modifies the shape of the eigenfuctions, especially that of the higher overtones, and the radiative damping is also depth-dependent. It is not easy to predict how this will modify the instability criterion above ; those colleagues who have a pulsation code at their disposal should be encouraged to reexamine the properties of the g^+ modes by introducing a realistic (i.e. non zero) superadiabatic gradient in the convective core of their models.

1.2. Convective forcing

1.2.1. Convection described by linear modes. It is possible to express the forcing term by using the linear modes to describe the convective motions, such as

$$\underline{u}(x,y,z,t) = \Sigma' \underline{v}(x,y,z) \exp \sigma t \quad (4)$$

However, the exponential growth of the unstable modes reduces this approach to a formal exercise of limited applicability.

In a rotating star, the situation is different. The convective modes are overstable, at least in the linear approximation (Cowling 1951), hence

$$\underline{u} = \Sigma \, \underline{v}(x,y,z) \exp (\sigma t + i\omega t) \quad . \quad (5)$$

Osaki (1974) made the assumption that this oscillatory behavior persists in the fully developed convection ; he noticed moreover that the dominant term in F is due to the Coriolis force in a convective core, and that it is therefore linear in $\underline{u}$. These oscillatory convective motions may thus couple linearly with the stable gravity modes (g^+), and Osaki proposed this as another explanation for the Beta Cephei instability.

Shortly after, Kato (1974) suggested that this two-mode coupling should be treated as a single mixed mode, with substantial amplitude both in the convective core and in the envelope, and this idea has been applied recently by Lee and Saio (1986). They use a method which is not restricted to small rotation rates, as in Osaki's work, and they find indeed such mixed modes for a set of values of the rotational frequency. Those depend on the superadiabatic gradient in the convective core, which the authors treat at this stage as an independent parameter ; the next step would be to calculate it, taking into account the stabilizing effect of the rotation on convection. Although this theory is still not fully selfconsistent, and does not include the radiative damping, it demonstrates the existence of modes which behave much like g^+ modes in the radiative envelope and which draw energy from the convective core of the star. This is precisely what is needed to interpret the Beta Cephei variability, and the predicted pulsation frequencies seem also to agree well with the observations.

1.2.2. Turbulent medium. In this treatment, the convective region is described as a turbulent medium, whose statistical properties are used to estimate the strength of the emission of waves. The theory has been worked out by Lighthill (1952) for a homogeneous medium, and it has been applied to the acoustic waves generation in the solar convection zone.

Lighthill's theory predicts that the fraction of the kinetic energy which is converted into acoustic energy is of the order of the 5th power of the Mach number characterizing the convective motions. Since that number is extremely small in the convective core of an early type star (less than 10^{-3}) we may conclude that little acoustic energy will originate from it.

The emission of gravity modes by a stratified turbulent medium has been studied by Stein (1967), using a similar procedure. He found that the efficiency of the transfer of kinetic energy from the turbulent motions to gravity waves is extremely high, so high in fact that the weak-emission theory does no longer apply. The problem was reexamined by Press (1981), who essentially confirmed Stein's result. But when calculating the total energy emitted in the form of gravity waves, one has to take into account the exact profile of the g^{+} modes in the convective core (which are of the evanescent type there), and this has not been done properly yet to our knowledge.

Furthermore, the energy spectrum of the gravity modes is likely to peak at a period which is of the order of the turnover time of the convective eddies (typically from one week to one month for early-type stars) ; the corresponding modes are of high spatial order, and therefore subjected to severe damping in the outer region of the radiative envelope. It remains to be verified if enough of this energy will reach the surface to produce observable effects there.

Both Lighthill's theory and its extension to gravity modes by Stein suffer from the fact that the nonlinear source term in Eq. 1 is estimated in a very crude way, using a scaling procedure which is far from straightforward - Goldreich and Keeley do not hesitate to call it "something of an art". Among others, one assumes some statistical properties for the convective motions ; in practice, a Kolmogoroff power spectrum, which stricly applies only to isotropic homogeneous turbulence. For that reason, those theories can only yield order of magnitude estimates of the wave flux. More precise methods are therefore highly desirable.

1.2.3. Numerical simulation. The most promising approach to study the coupling of convection and waves appears to be numerical simulation, since it permits, in principle at least, to avoid the simplifying assumptions made in the other approaches. One may wonder, therefore, why it has not been used so far.

The reason is simple : the simulations of thermal convection have still not reached a level of realism which would make them suitable for the calculation of the waves emitted by the convective motions. The core memory of the most powerful computers available until recently limits the spatial resolution to 64x64x64 in 3 dimensions, which implies that such simulations are very sensitive to the approximations made for the sub-grid scales.

The resolution is much higher when keeping only two dimensions, as was done for example by Toomre et al. (1985). They consider the same type of stratification as Souffrin and Spiegel, with an unstable zone lying above a stable one, and they find indeed that this type of convection, which penetrates deeply into the stable layer, generates gravity waves of large amplitude, much like those observed in the laboratory when performing the ice-water experiment (Whitehead and Chen 1970). However, the condition of two-dimensionality puts a severe constraint on the dynamics, favoring the formation of large and powerful eddies, and the conclusions reached by Toomre et al. are not easily transposable to the real, three-dimensional case.

Fortunately, the situation is changing very rapidly with the avent of a new generation of supercomputers, which have a parallel architecture and endowed with much larger memories. Numerical simulations will soon been performed that will permit a reliable description of the waves generated in stellar convection zones.

In the meanwhile, the picture is far from clear. My personal feeling, at present, is that the mechanism proposed by Osaki, in the recent description of Lee and Saio, looks very promising. If this is confirmed by further work, the rotation of the star would play a major role in permitting the transfer of energy from the convective core to waves, which in this case are modified g^+ modes, whose direct or indirect effects may be observed on the surface of early type stars.

But the rotation may also intervene directly and be itself responsible for instabilities, as we shall see next.

2. INSTABILITIES DUES DO THE ROTATION

Rotating flows are liable to a great number of instabilities, and we do not intend to describe them here in detail. We refer instead the reader to recent reviews on this subject, which had mainly the astrophysical applications in mind : Tritton and Davies (1981), Knobloch and Spruit (1982), Zahn (1983) and Zahn (1984).

We shall also restrict here our attention to those instabilities which arise even when the rotational kinetic energy is small compared to the gravitational potential energy ; in other words, we assume that the star is far from being disrupted by the centrifugal force. We leave thus aside the very interesting case of the fastest rotators, in which the centrifugal force may overcome the gravity, leading to large departures from spherical symmetry, and in the most extreme situation, to a disk in Keplerian rotation.

Furthermore, we shall focus only on those regions of the star where the entropy stratification is stable (i.e. radiative zones), since the instabilities related to the convection have been considered above.

For the present purpose, let us recall that the instabilities of rotating flows may be classified into two broad families, according to the source of energy which feeds them. The first is that of the Rayleigh-Taylor type instabilities, which draw their energy from the potential energy of the medium, built up by the entropy stratification or/and the centrifugal force. The second family is that of the shear instabilities, whose energy originates from the kinetic energy of the mean flow. Let us recall the main properties of some typical instabilities of each of these families.

2.1. Baroclinic instabilities

Those instabilities are, in the first family we just mentioned, the most likely to occur in the general case of a differentiallly rotating star, if the rotation rate is not constant on cylinders. Then the effective gravity (gravitational force + centrifugal force) does not derive from a potential, and the surfaces of constant pressure (i.e. the level surfaces) do not coincide with the surfaces of constant density. If the medium has a uniform chemical composition, the angle between the level surfaces and those of constant entropy is given by

$$\sin \alpha = N_T^{-2} \frac{\partial}{\partial z} (\varpi \Omega^2) \qquad (6)$$

where z is the coordinate along the rotation axis, $\bar{\omega}\Omega^2$ the centrifugal force, and N_T is the buoyancy frequency, which measures the strength of the entropy stratification :

$$N_T^2 = \frac{g}{H_p} [\nabla ad - \nabla rad] \qquad (7)$$

using classical rotations (H_p is the pressure scale height).

A naive gedankenexperiment, inspired by that which often used to establish the Schwarzschild criterion, would lead to the conclusion that such a baroclinic state is always unstable : a blob displaced sideways but upwards in the wedge between the two surfaces would become hotter than its environment, hence lighter, and it would thus keep moving this way. But such a displacement would not necessarily ensure the conservation of angular momentum, and therefore one must proceed with a more careful analysis.

Let us consider first the case where the perturbation is axisymmetric, i.e. where a whole annulus is expanded or contracted symmetrically around the rotation axis. Then, in the adiabatic case, neglecting thermal diffusion and viscous dissipation, the instability occurs only if the specific angular momentum decreases from pole to equator on an isentropic surface. As a matter of fact, a perturbation along such an isentropic surface does not experience the buoyancy force, and one recovers therefore the classical Rayleigh criterion which applies to a homogeneous rotating medium.

Departures from adiabacy allow other types of baroclinic instabilities, because in the stellar matter heat diffuses on a much faster rate than momentum (for the radiative conductivity is much larger than the viscosity). Goldreich and Schubert (1967) and Fricke (1968) were the first to describe this instability in the astrophysical context, hence its name, the Goldreich-Schubert-Fricke instability. More recently, Knobloch and Spruit (1983) have introduced a third agent, that diffuses sensibly on the same rate than momentum, namely the chemical composition. Their generalization of the theory leads to the conclusion that a baroclinic star of very moderate differential rotation is liable to linear, axisymmetric instabilities, whose growth rate is related to the local thermal time scale. But there is still some debate about the nonlinear fate of these multidiffusive instabilities, and hence about their efficiency ; it is likely that they will be obliterated by instabilities proceeding on a faster, dynamical time scale, should those occur.

When the perturbations are allowed to be non-axisymmetric, the constraint imposed by the conservation of angular momentum is much weaker. But the theory has only been worked out in some detail for an ideal fluid (no thermal diffusion, no viscosity) and in the linear limit. The instability criterion involves both the strength of the differential rotation (in the vertical direction) and its profile ; for a complete account on the subject, see Pedlosky (1979). Some specific profiles have been examined by Spruit et al. (1983) and by Zahn (1983) : their conclusion is that in stellar interiors, this instability is likely to occur only in the vicinity of convection zones, where the stabilizing effect of the entropy stratification is sufficiently weakened.

What remains to be verified, is whether baroclinic instabilities of finite amplitude are not as likely to occur as the shear instabilities which will be examined next. Here also, numerical simulation may provide the best tool to investigate this rather complex problem.

2.2. Shear Instabilities

Shear instabilities extract vorticity from a mean laminar flow, to build turbulent vortices. As stated earlier, they draw their energy from the kinetic energy of the mean flow.

Let us recall their main properties (for the relevant bibliography, see Zahn 1983).

- A laminar flow becomes unstable as soon as the Reynolds number which characterizes it becomes larger than some critical value. This critical number depends somewhat on the profile of the flow and on the boundary conditions, but typically it is always of order 100 or 1000. Depending on the profile and/or on the boundary conditions, the instability may be a linear one, or else of finite amplitude.
- The Coriolis force by itself is unable to inhibit the instability, as it is now well established by theoretical work and by laboratory experiments. In other words, a differentially rotating flow will become unstable under the same conditions as a plane-parallel flow.
- The only stabilizing mechanism (to be precise : in the absence of a magnetic field) is a density or entropy stratification (respectively in a incompressible or compressible medium). But even this stabilizing effect operates only in the vertical direction, and it is seriously weakened by thermal dissipation ; only a chemical composition gradient is in general strong enough to prevent the shear instability, provided

$$N_\mu^2 \,/\, (du/dz)^2 \gtrsim 1/4 \qquad (8)$$

where N_μ is the buoyancy frequency associated with the vertical gradient of molecular weight :

$$N_\mu^2 = \frac{g}{H_P} \frac{\partial \ln \mu}{\partial \ln P} \quad . \qquad (9)$$

The conclusion which must be drawn immediately is that a horizontal shear flow cannot be stabilized and therefore that, in a star, the slightest departure from uniform rotation, on equipotential surfaces, will provoke shear instabilities. Indeed, the Reynolds number characterizing such a flow will be supercritical even for an extremely small differential rotation, due to the size of the star.

Those instabilities will generate turbulent motions which are predominantly horizontal, thus two-dimensional ; only the smallest vortices will not feel the Coriolis force and will thus turn into three-dimensional turbulence. It is this thre-dimensional tail of the turbulent spectrum which will be therefore responsible for turbulent diffusion of momentum and chemical elements in the vertical direction.

From this we may conclude that a differentially rotating star will be the seat of instabilities, which transform the kinetic energy of the differential rotation into turbulent energy, a fraction of which may in turn be transformed into waves that can reach the surface and provoke there observable effects. To estimate the magnitude of such effects, which is of course related to the strength of the turbulent motions, we now must address the energy problem.

2.3. The energy problem

If there is no mass loss (or gain), the mean angular momentum of a rotating star is conserved. Since the shear instabilities extract kinetic energy from the mean flow, this flow will evolve towards uniform rotation, which is well known to be, for a fixed total angular momentum, the state of lowest rotational energy and the instabilities described above will then die out.

We may thus ask the question whether differential rotation may be created or maintained by some mechanism. Such mechanisms are well known :

- During the course of its evolution, a star expands or contracts, and these adjustments do not proceed in general in a homologous way. Differential rotation will thus results, presumably mainly in the vertical direction.
- If the star is a component of a binary system, the tidal torque exerted by its companion induces differential rotation, since the outer layers are more sensitive to the tide. The energy supply may be huge in this case, since it derives from the orbital motion.
- Waves, such as those emitted by the convective core and which we have examined earlier, can carry angular momentum and deposit it in the differential rotation, due to some dissipative effect. This question has been examined by Ando in a series of very interesting papers (1982, 1983, 1986).
- Another possibility is meridional circulation, which is due to the

thermal imbalance induced by the centrifugal distorsion of the star. Such a circulation advects angular momentum, and hence creates or maintains differential rotation. This mechanism has been explored with the goal of estimating the turbulent diffusitivity in the vertical direction (Zahn 1983). The flow diagram of the energy is sketched below (Fig. 1).

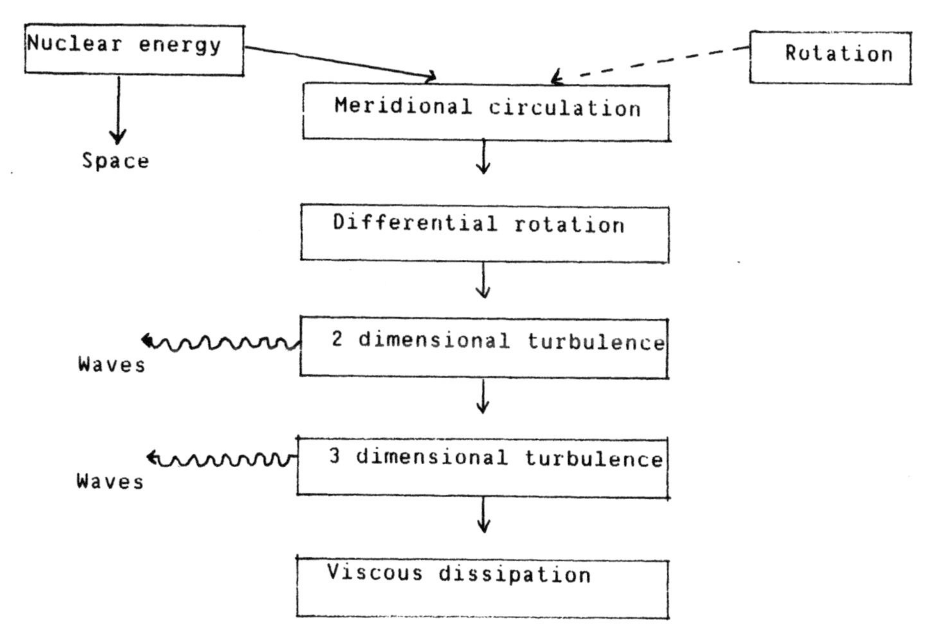

Both in Ando's scheme and in this one, the origin of the kinetic energy of the flow is the nuclear energy, and its supply lasts therefore as long as the star burns its nuclear fuel.

The main difficulty, when elaborating on such a mechanism, is to calculate the efficiency of the transformation of thermal energy into mechanical energy. The second principle of thermodynamics is not easily applicable, since one cannot isolate systems with definite sources and sinks. If one uses instead a phenomenological approach, one gets the crude estimate, for the latter scheme, that this efficiency η is of the order of the ratio of the centrifugal force to the gravity, i.e.

$$\eta = \varpi \Omega^2 / g \quad . \tag{10}$$

The turbulent energy created per unit time is therefore a substantial fraction of the total luminosity, at least in rapid rotators such as early type main sequence stars, since one can show that it il will be of order $\eta^2 L$.

The turbulent viscosity deduced from this scheme is in reasonable agrement with that postulated by Schatzman and Maeder (1981) to interpret the abundance of He^3 in the solar wind (see also Lebreton and Maeder 1986). It also provides an explanation for the C^{13} and nitrogen enrichment in stars evolving off the main sequence, and for the blue stragglers which are observed in some clusters, and which in this picture would be nearly homogeneous mixed stars, due to their fast rotation (Maeder, private communication).

Estimating what fraction of that mechanical energy may take the form of waves and be thus deposited in the outer layers of the star is an even more difficult task, and it has not been undertaken yet. The waves which are the most likely to be generated by the two-dimensional turbulence are the so-called Rossby waves, which may easily couple with the observed gravity and toroïdal oscillations.

3. CONCLUSION

We have seen that, in early type stars, both the convection (presumably in the convective core) and the differential rotation can cause instabilities which transform thermal energy (from nuclear origin) into mechanical energy, which may either be observed directly in form of oscillations or surface turbulence, or may be used to create chromospheres, coronae and winds.

However the theory is presently still in a state where it does not enable us to make quantitative predictions, to be compared directly with the observations. If one is in a pessimistic mood, one might even lament that the most relevant mechanisms are not identified with certainty. But work is in progress, and increasing contact is made between theory and observations.

Let me conclude by suggesting some observational tests which would be most helpful in guiding theoretical research, in my view at least :

- more precise estimates of the energy involved in the various instabilities that are observed ;
- clues on whether the oscillations (or more generally the phenomena) are periodic or aperiodic in time ;
- evidence for (or against) horizontal mass motions, or pole-equator temperature differences ;
- evidence for vertical transport, which may be revealed by abundance anomalies or through deviations from the standard location of the star in the Hertzsprung-Russel diagram.

REFERENCES

Ando,H. 1982, Astron. Astrophys., 103, 7
Ando,H. 1983, Publ. Astr. Soc. Japan, 35, 343
Ando,H. 1986, Astron. Astrophys., 163, 97
Cowling,T.G. 1951, Astrophys. J., 114, 272
Fricke,K.J. 1968, Z. Astrophys., 68, 317
Goldreich,P.A. and Schubert,C. 1967, Astrophys. J., 150, 571
Goldreich,P.A. and Keeley,D.A. 1977 Astrophys. J., 212, 243
Kato,S. 1974, Publ. Astron. Soc. Japan, 26, 341
Knobloch,E. and Spruit,H.C. 1982, Astron. Astrophys., 113, 261
Knobloch,E. and Spruit,H.C. 1983, Astron. Astrophys., 125, 159
Lee,U. and Saio,H. 1986 (preprint)
Lebreton,Y. and Maeder,A. 1986, Astron. Astrophys., in press
Lighthill,M.J. 1952, Proc. Roy. Soc. London, A, 211, 564

Osaki,Y. 1974, Astrophys. J., 1989, 469
Pedlosky,J. 1979, Geophysical Fluid Dynamics, 482 (Springer ; Berlin, Heidelberg, New-York)
Press,W.H. 1981, Astrophys. J., 245, 286
Schatzman,F. and Maeder,A. 1981, Astron. Astrophys., 6, 1
Souffrin,P. and Spiegel,E.A. 1967, Ann. Astrophys., 30, 985
Stein,R. 1967, Solar Phys., 2, 385
Toomre,J., Hurlburt,N. and Massaguer,J.M. 1985, Proc. Nation. Solar Obs. Workshop 1983, 222 (Keil edit.)
Tritton,D.J. and Davies,P.A. 1981, Hydrodynamic Instabilities and the Transition to Turbulence, 289 (Lecture Notes in Physics, vol. 71, Springer ; Berlin, Heidelberg, New York)
Whitehead,J.A. and Chen,M.M. 1970, J. Fluid Mech., 40, 549
Zahn,J.-P. 1983, Astrophys. Processes in Upper Main Sequence Stars, 253 (Publ. Geneva Observatory)
Zahn,J.-P. 1984, Proc. 25th Liège Intern. Astrophys. Coll., 407 (Univ. Liège, edit)

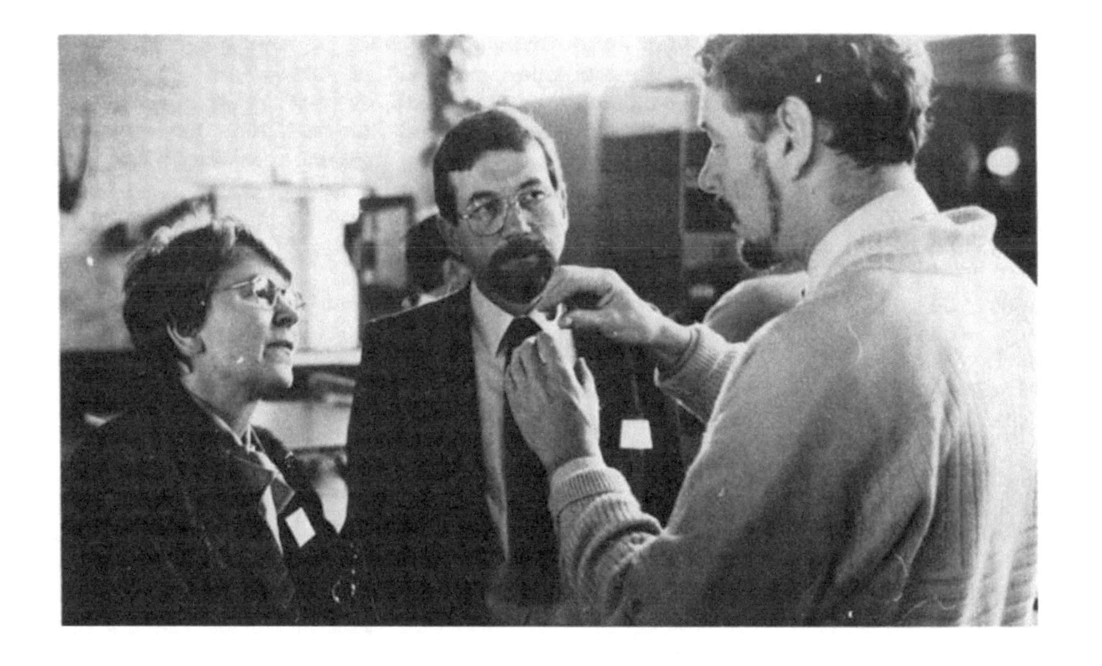

Jean-Paul Zahn and Françoise Praderie listening to an explanation by Henny Lamers. Do they believe it?

DISCUSSION

ULMSCHNEIDER: To make the picture more complicated I want to point out that for the acoustic energy generation the presence of magnetic fields can make order of magnitude differences. The quadrupole type Lighthill generation mechanism becomes a monopole generation mechanism in the presence of a magnetic fluxtube. In late type main sequence stars the chromospheric heating is almost exclusively in magnetic flux tube regions. I don't know how important magnetic fields are for early type stars.

ULMSCHNEIDER: Gravity wave generation has been computed for the sun by Bob Stein and published 1968 in Solar Physics. He however later concluded that gravity waves do not exist in the convection zone as the wave frequencies there become imaginary i.e. they are damped.

ZAHN: Gravity modes penetrate somewhat into a convection zone, and may therefore be excited by the convective motions. More likely, it is the penetrative convection which generates the gravity waves, as it occurs for instance in the laboratory (cf. the ice-water experiment).

CASSINELLI: I often hear that early type stars cannot have magnetic fields other than in primordial fields. Wouldn't the differential rotation and turbulence that you have discussed lead to the generation of magnetic fields through a dynamo process? As is shown on my poster paper, there is evidence from the X-ray spectra of early type stars for the presence of gas with temperatures around 20 million degrees (in τ Sco BOV, ζ Oph 09.5V and ζ Ori 09.5Ia). The presence of this hot gas could be understood if there are small magnetic loops with fields of about 30 to 100 Gauss on the stellar atmosphere. It is difficult to explain the hot gas with shocks or wind instabilities.

ZAHN: You are right. Differential rotation and turbulence may well produce magnetic fields through a dynamo mechanism, as we believe that it is the case in the solar convective zone. It is true that the turbulence velocities are probably much weaker than in a convective region, but their two-dimensional character can compensate this.

PRADERIE: In the paper of Mangeney and Praderie, (1984, A+A), we succeeded to correlate the X luminosity L_X in main sequence stars, from type O to M, with an effective Rossby number (Ro = inertia force/Coriolis force). The point is that the relation is unique for the whole main sequence. A word of caution is necessary as to the meaning of this relation for hot stars. The number Ro was computed with envelope models due to Maeder, which indeed have a small subphotospheric convective zone up to 20 $M_{\odot}$ or more. But we don't know at present if a magnetic field can be produced in such superficial convective zones in hot stars, or if Ro is the right number to characterize the superficial appearance of a magnetic field, which could be primordial and brought to the surface by rotation, as Uchida suggests.

MAEDER: Is there any numerical simulation of convection with velocities close to the sound velocities, as expected in red supergiants?

ZAHN: Not to my knowledge. Most numerical simulations have been carried out in the anelastric approximation, which filters out the sound waves and strictly applies only to small Mach numbers. Fully compressible simulations (such as by Graham in 1975) have been restricted so far to small Rayleigh numbers, and therefore to small velocities.

APPENZELLER: As we learned from your talk, rotation produces a thermal imbalance, driving circulations, which in turn may produce turbulence, which then may energize non-radial oscillations. As all stars rotate, circulations are probably very common and the above sequence could in principle work in about any star. Can you give some general rules for the conditions under which this process is likely to be important and will really drive observable oscillations?

ZAHN: The amount of energy released by the thermal unbalance due to rotation depends strongly on the rotational velocity. This may explain the large differences in behaviour that we observe from star to star.

DE LOORE: Can you comment on the relative importance of the various components you showed in your diagram, meridional circulation, 2D, 3D-turbulence?
How important is this for the stellar structure? (Atmosphere, internal regions.)

ZAHN: I have tried to address this question in more detail in the written version of my review.

OSAKI: I have a comment about the Souffrin-Spiegel mechanism for the excitation of nonradial g^+-modes by convection zones. You have mentioned that this mechanism may be rather unlikely in real stars. However, we (Shibahaski and Osaki, 1981, PASJ., 33, 427) have found in yellow supergiant models violently unstable non-radial f-modes which are presumably driven by the Souffrin-Spiegel mechanism.

OSAKI: You have discussed convection and rotation separately. Would you comment on the possibility of wave generation by unstable temperature stratification when the restoring force due to rotation exists?

ZAHN: You are the expert in this field, and you should answer this question!

OWOCKI: I always learned (e.g. in graduate school) that early-type star envelopes were relatively simple; that is radiative vs. convective. Now we're talking here about convection, waves, pulsations etc. in these stars. Please summarize what has happened in the last 5 years that has so revolutionized this classical view. Are these stellar envelopes now thought to be unstable, e.g. by the Schwarzschild criterion? Or is the origin of convection these more complicated mechanisms you have described?

ZAHN: These stellar envelopes are convectively stable, i.e. they satisfy the Schwarzschild criterion for stability. Moreover, other instabilities may occur, for instance due to the rotation. Their efficiency is still a matter of debate, and observational tests are highly desired. (Abundance anomalies, surface velocity fields, mechanical fluxes in the atmosphere, very precise location of a star in the HR-diagram, etc.)

CASSINELLI: I would like to respond to a question of Stan Owocki about spots on hot stars.There is an important difference from the cool star case. In cool stars much of the energy transport in the subphotospheric region is by convection and the strong fields in spots could inhibit the transport. However, in hot stars the energy transport is radiative, which would not be affected directly by the fields. So even if there are strong fields on the surface, dark spots would not be seen. If the fields led to a concentration of specific materials at zones on the surface as in Ap Stars, there could be changes in the opacity that could lead to spots. However, for the luminous hot stars that we have been discussing that would not occur, and so dark spots will not be observed.

BAADE: 1) It appears to me that the detection of non-radial pulsation with relatively high non-radial order ($m \simeq 10$; values up to 16 have been claimed) implies a limit amplitude of turbulent motions.
2) I now have clear evidence of a feature traversing the line profiles of the Be star μ Cen (B2). A possible explanation is by a wave which is retrograde even in the inertial frame. Would the instability mechanisms discussed by you have problems accommodating such a case?

ZAHN: 1) You probably refer to the direct observation of high-order modes through Doppler-imaging. In this case, as you state, too large turbulent velocities would render the bumps in the profile invisible.
2) A wave travelling in the retrograde sense in the inertial frame cannot be a R-mode. But p and g-modes can drift faster than the rotational speed, and if they travel in the retrograde sense, they will appear to do so also in the inertial frame.

LAMERS: During the previous reviews and discussions the observers have shown evidence for an overwhelming range of variations of different types, periods etc. Yet, at the end of your talk you showed a 'shopping list' of features which we should observe. Did we make the wrong observations up to now?

ZAHN: You didn't, of course. All observations are most welcome. My list suggests some tests which are very important for the validation of current theories.

MAEDER: Regarding the efficiency of turbulent diffusion induced by differential rotation, we can notice that it is generally considered that the effect needs a long main sequence lifetime and thus is likely to occur in low mass stars. However, the diffusion coefficient contains the viscosity $\nu \sim T^4/\kappa\rho^2$ which is very large in massive stars and the process is likely to reappear in very massive stars.

STICKLAND: You and Dr. Osaki mentioned, but did not elaborate on, tidally induced instabilities. If you have a binary system like Iota Orionis where e = 0.46 and the period is 29 days, you must have an impulse lasting a day or so after which the stars can relax. What would you expect to see on the surface of the components?

ZAHN: There would be the usual effects of interaction such as synchronization, but the surface phenomena would be difficult to predict.

PRADERIE: If it is important to have direct determination of differential rotation, could one imagine an observational program looking at rotational splitting in well chosen modes, so that one probes different depths in hot stars, as was done succesfully in ZZ Ceti type dwarfs, and also in the Sun? This requires very high temporal resolution observations, and would be suited for a variability devoted satellite experiment.

BAADE: Claims of detected rotational m-mode splitting have been made only for ß Cephei stars. The main argument were 3 equidistant frequencies in some stars. But it has been pointed out that this is exactly what one should not expect in a star with finite rotation. In purely nonradially pulsating stars which often have much longer periods, the matching of the spectroscopic and photometric periods not seldom is still unsatisfactory even for the strongest mode. All this plus the apparently strong increase of the amplitudes with wavelength and the aliassing problem with periods comparable to one day or one night call for observations from space. Only then there is a chance to detect sufficiently many modes.

André Maeder during one of his many contributions to the workshop

SHOCK WAVES IN LUMINOUS EARLY-TYPE STARS*

John I. Castor
Lawrence Livermore National Laboratory
University of California
Livermore, CA 94550
U. S. A.

ABSTRACT. Shock waves that occur in stellar atmospheres have their origin in some hydrodynamic instability of the atmosphere itself or of the stellar interior. In luminous early-type stars these two possibilities are represented by shocks due to an unstable radiatively-accelerated wind, and to shocks generated by the non-radial pulsations known to be present in many or most OB stars. This review is concerned with the structure and development of the shocks in these two cases, and especially with the mass loss that may be due specifically to the shocks. Pulsation-produced shocks are found to be very unfavorable for causing mass loss, owing to the great radiation efficiency that allows them to remain isothermal. The situation regarding radiatively-driven shocks remains unclear, awaiting detailed hydrodynamics calculations.

1. GENERAL PROPERTIES OF SHOCKS

In order to understand how shock waves behave in a stellar atmosphere, it is first necessary to recall the Rankine–Hugoniot shock jump conditions, and then consider the various relaxation processes that take place near the shock. What we call a 'shock' when we view it in the large is really a complex non-equilibrium flow when examined in detail. The details must be considered before choosing the correct way of embedding the shock in the larger problem.

1.1. Jump Conditions

The effects of a shock are all consequences of the jumps in density, pressure, temperature and velocity that are dictated by the laws of conservation of mass, momentum and energy. If ρ, p, T, and v are, respectively, the density, pressure, temperature and flow velocity relative to the shock, then the jump conditions for a strong shock are

$$\frac{\rho_1}{\rho_0} = \frac{v_0}{v_1} = 4, \quad p_1 = \frac{3}{4}\rho_0 v_0^2, \quad and \quad T_1 = \frac{3}{16}\frac{\mu v_0^2}{\mathcal{R}}.$$

* This work was performed under the auspices of the U. S. Department of Energy by the Lawrence Livermore National Laboratory under Contract No. W-7405-ENG-48.

H. J. G. L. M. Lamers and C. W. H. de Loore (eds.), Instabilities in Luminous Early Type Stars, 159–174.

(μ is the mean atomic weight, $\mathcal{R}$ is the gas constant, and subscripts 0 and 1 refer to pre- and post-shock, respectively.) A shock is strong when the upstream flow velocity relative to the shock, v_0, is large compared with the upstream sound speed $a = \sqrt{p_0/\rho_0}$. Many of the interesting effects of shocks are due to the elevated temperature of the post-shock gas; the numerical value of T_1 is given by

$$T_1 = 1.51 \times 10^5 \left(\frac{v_0}{100\,\mathrm{km\,s^{-1}}}\right)^2 \mathrm{K}.$$

Since the speeds that have been suggested for shocks in OB stars are of order 200–600 km s^{-1}, post-shock temperatures in the range $T_1 = 6 \times 10^5$–5×10^6 K are expected.

1.2. Internal Structure of Shocks

The 'discontinuity' of flow variables at a shock is an idealization—the atomic nature of the gas, and the transport properties due both to atoms and radiation, smooth out the jumps and give them a finite width. Atomic transport gives shocks a width about equal to the gas-kinetic mean free path in the shocked gas (with a correction for the electron/proton mass ratio). (See Zel'dovich and Raizer [1967], § VII.2.) This thickness corresponds roughly to a particle column thickness (*i.e.*, $N = \int n\,dx$) $\approx 1 \times 10^4 T_1^2/\ln\Lambda$, where $\ln\Lambda$ is the usual Coulomb logarithm. For $T \approx 10^6$ K, this thickness is $\approx 10^{15}\,\mathrm{cm^{-2}}$. As we will see shortly, this is quite narrow compared with the broader parts of the shock's internal structure. A very useful approximation is that there is a true discontinuity—the 'gas-dynamic' shock—embedded in a broader region of radiation transport effects and excitation and ionization relaxation.

It may be helpful to picture the whole structure of the shock as containing four regions, from upstream to downstream: (A) cold unshocked gas; (B) hot shocked gas—region of ionization run-up; (C) radiative cooling, ionization about in balance; and (D) cold dense gas. The 'gas-dynamic' shock separates regions (A) and (B). Some of the radiation produced as the gas cools in region (C) may be absorbed in region (A), leading to a 'radiative precursor' (Zel'dovich and Raizer [1967], §§ VII.14–18). The length scale of this precursor, if the gas flows into the shock too quickly for it to be able to come to thermal equilibrium with the precursor radiation, is a mean free path of the predominant radiation. This radiation is mostly in the He II Lyman continuum, and the mean free path corresponds to a column N of $\approx 10^{19}\,\mathrm{cm^{-2}}$, or more if the predominant photon energy is above 100 eV. This scale is great enough that it belongs to the outer structure of the stellar atmosphere, rather than to the internal structure of the shock. That is, for our purpose region (A) can be considered transparent, and whatever preheating and preionization occur have taken place before the gas flows into region (A).

The Mach number (v/a) of the flow is small throughout regions (B), (C) and (D), with the result that the pressure is nearly constant ($= \rho_0 v_0^2$). The temperature in region (D) has cooled to the level determined by radiative energy balance with the ambient radiation field, which also determines the temperature in region (A). (Regions [A] and [D] view the same radiation unless region [A] is opaque.) Thus the jump from region (A) to region (D) can be called an 'isothermal shock'. Although there is no net temperature jump, there is a large density jump

$$\frac{\rho(\mathrm{D})}{\rho_0} = \left(\frac{v_0}{a_0}\right)^2 \gg 1,$$

an equal pressure jump, and a reciprocal velocity jump. Notice that the density jump can become arbitrarily large instead of being limited to 4, as in an adiabatic shock.

The processes that occur in regions (B) and (C) are these: The gas enters region (B) quite hot, since the pre-shock kinetic energy has been converted to enthalpy. But the state of ionization and excitation of the material is unchanged by the shock. This situation is very much out of equilibrium, and a process of relaxation begins, in which thermal energy is used up in ionizing and exciting the atoms. As long as the degree of ionization of the atoms is low, the ionization rate is very rapid; as the ionization increases, the rate slows, and a steady state of ionization balance would be reached if time and space permitted and if radiative processes did not intervene. This is about the state of the matter as it exits region (B) for region (C). The ionization and recombination rates around the state of ionization equilibrium are comparable with the rates of emission of radiation, such as free-bound emission and collision-induced resonance line emission. These processes convert thermal and excitation energy into radiation, which leaves the region of the shock. This cooling is what occupies region (C). Since ionization rates are somewhat larger than the cooling rate, the material stays relatively near the condition of ionization equilibrium as it cools. Region (C) ends when the radiative cooling rate is balanced by the heating rate due to the ambient radiation.

There are a variety of characteristic column thicknesses N associated with regions (B) and (C). The thickness of the ionization layer for an ion depends on its ionization potential, χ, with the layer being very thin if χ is small, and thick if χ is large:

$$N \approx \begin{cases} 10^{14}\left(\dfrac{v_0}{100\,\mathrm{km\,s^{-1}}}\right)^4 \mathrm{cm}^{-2}, & \chi \approx kT; \\ 4\times 10^{17}\left(\dfrac{v_0}{100\,\mathrm{km\,s^{-1}}}\right)^4 \mathrm{cm}^{-2}, & \chi \approx 5kT. \end{cases}$$

The typical value of χ when the degree of ionization is highest is about $5kT$. The column thickness of the ionization region for the low ionization potential case is similar to, and perhaps smaller than, the thickness of the gas-kinetic shock. This means that low ionization potential species may be ionized within the shock. However the peak ionization is attained only at a downstream distance that is many times greater than the shock width.

The processes that contribute to cooling of the gas are many, and a proper cooling calculation must include many different ionic species with all their possible resonance excitations. One such set of calculations was made by Raymond, Cox and Smith (1976). The cooling rate per unit volume is expressed as $n^2\Lambda$, where n is the total number density of nuclei and Λ (not to be confused with the Coulomb logarithm) is a function of T and also weakly a function of the radiation environment and the past history of the material. (See Fig. 1 of Raymond, Cox and Smith.) Λ has a broad maximum near 10^5 K, where the cooling is dominated by collisional excitation of abundant lithium-like ions. Above about 3×10^5 K Λ declines, with bumps due to other collisional excitations. This decline can very roughly be fitted to a power law: $\Lambda \propto T^{-1/2}$. From this formula it is easy to calculate the column thickness needed for cooling to remove all the enthalpy of the shocked gas. This result has been given by Krolik and Raymond (1985):

$$N_{\mathrm{cool}} = \text{total column to return to ambient } T \approx 7\times 10^{17}\left(\frac{v_0}{100\,\mathrm{km\,s^{-1}}}\right)^4 \mathrm{cm}^{-2}.$$

This is comparable with the thickness needed to reach peak ionization, so we conclude that the mean ionization starts downward, as the temperature declines, just as the ionization and recombination rates are coming into balance. We can regard N_{cool} as the thickness of region (C), and, to a fair approximation, of the entire internal structure of the shock.

The bulk of the emission from the shock is produced in region (C). The net flux emitted is, in total, about equal to the kinetic energy flux into the shock, $\frac{1}{2}\rho_0 v_0^3$. How this flux is distributed in the spectrum depends on the actual temperature structure in the shock, which in turn is primarily sensitive to v_0. As the observations of x-rays from OB stars with the *Einstein Observatory* have shown (Cassinelli and Swank [1983]), the source of the x-rays must be material with a temperature of order 3×10^6 K, which requires $v_0 \approx 450\,\mathrm{km\,s^{-1}}$. The corresponding N_{cool} is of order $10^{20}\,\mathrm{cm}^{-2}$, which exceeds the radiation mean free path estimate given earlier; however, the typical photon energy is about 300 eV at this temperature, which increases the He II Lyman continuum mean free path to the equivalent of $N \approx 7 \times 10^{21}\,\mathrm{cm}^{-2}$.

2. SHOCKS FROM PERIODIC PULSATION

The first mechanism for shock production that I want to consider is the one in which the root instability originates in the stellar interior, so that the whole star is pulsationally unstable and can be supposed to pulsate in some normal mode with a well-defined period, P. From the atmosphere's point of view, it is being driven by an oscillating piston characterized by P and the velocity amplitude, U. If P is comparable with the period of the radial fundamental mode, about 4 hours for main-sequence OB stars, then the atmosphere is being driven at a period below its acoustic cut-off frequency (see Lamb [1945], § 309). Thus the motion of the atmosphere tends to be a standing wave, and lacks the running-wave character that most easily leads to shock formation. If the piston velocity is moderate, however, non-linear effects still produce a shock at a certain height above the piston that increases with P and decreases with U.

Once a shock forms, its strength, measured by the jump in velocity across it, increases as the shock runs upward through material of lower and lower density. When the shock is sufficiently strong, the density and velocity of the material into which the shock runs will be affected considerably by the *previous* similar shock that passed through that material. The passage of each shock delivers an upward impulse to the material, and the effect is a 'shock levitation' of the atmosphere, partially offsetting gravity. As a result, the scale height of the atmosphere is expanded, which also diminishes the tendency of the shock strength to increase with height. In this way, the shock strength finds a stable limiting value, which is a definite function of height.

The picture just described is possible only so long as the shock is effectively isothermal—the shocked gas can cool within a time shorter than, or at most comparable with, the period. The cooling time behind the shock increases as the pre- (and post-) shock density declines, so the isothermal condition inevitably fails above some height in the atmosphere. Above this point the temperature remains near the post-shock temperature, T_1, and the shocks, now weak, provide the energy deposition that maintains the temperature. This temperature is comparable with the 'escape temperature' at which the sound speed equals half the escape velocity, thus a stellar wind of the kind described by Parker (1958) results.

2.1. The Height of Shock Formation

An accurate calculation of the height at which a shock is formed when an atmosphere is driven by a periodically oscillating piston can only be done numerically. A simple estimate of the scaling of this height can, however, be made by the following argument. Assuming for the moment, as will be verified shortly, that the sound-travel time up to the height of shock formation is a small fraction of the period, then the shock should form relatively soon after the beginning of the outward-acceleration phase of the piston motion. One possible idealization, therefore, is to consider a piston moving into an atmosphere at rest with a position $R(t)$ that is constant for $t \leq 0$ and has a constant positive acceleration for $t > 0$. This has the objection that the atmosphere will respond to the piston acceleration (the standing-wave effect) so that the matter has just the same acceleration as the piston, and no shock is produced. That is, the shock creation is tied to the *change* in piston acceleration. This leads to a model like the previous one, but with a constant positive d^3R/dt^3 for $t > 0$.

Let a be the isothermal speed of sound. Then at each instant t_0 a sound wave of speed a leaves the piston at $R(t_0)$. These sound waves cover the region above the piston, and, since the later-departing waves have a larger absolute velocity, they in fact cross each other. Thus the wave trajectories form an *envelope* in the r–t diagram, with a cusp where the shock forms. (See § 49 of Courant and Friedrichs [1948].) The cusp forms in this simple model at the height

$$h \approx \frac{1}{3}\sqrt{\frac{a^3}{d^3R/dt^3}}.$$

If now d^3R/dt^3 is related to U using the assumption of sinusoidal oscillation, and taking H to be the static scale height of the atmosphere, a^2/g, the height expressed in scale heights is

$$\frac{h}{H} \approx \frac{1}{6\pi}\frac{Pg}{\sqrt{aU}}.$$

The velocity amplitude is thought to be comparable to a, and in this case h/H ranges from a few to 15 for a plausible range of P. (The sound-travel time is then about 1/3 radian of pulsation phase, justifying the earlier assumption.)

2.2. The Growth of Shock Strength with Height

For cases of interest, the shock first forms in a layer dense enough that the shock is quite isothermal, *i.e.*, the picture advanced in § 1.2 is applicable. In general, the evolution of the shock as it moves upward is governed by the pre-shock density and velocity, and all the conditions in the post-shock region. If the Mach number is large, however, the dependence on post-shock conditions becomes weak, and an approximation due to Whitham (1958) can be used. The pre-shock conditions are taken as input, and the jump conditions are solved for the post-shock flow variables with the Mach number as a parameter. These relations are substituted into the differential equation that is valid along an outward-running characteristic (sound wave), *neglecting the difference between the path of the outward characteristic and the shock*. The result is a differential equation for the Mach number:

$$\left(1+\frac{1}{M}\right)^2\frac{dM}{dr} = -\frac{d\ln\rho_0}{dr} - \frac{g}{a\left(v_0+\left(M-\frac{1}{M}+1\right)a\right)}.$$

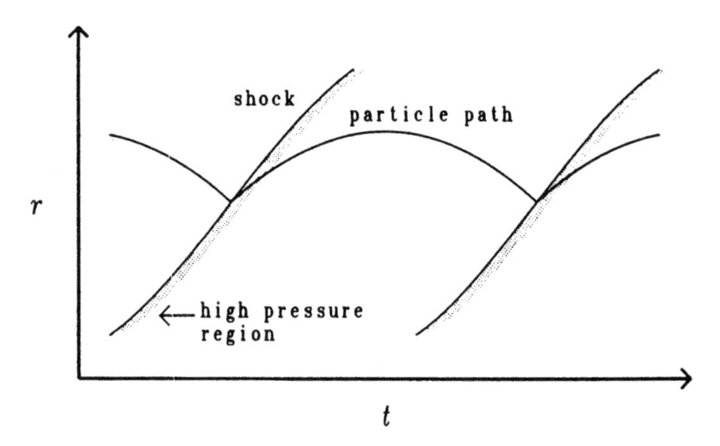

Figure 1. Space–time diagram showing the periodic shock trajectory and the path of a parcel of material. The shading indicates the high-pressure post-shock region.

If the density distribution is close to hydrostatic, then the first term on the right side is $\approx 1/H$ and is much larger than the second term, provided $M \gg 1$. This simple relation results:

$$M \approx \text{constant} - \ln \rho_0.$$

The interpretation of this relation is that it gives the Mach number of the shock as it moves upward in terms of the current density just in front of the shock. This relation breaks down when the density gradient starts to differ substantially from the hydrostatic one, which happens when the 'shock levitation' becomes significant. The dynamics then passes to the other limit, in which the Mach number no longer varies rapidly with height, and the two terms on the right of the equation above balance each other. In this case the density scale height has been extended by just a factor M.

If the shock were to become adiabatic before the levitation effect reduced the density gradient, then the development of the Mach number with height would be quite different. For an exponential distribution of density, the variation of Mach number is given by Zel'dovich and Raizer (1967, § XII.25):

$$M \propto \rho_0^{-1/\alpha},$$

where $\alpha = 4.90$ for $\gamma = 5/3$. The growth of Mach number in the adiabatic case is much stronger than in the isothermal case, once the Mach number is large. The two cases are illustrated by Castor (1970, Fig. 8).

2.3. The Limiting Strength for Periodic Shocks

The shock growth decribed in the previous section can be thought of as a transient effect before the shock attains the strength that gives full levitation of the atmospheric material. 'Full levitation' means that all the outward force on a parcel of material is exerted either in one shock jump each period, or in a relatively thin high-pressure zone behind the shock, and that for the rest of the period the parcel is essentially in free fall. (See Fig. 1.) This simple situation is amenable to both numerical and analytic treatment, and has been studied by Hill (1972), Hill and Willson (1979), Willson and Hill (1979), Willson and Bowen (1985), and Bertschinger and Chevalier (1985).

High Mach number allows the pressure-gradient force to be neglected altogether (apart from the shock jump), which leads to even greater simplification, the ballistic limit. This is justified when $Pg \gg a$, since the flow velocity scales as Pg. Values of Pg/a range from 40 to 150 for the non-radial pulsations observed in OB stars (from the periods quoted by Smith [1986]), so the approximation should be excellent.

In the ballistic model, the parcel of material is exactly in free fall between shock passages. Periodicity dictates that the pre-shock and post-shock velocities of the parcel be numerically equal ($u_0 = -u_1$) and the large Mach number also implies that the shock velocity u_s be $\approx u_1$. In order that the period be compatible with these initial and final velocities, the following condition must apply:

$$\frac{PV_{esc}(r)}{2r} \equiv 38.3Q\left(\frac{R_*}{r}\right)^{\frac{3}{2}} = \frac{\beta}{1-\beta^2} + \frac{\sin^{-1}\beta}{(1-\beta^2)^{\frac{3}{2}}},$$

where $V_{esc}(r)$ is the local escape velocity, Q is the pulsation constant in days, R_* is the stellar radius, and β stands for $u_s(r)/V_{esc}(r)$. This equation provides u_s, and indirectly all the other shock properties, as a function of r.

From the point of shock formation, the shock strength increases as it moves upward, according to the relations in § 2.2, until it reaches the limiting value just determined, whereafter the strength slowly decreases with r, following the formula above. Of course, periodic motion as described here precludes the possibility of any mass loss, and, indeed, that is the result of numerical calculations (*e.g.*, those of Hill [1972]), so long as isothermal conditions obtain.

3. MASS LOSS DUE TO PULSATION SHOCKS

As I just noted, shock-induced mass loss is tied to the breakdown of the isothermal shock approximation. Specifically, if the cooling time of the post-shock gas is longer than the pulsation period, then a parcel of material steadily gains heat as it is successively shocked, so that periodicity is impossible. This extra heat is used to do work lifting the parcel upward, producing a net outward flow or mass loss. (See Wood [1979].) The key question is the height at which the isothermal approximation breaks down; the next question is how the transition to a wind then occurs.

3.1. The Post-shock Cooling Time and the Transition to a Wind

The same fit to Λ *vs.* T quoted earlier from Krolik and Raymond (1985) can be used to find the flow time through region (C),

$$t_{cool} \approx 2\times10^{-14}\frac{(v_0/100\,\mathrm{km\,s^{-1}})^3}{\rho_0}\,\mathrm{s} \quad \text{for } 100 \le v_0 \le 1000\,\mathrm{km\,s^{-1}}.$$

The dynamic range of the density is much greater than that of the shock speed, v_0, so the critical condition $t_{cool} = P$ essentially fixes ρ_0. With P in the range 10^4–10^5 s and v_0 in the range 100–300 km s^{-1}, ρ_0 lies between 10^{-19} and 10^{-17} g cm^{-3}. This is quite a low density, due to the fact that the radiating efficiency of a moderately hot, ionized plasma is excellent.

We may define r_{ad}, the adiabatic radius, as the place where $t_{cool} = P$. Above the adiabatic radius the shock may be treated as adiabatic. Since the pre-shock temperature is now high, the shock is no longer strong, (*i.e.*, $M \approx 1$). The nature of the transition to a wind depends on the value of β at r_{ad}. If $\beta \approx 1$, then the temperature at r_{ad} is already comparable to the Parker temperature and the sonic point of a Parker wind will be at or near r_{ad}. If $\beta \ll 1$, however, the temperature at r_{ad} is less than the Parker temperature by about a factor β^2. In the latter case the sonic point lies a modest distance outside r_{ad}, and the intervening region is approximately hydrostatic with a temperature that increases outward determined by a balance between shock heating and adiabatic expansion. Simple estimates of the mass loss rates that result in these two cases give

$$\dot{\mathcal{M}} \approx \begin{cases} 4\pi r_{ad}^2 \rho_0(r_{ad}) u_s(r_{ad}), & \beta \approx 1; \\ 4\pi r_{ad}^2 \rho_0(r_{ad}) \dfrac{(u_s(r_{ad}))^2}{V_{esc}(r_{ad})}, & \beta \ll 1. \end{cases}$$

It should be noted that the mass loss rates scale directly with the density at the adiabatic radius, and thus inversely with the cooling efficiency, Λ.

As an example, consider a non-radially pulsating B star with the properties

$$\mathcal{M} = 15\,\mathcal{M}_\odot, \quad R_* = 6.4\,R_\odot, \quad V_{esc} = 944\,\mathrm{km\,s^{-1}},$$

and with a pulsation period equal either to 3 hours (in an $l = 8$ mode, say, with $Q = 0.03\,\mathrm{d}$) or to 12 hours ($l = 2, Q = 0.12\,\mathrm{d}$). The results of applying the ballistic theory are that $\beta = 0.43$ and 0.75, and $u_s = 400$ and $700\,\mathrm{km\,s^{-1}}$, for $P = 3$ hours and 12 hours, respectively, assuming $r = R_*$. However, it can be seen that the limiting amplitude is never attained while the shock is isothermal. If it is assumed that the density at the photosphere (the 'piston' location) is $10^{-9}\,\mathrm{g\,cm^{-3}}$ and that $U = a \approx 14\,\mathrm{km\,s^{-1}}$, then the densities at the height of shock formation in the two cases are $\rho \approx 2 \times 10^{-11}$ and $5 \times 10^{-17}\,\mathrm{g\,cm^{-3}}$. (The low value of ρ for $P = 12$ hours is due to the difficulty of forming a shock from a low-frequency wave.) If these densities are used in the shock-growth relation between density and Mach number, it is found that the velocities 400 and 700 $\mathrm{km\,s^{-1}}$ are attained only when ρ_0 is less than the critical value at r_{ad}; the shocks become adiabatic in the growth region. A simultaneous solution of the shock growth relation and the adiabatic radius condition leads to these data at r_{ad}:

$$\rho = 1 \times 10^{-17}\,\mathrm{g\,cm^{-3}} \quad u_s = 190\,\mathrm{km\,s^{-1}} \quad \text{for } P = 3\,\mathrm{h},$$
$$\rho = 2 \times 10^{-19}\,\mathrm{g\,cm^{-3}} \quad u_s = 80\,\mathrm{km\,s^{-1}} \quad \text{for } P = 12\,\mathrm{h}.$$

The estimates for the mass loss rate turn out to be $\dot{\mathcal{M}} \approx 2 \times 10^{-12}\,\mathcal{M}_\odot\,\mathrm{y^{-1}}$ for $P = 3$ hours, and $\dot{\mathcal{M}} \approx 1 \times 10^{-14}\,\mathcal{M}_\odot\,\mathrm{y^{-1}}$ for $P = 12$ hours.

These rates of mass loss are considerably less than the smallest rates that might be observed in OB stars, and therefore this mass loss mechanism, for these stars, does not appear to be significant. Two factors are responsible for this: One is the high cooling efficiency, which forces the density to be very low before adiabatic shocks are possible. The second factor, which is important for $P = 12$ hours, is that the long period means that the density is already quite low where the shock forms. These factors are much more severe for OB stars than for red giant variables, such as those studied by Willson and Wood, for which shock-driven mass loss appears to be quite important.

4. RADIATION-DRIVEN SHOCKS

The luminous early-type stars all have winds that are thought to be driven by the force due to resonance-line scattering (Lucy and Solomon [1970], Castor, Abbott and Klein [1975]). That these winds are unstable was suggested by Lucy and Solomon (1970); the later work on the radiatively-driven instability is reviewed by Rybicki elsewhere in these proceedings.

There is also an abundance of observational indicators of instability: x-ray emission seen with the *Einstein Observatory*; strong UV absorption by 'superionized' species like O VI and N V; non-thermal radio emission (Abbott, Bieging and Churchwell [1984]); and variable features in the UV absorption lines (Henrichs 1986). These may very well all be a result of shock waves passing through the stellar wind, and the shock waves may very well be due to the instability of the radiative driving. In this section I will discuss two such models that have been proposed.

4.1. Lucy's Periodic Shock Model

Lucy (1982) proposed a model involving a train of shock waves moving outward, spaced in time by an amount τ which turns out to be a few tens of seconds. This imparts a 'saw-tooth' structure to the flow velocity (Fig. 2). On the assumption that the cooling length is negligible compared with the shock separation $l = V\tau$, isothermal hydrodynamics is used. The shock velocity V and the shock jump U are supposed to be slow functions of r. The flow velocity in the shock frame, w, is supposed to be a strong function of the distance behind the shock front, but a slow function of r. The result is that in the accelerating frame of the shock there is a nearly-steady inward flow, with density obeying the continuity equation $\rho w \approx$ constant, and with w obeying a momentum equation that includes the outward radiation force and the inward forces of gravity and the reaction to the shock acceleration, in addition to the pressure-gradient force.

Since w must be subsonic just behind the shock, and is definitely supersonic in front of the next shock, the flow in the shock's frame must pass through a sonic point. This is possible only if the radiation force decreases inward, so that the net force is outward on the outer side of the sonic point (*i.e.*, just behind the shock), and inward on the inner side of the sonic point. This modulation of the radiation force can come only from the 'velocity shadowing' effect of the shock(s) at smaller r. In particular, the radiation force can change substantially near the sonic point $w = a$ only if the drop in V to the previous shock, $w_\dagger$, is also comparable with a. Lucy argued that stabilizing effects exist to ensure that this will be the case: A shock will 'clone sisters' if the gap to the next shock is too great, since in that case it will act like an isolated shock, which is known to be unstable. A shock will 'eat its sisters' if they get too close, since all the intervening matter will fall into the 'velocity shadow' and the outer shock will be unable to stay ahead of the inner one.

The details of the model are fairly easily worked out. The actual value of $w_\dagger$ is a parameter, which must be comparable with a. The value of the shock time interval is then determined, $\tau = w_\dagger/(V dV/dr)$. This takes values in the range 10–100 s. The shock strength and temperature are then given by

$$U \approx 300\sqrt{\frac{w_\dagger}{20}\frac{V}{2000}}\,\mathrm{km\,s^{-1}}, \quad \text{and} \quad T_1 \approx 116\frac{w_\dagger}{20}\frac{V}{2000}\,\mathrm{eV}.$$

The shock velocity V and the mean density obey the equations of steady flow, as if there were no shocks and V were the flow velocity, except that the radiation force becomes a

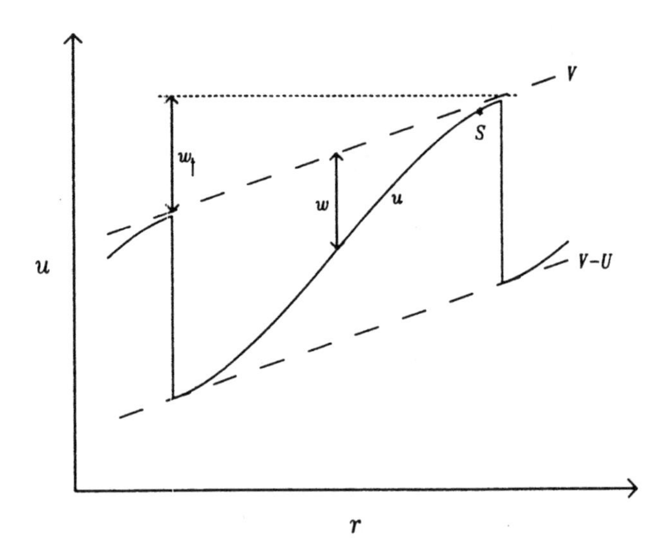

Figure 2. A small section of the velocity distribution u *vs.* r. $V(r)$ is the shock velocity; $U(r)$ is the velocity jump at the shock; $w = V - u$ is the inward flow velocity in the shock frame; $w_\dagger$ is the step in V between successive shocks. The point marked 'S' is the sonic point of the flow in the shock's frame.

suitable average over the inter-shock region. Lucy argued that the radiation force may also be about the same as it would be without shocks, so that the mean flow is in fact identical to that calculated by Castor, Abbott and Klein (1975).

In order for the x-ray output of the shocks to agree with the *Einstein* observations, T_1 should be about 300 eV, which requires $w_\dagger$ to be $60\,\mathrm{km\,s^{-1}}$, 3–4 times larger than the sound speed. It is difficult to understand why this should be the stable value of $w_\dagger$. Other possibilities were discussed by Cassinelli and Swank (1983). The shock period τ is quite short—shorter than any natural time scales of the stellar photosphere or interior—and therefore if the shocks originate from some noise source in those regions there must be considerable 'cloning' to reduce the period to the required value. And, of course, the periodic structure cannot exist if the mechanisms that stabilize $w_\dagger$ do not work.

The column density between shocks in Lucy's model is

$$N_1 \approx 5 \times 10^{20} \left(\frac{\dot{\mathcal{M}}}{10^{-6}}\right) \left(\frac{w_\dagger}{20}\right) \left(\frac{2000}{V_\infty}\right)^2 \left(\frac{R_*}{10^{12}}\right) \mathrm{cm}^{-2}.$$

Equating this to N_{cool} for a shock speed equal to U gives

$$\dot{\mathcal{M}} = \dot{\mathcal{M}}_{\mathrm{min}} \approx 1 \times 10^{-8} \left(\frac{V_\infty}{2000}\right) \left(\frac{w_\dagger}{20}\right) \left(\frac{R_*}{10^{12}}\right) \mathcal{M}_\odot\,\mathrm{y}^{-1}.$$

If $\dot{\mathcal{M}}$ is below $\dot{\mathcal{M}}_{\mathrm{min}}$ the shocked gas in Lucy's model never cools between shocks. Such a hot wind ($T \approx 10^6\,\mathrm{K}$) is possible, and it can still be driven by radiation, but the force is reduced by the high temperature—the driving ions are largely stripped—so less mass loss is produced for a given star than if the wind were cool.

4.2. The Krolik and Raymond 'Shell' Model

In a recent paper, Krolik and Raymond (1985) have proposed a model of radiatively-driven shocks that has some different aspects from Lucy's. They consider a single shock (although there may be others some distance away), and treat in detail the ionization, recombination and cooling behind the shock, as described in § 1.2. From the resulting structure of velocity, temperature and density of various ions, they calculate how much momentum is absorbed within the shock from the photospheric radiation field. The shock is considered to have a definite column thickness, N, which is an unknown of the problem. A simple dynamical model is then used to estimate how N and the shock velocity evolve with time.

This model is, in effect, one of a pancake-like shell of shocked gas that is confined in front by the ram pressure of the pre-shock material, and driven from the rear by a radiation pressure which is the momentum deposition calculated for the shock. This is basically an *episodic* rather than a *periodic* model. It is not unlike Lucy's model, but for two key differences: In the Krolik and Raymond model all the gas swept up by the shock remains confined in the shell, while in Lucy's model gas streams out the back of the shock to balance the gas entering at the front. Lucy's model also accounts in a more consistent way for the dynamics of the post-shock flow. The hydrodynamic boundary conditions at the back of the shell in the Krolik and Raymond model are unclear.

The numerical estimates obtained by Krolik and Raymond for the typical shock strength and x-ray emission are quite similar to the requirements of the *Einstein* data. Since the spacing of shocks is not constrained in this model, in contrast to Lucy's, the shock strength can become greater, giving the desired shock temperature.

Further work by Krolik and Raymond will account for the global dynamics of the wind including such shells of shocked gas.

5. SUMMARY

The discussion I have given above of pulsation-driven shocks and radiation-driven shocks has raised and partially answered several of the interesting questions about shocks in OB stars:

- Are shocks formed by pulsation?
  The answer is "yes" if the mass flux ρu_s of the shock is $\gg$ the radiatively-driven mass flux; otherwise the radiation force overwhelms pulsation as the cause of shocks. For the B star of § 3.2 the answer is "yes" for $P = 3$ h, and "no" for $P = 12$ h.
- Are the periodic shock dynamics unaffected by radiation pressure?
  "Yes", but only if the shock mass flux at the *adiabatic radius* is $\gg$ the radiatively-driven mass flux. This limits the radiatively-driven $\dot{\mathcal{M}}$ to $\leq 10^{-11}\ \mathcal{M}_\odot\,\mathrm{y}^{-1}$.
- Is there pulsation-produced mass loss, without assistance from radiation?
  Only if the radiatively-driven $\dot{\mathcal{M}} \leq 10^{-12}\ \mathcal{M}_\odot\,\mathrm{y}^{-1}$.
- Do radiatively-driven shocks have a sawtooth or a shell structure?
  - If Lucy's shock cloning and eating mechanisms work: a sawtooth structure results, with or without pulsation.
  - Without Lucy's mechanisms: shells result with pulsation, and no shocks result without pulsation.

Further insight into the morphology of radiatively-driven shocks awaits the detailed hydrodynamic modeling of stellar winds now in progress, such as the effort by Owocki, Rybicki and myself, some preliminary results of which are described elsewhere in these proceedings.

6. REFERENCES

Abbott, D. C., Bieging, J. H. and Churchwell, E. 1984, *Astrophys. J.*, **280**, 671.

Bertschinger, E. and Chevalier, R. A. 1985, *Astrophys. J.*, **299**, 167.

Cassinelli, J. P. and Swank, J. H. 1983, *Astrophys. J.*, **271**, 681.

Castor, J. I. 1970, "Shock Waves in Population II Variable Stars", in *The Evolution of Population II Stars*, ed. A. G. D. Philip, Dudley Observatory Report No. 4, 147.

Castor, J. I., Abbott, D. C. and Klein, R. I. 1975, *Astrophys. J.*, **195**, 157.

Courant, R. and Friedrichs, K. O. 1948, *Supersonic Flow and Shock Waves* (New York: Interscience Publishers, Inc.).

Henrichs, H. 1986, *Publ. Astr. Soc. Pacific*, **98**, 48.

Hill, S. J. 1972, *Astrophys. J.*, **178**, 793.

Hill, S. J. and Willson, L. A. 1979, *Astrophys. J.*, **229**, 1029.

Krolik, J. H. and Raymond, J. C. 1985, *Astrophys. J.*, **298**, 660.

Lamb, H. 1945, *Hydrodynamics* (New York: Dover Publications).

Lucy, L. B. 1982, *Astrophys. J.*, **255**, 286.

Lucy, L. B. and Solomon, P. 1970, *Astrophys. J.*, **159**, 879.

Parker, E. N. 1958, *Astrophys. J.*, **128**, 664.

Raymond, J. C., Cox, D. P. and Smith, B. W. 1967, *Astrophys. J.*, **204**, 290.

Smith, M. A. 1986, *Publ. Astr. Soc. Pacific*, **98**, 33.

Whitham, G. B. 1958, *J. Fluid Mech.*, **4**, 337.

Willson, L. A. and Hill, S. J. 1979, *Astrophys. J.*, **228**, 854.

Wood, P. R. 1979, *Astrophys. J.*, **227**, 220.

Zel'dovich, Ya. B. and Raizer, Yu. P. 1967, *Physics of Shock Waves and High-Temperature Hydrodynamic Phenomena* (New York: Academic Press).

DISCUSSION

CASSINELLI: You conclude that pulsation alone cannot drive very large mass loss rates (> 10^{-11} $M_\odot$/yr). Wolf Rayet stars have mass loss rates much larger than this and Maeder has explained these are being pulsationally driven, because the stars are at the vibrational instability limit. Are you saying therefore that the Maeder mechanism cannot explain the large $\dot{M}$ of the WR stars?

CASTOR: I think that the pulsations will not produce the mass loss unassisted by radiation pressure. The pulsation could affect the structure of a radiatively driven wind - for example, by levitating the material to a larger radius, where the radiative acceleration is more effective.

MAEDER: It is a very interesting question that was asked by Cassinelli as to whether pulsations are able to sustain the shocks. In order to answer it, it would be necessary to know the amplitude of pulsations. Unfortunately, they are not yet observed. Maybe we may have an indication from non linear calculations of pulsations concerning this very important point.

APPENZELLER: Concerning the efficiency of pulsation driven mass loss: If you replace the B star which you used as an example by a model star very close to the Eddington limit and hence a very low 'effective escape velocity', could you then get a significant mass loss from the pulsations?

CASTOR: In fact, the mass loss rate would be somewhat reduced, since the shock strength would go down with v_{esc}, hence the density required for marginal cooling would also go down. Much of a change in v_{esc} by the factor $(1 - L/L_{Edd})$ would also make a radiation-driven wind very likely!

HENRICHS: What is the propagation velocity of the shells with swept-up matter with respect of the terminal velocity of an unshocked wind in Krolik & Raymonds model?

CASTOR: The shells move into the wind in front of them at about 300-600 km/s. How this relates to the terminal velocity is not yet worked out.

ULMSCHNEIDER: When you have time-scales of 22 sec then the time dependence of ionization will be very important and could greatly change the post shock cooling.

CASTOR: The time scales for ionization balance and cooling are just the ones I compared with the shock spacing. The Krolik and Raymond cooling calculations, which are consistent with the cooling time I quoted, allow for non-equilibrium ionization.

CASSINELLI: Would you please comment on the presence of the very hot gas (T = 15 to 20 million degrees) in several early type stars: τ Sco, ζ Oph and ζ Ori?

CASTOR: One possibility is that the wind truly 'runs into a brick wall', with a shock equal to the total wind speed. This raises problems of how to place a structure like this in the wind; it does not seem likely. More probably, this emission comes from matter confined in magnetic loops.

NOËLS: You have shown that stellar pulsations can produce only very weak mass loss rates. In presence of a strong stellar wind already existent, do you think a pulsation could produce an enhancement of the wind?

CASTOR: If Lucy's model is correct, the effect is slight. But the door is open, with shock models not yet worked out, for pulsation to have some effect.

LAMERS: In Lucy and White's original model the clumps move like bullets through the slower moving ambient medium. The observations however show the opposite: the shells or puffs move slower than the wind. So probably the shocks occur at the lower side of the blobs rather than at the top. Do you think that this will change the general conclusion of this model in terms of the generation of X-rays and the general structure of the wind?

CASTOR: The X-rays should be about the same, but it seems difficult to me to drive the ambient gas more strongly than the clumps. I think there are problems with the Lucy-White clump model in its kinematic properties relative to the narrow components.

HENRICHS: How would you put the 'narrow absorption components' in terms of Lucy's model and Krolik-Raymond's model?

CASTOR: The narrow components are not predicted by Lucy's model. They can very possibly be identified with the Krolik and Raymond shells. Whether the proper relation of velocity and ionization state of the shell to the background is found depends on aspects of their model that have not been worked out.

HENRICHS: An interesting property of the 'discrete' (or narrow) absorption components is that their column density is independent of the wind causing the 'underlying' P Cygni profile. This result is valid from O3 to B6 for all luminosity classes. Can this be understood in terms of the Krolik-Raymond model?

CASTOR: The Krolik-Raymond model sets the column density of the shell equal to the total column swept up by the shock; it therefore is tied to the column density of the background, unlike the relation you quote. But Krolik and Raymond did not allow matter to fall off the back of the dense shell. If they had, a fixed column density might have been found.

DE JAGER: In a poster we showed that:
1) V_{turb}/V_{sound} increases if a star approaches the Humphreys-Davidson limit, and is ≃ 1 for L ≃ 0.3L $_{limit}$.
2) $\dot{M}$ increases with V_{turb}/V_{sound}.
3) If all turbulent energy is assumed to be used to feed the stellar wind, then the thus calculated rate of mass loss agrees well with the observed values.

CONTI: Have you considered the data in the literature concerning macroturbulence and potential correlations with mass loss rates?

DE JAGER: I hesitate to use macroturbulence data from literature because many of these data are not very well defined. I suspect that for many hypergiants and supergiants the communicated V_M-data are a rather confuse mixture of real large-scale stochastic motions, rotation and nonlinear pulsations.
In contrast to that microturbulence is much better defined.

CASSINELLI: Does the turbulence drive the outflow by first forming a coronal zone and with the flow occurring by a Parker mechanism?

DE JAGER: That is a good question to which the answer is not yet evident to me. The only thing I said is that the energy contained in the microturbulent photospheric motions of the stars, if fully dissipated, is sufficient to drive the stellar winds. But how this energy is transferred is not yet clear.

John Castor giving his review on shock waves in luminous early type stars

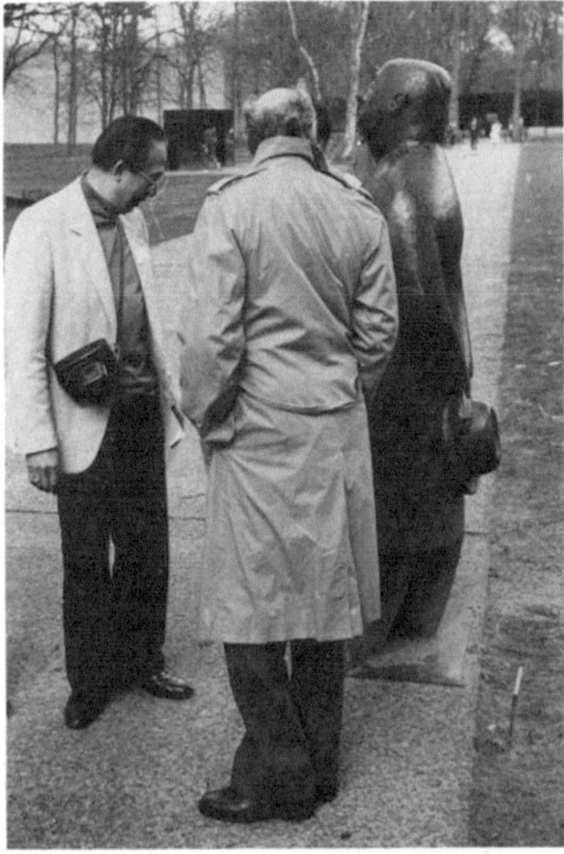

Above:
A break at the bicycle-trip: a visit to the "Hubertus Castle".

Below:
Pik Sin Thé and Bert de Loore admiring a statue in the Kröller-Müller museum.

RADIATION DRIVEN INSTABILITIES

George B. Rybicki
Harvard-Smithsonian Center for Astrophysics
60 Garden Street
Cambridge, MA 02138
USA

ABSTRACT. Various radiation driven instabilities have been shown to operate in the atmospheres and winds of early-type luminous stars. The strongest of these occur in the supersonic parts of the winds, where as many as one hundred *e*-folds of linear growth can occur during a typical outflow time. The nonlinear growth of such instabilities can possibly account for the observed superionization and X-ray emission in these stars. Developments in the linear theory of these instabilities is reviewed.

1. INTRODUCTION

The force due to radiation is an important physical mechanism in the early-type luminous stars. As de Jager (1980, 1984) has emphasized, the weakening of the effective gravity in such stars by radiation pressure makes them particularly vulnerable to instabilities, especially when combined with the destabilizing effect of the very strong turbulent motions, which are inferred from the broadening the photospheric lines. These combined effects may explain the observed upper luminosity limit for early luminous stars (Humphreys, this volume).

In addition to its importance in setting the upper limit for the luminosity, the radiation force also plays a role in many other specific instabilities in the atmospheres and winds of such stars. It is the purpose of this review to consider the various types of radiation driven instabilities that have been proposed, especially those that are relevant in the atmosperes and winds of early-type luminous stars. As will be shown, the strongest of these instabilities are associated with the supersonic part of the wind, and most of our discussion will be concentrated there.

The winds of early luminous stars are known to be very strong, with mass loss rates ranging from 10^{-7} to 10^{-4} $M_{\odot}$/yr, and with terminal wind speeds ranging from 1000 to 3000 km/s. The only known way to account for these extreme values (especially the wind speeds) is by radiation driving due to spectral lines. Theories of steady-state winds based on radiation driving have been developed by Lucy and Solomon (1970) and by Castor, Abbott and Klein (1975). These theories have been successful in matching many of the observed properties of such winds.

There are a number of observations that do not fit easily into a steady-state theory, however. These are of three types: First, many of these stars show variability in their winds on a variety of times scales (de Jager et al. 1979; see also in this volume the

H. J. G. L. M. Lamers and C. W. H. de Loore (eds.), Instabilities in Luminous Early Type Stars, 175–190.

articles by Baade, Lamers and Vreux). In particular, variable discrete components are often found in the absorption portions of the P Cygni profiles (Abbott 1985, Heinrichs 1986). Second, the X-ray emission from these stars (Harnden et al. 1979; Seward et al. 1979) and the high stages of ionization observed (Lamers and Morton 1975) require wind temperatures of the order of 2 to 5 $\times 10^5$ K, much higher than the radiative equilibrium temperatures of the steady-state theories. Third, the steady-state theories give velocity laws inconsistent with the observed shapes of the P Cygni profiles, especially the extended, dark absorption troughs (Lucy 1982). It has been suggested that instabilities in the winds of early luminous stars might resolve all three of these difficulties.

In this paper we review the linear theory of radiation driven instabilities, with particular attention to the strongest type, the line-shape instabilities. Previous reviews on this and related topics have been given by Hearn (1985) and Lucy (1986). The important area of the nonlinear development of the instabilities will not be considered here, as it is treated in the article by Castor (this volume).

2. THE DYNAMICAL EFFECTS OF RADIATION

The radiation force per unit mass, also called the *radiative acceleration*, can be written

$$\mathbf{g} = \frac{1}{c} \int \kappa_\nu \mathbf{F}_\nu \, d\nu, \tag{2.1}$$

where κ_ν is the opacity per unit mass, $\mathbf{F}_\nu$ is the monochromatic radiative flux vector, and c is the speed of light. This formula assumes that the opacity is purely absorptive or that any scattering has front-back symmetry, so that no momentum is transferred upon re-emission. These assumptions are generally valid for the opacities found in early-type stars.

The radiative acceleration can in general be exceedingly complex. However, deep in the stellar interior $\mathbf{F}_\nu$ takes the simple Rosseland form,

$$\mathbf{F}_\nu = -\frac{4\pi}{3\rho\kappa_\nu} \nabla B_\nu, \tag{2.2}$$

where ρ is the mass density and B_ν is the Planck function. Substitution into Eq. (1) then yields

$$\mathbf{g} = -\frac{1}{\rho} \nabla P_{rad}, \tag{2.3}$$

where P_{rad} is the isotropic, thermal radiation pressure

$$P_{rad} = \frac{1}{3} a T^4, \tag{2.4}$$

and a is the radiation constant.

One notes from this result that the nature of the opacity is irrelevant to g in the deep interior, since it has canceled out of the final formula. If, as is the usual case, one includes the radiation pressure as part of the equation of state, then the dynamical effects of radiation in the interior are included automatically.

In the atmosphere of the star the use of equation (2.4) is not appropriate, but at least one source of opacity, that due to electron scattering, can be treated in a very simple way. Since this opacity is frequency-independent, $\kappa_\nu = \kappa_{es}$, its contribution to the radiative acceleration can be written

$$\mathbf{g}_{es} = \frac{1}{c}\kappa_{es}\mathbf{F}, \tag{2.5}$$

where **F** is the total radiative flux.

In spherical geometry the magnitude of the flux **F** can be related to the luminosity L through $F = L/(4\pi r^2)$. Since the flux, like the Newtonian gravity $g_{grav} = GM/r^2$, obeys an inverse square law, it is convenient to define an effective gravity

$$g_{eff} = g_{grav} - g_{es} = g_{grav}(1 - \Gamma), \tag{2.6}$$

with $\Gamma \equiv \kappa L/(4\pi GMc)$. The radiative force due to electron scattering can then be taken into account by using g_{eff} in place of g_{grav} in the momentum equation.

For other sources of opacity in the atmosphere, the radiative transfer is much more complex, and a detailed treatment is required. Nonetheless, at the risk of some oversimplification, it may be useful to make a few general remarks about radiative forces in the atmosphere.

There is a tendency for spectral regions of high opacity to *saturate* and carry little radiative flux. This property is seen in its purest form in the stellar interior, as exemplified by equation (2.2), but it applies to some extent even in the atmosphere. Because of this, a large opacity does not necessarily lead to a large radiative force.

Continuous and spectral line opacities both give rise to radiative forces, but there is a crucial distinction between them when strong velocity fields are present, such as in a wind. When gas velocities in the atmosphere are larger than the thermal velocities of the ions, then the Doppler effect can substantially affect the radiative transfer in the lines, but will have virtually no effect on the continuum. Since ion thermal velocities are roughly of the same order as the sound speed, the treatment of the line transfer depends on the Mach number of the flow. If the flow is very subsonic radiative forces can be adequately treated by assuming a static atmosphere. In supersonic regions the velocities can shift spectral lines out of their own absoption profiles, and thus subject them to much larger radiation fields. This desaturation of the radiation field tends to make spectral line opacity the principal source of radiative force in supersonic regions.

3. CLASSIFICATION OF INSTABILITIES

A formal linear stability analysis of radiative driven instabilities should in principle be based on the hydrodynamical equations of mass, momentum and energy. However, since the main purpose here is to review radiation driven instabilities, the simplifying assumption will be made that the gas pressure obeys the polytropic law, $P \propto \rho^{-\gamma}$, where ρ is the density. The energy equation is then not needed, and the required hydrodynamical equations are just those of mass and momentum,

$$\frac{\partial \rho}{\partial t} + \mathbf{v} \cdot \nabla \rho + \rho \nabla \cdot \mathbf{v} = 0, \tag{3.1}$$

$$\frac{\partial \mathbf{v}}{\partial t} + \mathbf{v} \cdot \nabla \mathbf{v} = -\frac{1}{\rho}\nabla P - \frac{GM\mathbf{r}}{r^3} + \mathbf{g}. \tag{3.2}$$

where **v** is the gas velocity.

In order to investigate the linear stability of some zeroth order unperturbed solution, the various physical quantities are expressed as a sum of an unperturbed part (with subscript "0") and an infinitessmal perturbed part (preceded by "δ").

$$\begin{aligned} \rho &= \rho_0 + \delta\rho, \\ \mathbf{v} &= \mathbf{v}_0 + \delta\mathbf{v}, \\ \mathbf{g} &= \mathbf{g}_0 + \delta\mathbf{g}, \\ &\text{etc.} \end{aligned} \tag{3.3}$$

Substitution of these expressions into equations (3.1) and (3.2) and neglecting second order and higher terms in the perturbations yields the equations for the unperturbed steady state,

$$\mathbf{v}_0 \cdot \nabla\rho_0 + \rho_0 \nabla \cdot \mathbf{v}_0 = 0, \tag{3.4}$$

$$\mathbf{v}_0 \cdot \nabla\mathbf{v}_0 = -\frac{1}{\rho_0}\nabla P_0 - \frac{GM\mathbf{r}}{r^3} + \mathbf{g}_0. \tag{3.5}$$

Similarly, the equations for the pertubations are found to be,

$$\frac{\partial\delta\rho}{\partial t} + \mathbf{v}_0 \cdot \nabla\delta\rho + \delta\mathbf{v} \cdot \nabla\rho_0 + \rho_0\nabla \cdot \delta\mathbf{v} + \delta\rho\nabla \cdot \mathbf{v}_0 = 0, \tag{3.6}$$

$$\frac{\partial\delta\mathbf{v}}{\partial t} + \mathbf{v}_0 \cdot \nabla\delta\mathbf{v} + \delta\mathbf{v} \cdot \nabla\mathbf{v}_0 = -\frac{a^2}{\rho}\nabla\rho + \delta\mathbf{g}, \tag{3.7}$$

where the sound speed a is given by $a^2 = (\partial P/\partial\rho)_0$.

These perturbation equations are still very difficult to solve, so it is usual to make the WKB assumption that the solutions of interest are those in which the perturbations vary on a scale small compared to typical scale lengths of the unperturbed solution. The terms involving gradients of the unperturbed quantities can then be dropped, and the coefficients in the equations are considered to be locally constant. Solutions of the form

$$\left.\begin{matrix} \delta\rho \\ \delta\mathbf{v} \\ \delta\mathbf{g} \end{matrix}\right\} \propto e^{i(\mathbf{k}\cdot\mathbf{r}-\omega t)} \tag{3.8}$$

can then be assumed, which lead to the algebraic equations

$$-i\omega\,\delta\rho + i\rho_0\mathbf{k} \cdot \delta\mathbf{v} = 0, \tag{3.9}$$

$$-i\omega\,\delta\mathbf{v} + \frac{a^2}{\rho_0} i\mathbf{k}\,\delta\rho - \delta\mathbf{g} = 0, \tag{3.10}$$

in a frame of reference comoving with the mean flow.

Local analysis differs from the mode analysis of pulsation theory in that spatial eigenfunctions are constrained to be of the form $\exp(i\mathbf{k} \cdot \mathbf{r})$. Boundaries are assumed to play no role. This approach is valid as long as the important scale lengths of the perturbations are indeed small compared to scale lengths H of the mean flow, that is, $kH \gg 1$.

In order to complete the specification of these equations, the perturbation in the radiative acceleration $\delta \mathbf{g}$ must be expressed in terms of the other hydrodynamical variables in the problem. From equation (2.1) this can be written

$$\delta \mathbf{g} = \frac{1}{c} \int \kappa_{0\nu} \, \delta \mathbf{F}_\nu \, d\nu + \frac{1}{c} \int \delta \kappa_\nu \mathbf{F}_{0\nu} \, d\nu. \tag{3.11}$$

The problem is thus reduced to finding expressions for the perturbations in opacity and in radiative flux. The opacity perturbations $\delta \kappa_\nu$ are of two general types: those caused by perturbations in thermodynamic variables, such as density or temperature; and, in the case of spectral lines, those caused by velocity perturbations. The perturbations in the radiative flux $\delta \mathbf{F}_\nu$ pose a particularly difficult problem, since the radiative flux depends nonlocally on the state of the medium. To the extent that the opacity depends on processes that can be affected by the radiation field, such as heating, ionization or non-LTE effects, opacity perturbations are also determined nonlocally. Because of these dificulties, only a limited number of cases have been treated in the literature, those for which these perturbations take a particularly simple form.

Martens (1985) gave a classification of radiation driven instabilities based in part on the nature of the functional dependence of the opacity and radiation flux perturbations. The complete scheme was based on answers to the following questions: 1) Is the unperturbed state subsonic or supersonic? 2) Does the opacity depend on velocity? 3) Does the radiation field respond to perturbations? 4) Is the wavevector purely radial? Since each question offers a twofold choice, there are in principle sixteen types of instabilities, only about half of which have so far been investigated.

Hearn(1972, 1973) was the first to treat analytically the stability of a medium under the influence of a continuum radiation force where the opacity depended on temperature and density fluctuations. He found that sound waves could amplify under such circumstances, and he speculated that this instability might contribute to chromospheric heating.

However, the strongest of the instabilities studied so far seem to be those due to the Doppler shifting of spectral lines in supersonic winds, which Martens calls *line-shape* instabilities. These have growth rates of order one hundred e-folds in a typical outflow time. All other known instabilities have much more moderate growth rates of order one e-fold in a typical outflow time. Because the line-shape instability is so strong, it clearly dominates the growth of perturbations, and we therefore confine most of our subsequent discussion to it.

4. THE LINE-SHAPE INSTABILITY

One of the earliest attempts to treat a radiation driven instability was by Milne (1926). Although his goal was somewhat different (he was trying to explain the selective ejection of certain ions), the mechanism he proposed is immediately applicable to the case of a supersonic wind. Milne pointed out that the radiation force due to absoption in a spectral line was a very sensitive function of velocity, due to the Doppler shift of the material out of an absorption line. An element of material that moved faster than neighboring elements would therefore feel a stronger radiation field and would move still faster, leading to an instability.

Lucy and Solomon (1970) were the first to point out the possibility of such instabilities in the line driven winds of hot stars, and they also sugested that the resultant nonlinear growth into shocks could explain heating in the gas.

The first attempts to calculate the growth rates for this instability were made almost simultaneously by by MacGregor, Hartmann and Raymond (1979), by Martens (1979), and somewhat more accurately by Carlberg (1980). All of these treatments assume that the perturbations in velocity occure over a sufficiently small scale that they are optically thin and do not affect the radiation field; for convenience we refer to these treatments collectively as the *Optically Thin Approximation*, or *OTA*. This assumption implies that $\delta \mathbf{F}_\nu \equiv 0$, so equation (3.11) reduces to

$$\delta g = \frac{1}{c} \int \delta \kappa_\nu F_{0\nu} \, d\nu. \tag{4.1}$$

(We now confine our attention to perturbations solely in the radial direction, so all vectors are now treated as scalars.) MacGregor et al. (1979) used an approximate expression for the mean flux (essentially a step function in frequency), whereas Carlberg (1980) used the exact expression, taking the line profile properly into account. Martens (1979) used an approximate expression based on a similar calculation by Nelson and Hearn (1978). By expanding the velocity dependence of the opacity, these groups found that

$$\delta g = A \omega_0 \delta v. \tag{4.2}$$

Where A is a dimensionless constant and

$$\omega_0 = g_{thin}/v_{th}. \tag{4.3}$$

Here v_{th} is the thermal velocity and g_{thin} is the radiative acceleration that would act on an optically thin element exposed to the full stellar flux, namely,

$$g_{thin} = \frac{F_\nu \Delta\nu_D \chi_0}{\rho c}, \tag{4.4}$$

with $\Delta\nu_D$ denoting the line Doppler width, χ_0 the line strength, and ρ the density. The value for the dimensionless constant A found by MacGregor et al. and Martens differed from the more accurate value of Carlberg only by a factor of order unity. Substitution of the expression (4.2) into equations (3.9) and (3.10) leads to a dispersion relation with imaginary roots, implying the presence of instabilities. One might understand this instability as being due to the velocity and force perturbations being *in phase*, and thus doing work over the entire cycle of the wave.

The growth rates for these instabilities are approximately equal to $(\omega_0 A)^{-1}$. For parameters corresponding to typical hot star winds, this implies of the order of one hundred e-folds of linear growth in a typical outflow time. This strongly suggested the possibility of nonlinear dissipation in the wind.

About the same time Abbott (1980) investigated wave propagation in line driven winds, assuming that the scale of the perturbations was sufficiently large that they could be treated by the Sobolev approximation (Sobolev 1957, 1960). This led to the result that

$$\delta g(r) = U \delta v'(r), \tag{4.5}$$

where the prime denotes a derivative with respect to radius. The constant U has dimensions of velocity and is expressed as

$$U = \frac{\omega_0}{\chi_0}. \tag{4.6}$$

In terms of the Fourier components (3.8), this can be written

$$\delta g = ikU\delta v. \tag{4.7}$$

Substituting this approximation into equation (3.9) and (3.10), Abbott found a dispersion relation with real roots, implying that the perturbations traveled as waves, with zero growth rate, in contrast to the results of the OTA. The lack of instability here can be understood by the presence of the imaginary unit in equation (4.4), which implies that the velocity and force pertrubations are now 90 degrees *out of phase*, and thus no work is done during a cycle of the wave.

The contradictory results of these two different approaches left the instability question in doubt, until the work of Owocki and Rybicki (1984), which demonstrated that the two approaches corresponded to different assumptions as to the length scale of the perturbations. They developed a method that could treat uniformly the full range of possible pertrubation scales assuming the case of *pure absorption*, that is, no scattering in the lines. In this case they showed that the perturbations in radiation acceleration could be expressed as

$$\delta g(r) = A\omega_0 \delta v(r) - \int_{-\infty}^{r} N(r - r')\delta v(r')\, dr'. \tag{4.8}$$

The first term on the right hand side corresponds to the local effect of a velocity perturbation in moving the line profile into the stronger stellar continuum outside the absorption line, the same effect treated by the OTA. The integral term, with its kernel function N, expresses the nonlocal *shadowing effect* of such velocity perturbations on material further from the star. The one-sidedness of this shadowing term is a direct consequence of the assumption of pure absorption.

Owocki and Rybicki (1984) showed that for a wide class of line profiles, the kernel function N could be well represented as an exponential

$$N(r) = \omega_b \chi_b e^{-\chi_b r}, \tag{4.9}$$

where

$$\omega_b = A\omega_0; \qquad \chi_b = A\chi_0. \tag{4.10}$$

The shadowing described by this exponential falls off over a scale length $\lambda_b = 1/\chi_b$, which is called the *bridging length*. Owocki and Rybicki (1984) showed that for cases where the line profile was well represented by a Doppler profile the bridging length is approximately equal to the *Sobolev length*, defined as the length over which the mean flow velocity changes by a thermal velocity.

Using equation (4.9) in equation (4.8) gives what is called the *exponential shadowing approximation*. In terms of the Fourier components (3.8) this can be written,

$$\frac{\delta g}{\delta v} = \omega_b \frac{ik}{\chi_b + ik}, \tag{4.11}$$

This form has the property of reducing to equation (4.2) for small perturbation wavelengths ($k\lambda_b \gg 1$), and to equation (4.7) for large perturbation wavelengths ($k\lambda_b \ll 1$).

Using equation (4.11) in equations (3.9) and (3.10), Owocki and Rybicki (1984) obtained instability growth rates that increased continuously with the perturbation wavenumber, giving the Abbott (1980) results in the low wavenumber (long wavelength) regime and the OTA results in the high wavenumber (short wavelength) regime. The characteristic scale dividing the two regimes is the bridging length λ_b, which is of order of the Sobolev length. Thus the wind is very unstable to disturbances of scale less than the Sobolev length, which for typical conditions is of order one-hundreth of the stellar radius.

It might be remarked here that the existence of these two different regimes of stability was already understood by Milne (1926), who recognized that if an atom is "accompanied with a sufficiently large crowd of other atoms," the instability would be weakened due to the simultaneous shifting of the absorption line itself. The main advance of the preceding theory, besides making Milne's argument quantitative, is to establish the value of the bridging length that separates the two regimes.

The pure absorption treatment of Owocki and Rybicki (1984) was criticized by Lucy (1984), who pointed out that most driving lines in stellar winds would be scattering lines. He showed that the "line drag" effect of a moving element through the mean scattered radiation field would be of substantial magnitude, and could in certain cases cancel the instability. Lucy's work and subsequent work by Owocki and Rybicki (1985) have shown that the growth rate of the instability can be substantially reduced very near the stellar surface, but it quickly rises to about half the absorption line value at heights of order of a stellar radius. Since the number of e-folds of growth is so large, a reduction in the growth rate by one-half, say, does not change the conclusion that radiation driven winds are very unstable.

The question naturally arises why Lucy's line drag effect should be so important here, since radiation drag is usually regarded as a relativistic effect, proportional to v/c (see e.g., Peebles 1971, pp. 220–222). The reason is that the relevant radiation field at each point in the wind depends very strongly on frequency, varying on the scale of the Doppler width. An element of gas that moves one thermal velocity unit relative to the mean flow is thus subjected to a completely anisotropic radiation field in its own rest frame. Since the mean scattered radiation field is of the same order as the incident unscattered field, the element can experience a marked change in radiation force by moving at a relative velocity v_{th}. Thus the line drag effect is of order v/v_{th} rather than v/c.

A number of extensions and variations of the line-shape instability have been considered and should also be mentioned. Nelson and Hearn (1978) treated the case of perturbations at an arbitrary angle to the radius vector, a case clearly related to the classical *Rayleigh-Taylor Instability*. Since their results are based on the static solution to the transfer equation, they do not apply to the case of supersonic winds.

Lucy (1984) and Owocki and Rybicki (1985) considered the effects of the finite stellar disk (previous work had assumed a purely radially directed beam of radiation). They found that details of the photon drag effect near the stellar surface is very dependent on the limb-darkening law, so that, for example, limb-brightening can even lead to damping of the instability. Another effect treated by Owocki and Rybicki (1985) was the effect of line overlap, which they showed did not eliminate the instability, as was suggested by Lucy (1984).

Owocki and Rybicki (1986) investigated the propagation of pulses in radiation driven winds in order to determine whether the instability is *absolute* or *advective*. In an absolute instability an initial perturbation grows in place and its growth must be limited by nonlinear effects in the medium. In an advective instability an inital perturbation moves as it grows, and may or may not reach nonlinear size within the medium. Owocki and Rybicki showed that the line-shape instability was advective in the case of a pure absorption line, so that any growing disturbance would be swept out of the system. This raises the question of the source of new disturbances, perhaps through sound wave from the photosphere, or perhaps through the same mechanism that gives rise to the observed turbulence in these stars (de Jager 1980, 1984). A proper treatment of line scattering may well change this conclusion, however, and perhaps it is best not to speculate too much on this point.

Finally we mention the work of Kahn (1981), who also treated the stability of radiation driven stellar winds and claims to have found a new type of instability that amplifies nonradial perturbations. However, the "quasi-monochromatic" radiative transfer he used seems difficult to justify in a case, such as this, where the intensities vary over frequencies of the order of a Doppler width. Until this instability can be verified by a better treatment of the radiative transfer, its status is uncertain.

5. FINAL REMARKS

Much progress has been made in the linear theory of radiation driven instabilities, but there are a number of points that still need attention. Here is a list of what I regard as the most important remaining questions in this field:

1. There are a number of instabilities in the scheme of Martens (1985) that have not been investigated at all, or have not been investigated fully. In particular, the Rayleigh-Taylor instabilities badly need to be reconsidered, with attention paid to shadowing, scattering and finite disk effects. Only when this is done can the possible relation of such instabililties to the observed discrete components be realistically discussed.
2. The pure absorption calculations have shown that the instabilities are advective in the supersonic part of the wind. However, scattering has not yet been included, and with it the wind could turn out to be absolutely unstable. This is an important issue to decide, since it determines whether the instability is determined by a continual input of disturbances from some lower (possibly photospheric) source, or if the instability can grow by itself and be determined solely by nonlinear processes within the wind.
3. So far the theories have concentrated on the supersonic region for technical reasons. It would be of considerable interest to understand the role of the subsonic region of the wind, especially if the instability turns out to be indeed advective.
4. The relationship between the radiation driven instabilities and nonradial pulsations (see e.g., Castor 1986) is very intriguing, and should be pursued further. It is possible that the discrete components can only be explained in terms of such a joint theory.

REFERENCES

Abbott, D. C. 1980, *Astrophys. J.*, **242**, 1183.
____________1985, in *Progress in Stellar Spectral Line Formation Theory*, Eds., Beckman, J. E., and Crivellari, L., (Dordrecht: Reidel), p. 279.
Carlberg, R. G. 1980, *Astrophys. J.*, **241**, 1131.
Castor, J. I. 1986, *Publ. Astron. Soc. Pac.*, **98**, 52.
Castor, J. I. Abbott, D. C. and Klein, R. I. 1975, *Astrophys. J.*, **195**, 157.
de Jager, C. 1980, *The Brightest Stars*, (Dordrecht: Reidel).
____________1984, *Astron. Astrophys.*, **138**, 246.
de Jager, C., Lamers, H., Macchetto, F., and Snow, T. P. 1979, *Astron. Astrophys.*, **79**, L28.
Harnden, F. R., Branduardi, B., Elvis, M., Gorenstein, P., Grindlay, J., Pye, J., Rosner, R., Topka, K., and Vaiana, G. S. 1979, *Ap. J. (Letters)*, **239**, L65.
Hearn, A. G. 1972, *Astron. Astrophys.*, **19**, 417.

____________1973, *Astron. Astrophys.*, **23**, 97.
____________1985, in *The Origin of Nonradiative Heating/Momentum in Hot Stars*, Eds., Underhill, A., and Michalitsianos, A. G., NASA conf. publ. 2358, (Greenbelt, MD: NASA), p. 188.
Heinrichs, H. 1986, *Publ. Astron. Soc. Pac.*, **98**, 48.
Kahn, F. D. 1981, *Mon. Not. Roy. Astron. Soc.*, **196**, 641.
Lamers, H., and Morton, D. C. 1975, *Ap. J. Suppl.*, **32**, 715.
Lucy, L. B. 1982, *Astrophys. J.*, **255**, 278
____________1984, *Astrophys. J.*, **284**, 351.
____________1986, in *Radiation Hydrodynamics in Stars and Compact Objects*, eds., Mihalas, D., and Winkler, K.-H.A., (New York, Springer-Verlag), p. 75.
Lucy, L. B., and Solomon, P. M. 1970, *Astrophys. J.*, **159**, 879.
MacGregor, K. B., Hartmann, L., and Raymond, J. C. 1979, *Astrophys. J.*, **231**, 514.
Martens, P. C. H. 1979, *Astron. Astrophys.*, **75**, L7.
____________1985, in *The Origin of Nonradiative Heating/Momentum in Hot Stars*, Eds., Underhill, A., and Michalitsianos, A. G., NASA conf. publ. 2358, (Greenbelt, MD: NASA), p. 226.
Milne, E. A. 1926, *Mon. Not. Roy. Astron. Soc.*, **86**, 459.
Nelson, G. D., and Hearn, A. G. 1978, *Astron. Astrophys.*, **65**, 223.
Owocki, S. P. and Rybicki, G. B. 1984, *Astrophys. J.*, **284**, 337.
____________1985, *Astrophys. J.*, **299**, 265.
____________1986, *Astrophys. J.*, **309**, in press.
Peebles, P.J.E. 1971, *Physical Cosmology*, (Princeton, Princeton Univ. Press).
Seward, F. D., Forman, W. R., Giacconi, R., Griffith, R. B., Harnden, F. R., Jones, C., and Pye, J. P. 1979, *Ap. J. (Letters)*, **234**, L51.
Sobolev, V. V. 1957, *Soviet Astr. - A. J.*, **1**, 678.
____________1960, *Moving Envelopes of Stars*, (Cambridge: Harvard Univ. Press).

DISCUSSION

CASSINELLI: You said that as a result of discussions in the meeting you no longer think the instability can explain the discrete components. Would you explain why that is?

RYBICKI: The time-scales associated with discrete components cover a wide range, from hours to years. Wind instabilities can only easily explain the shorter scales, since disturbances are swept out on the time scale of hours. Also there is no obvious mechanism to achieve horizontal coherence of the instabilities over the disk, except perhaps excitation by non-radial pulsation.

HENRICHS: I want to correct your statement about the timescale of variability in the discrete components. At first sight it might seem indeed that in many cases they persist for months or years at the same velocity and with the same strength. But in all cases I know of, a study with higher time resolution (a day or so) shows that they are always variable, not dramatic, but significant. We have not been able, except in a very few cases, to follow the development of these short timescale variations on a typical flow timescale (a day).

RYBICKI: That is a very interesting point. It would be interesting to observe these variations on still shorter timescales to see if they can be fitted to reasonable propagation models.

LAMERS: You have shown that instabilities can grow in the wind.
a) Do you think that these instabilities (shocks, blobs) keep their identity when they travel with the wind?
b) Do the instabilities have a velocity relative the general outflow?

RYBICKI: The linear theory suggests that disturbances grow and move coherently, and they tend to move with the wind. However, the non linear theory may well produce structures that move at substantial velocities relative to the wind.

DAVIDSON: How susceptible is this topic to brute force numerical work - supercomputers, hydrocodes etc.? As a practical matter, what are the limitations and how far can you go?

RYBICKI: Castor, Owocki and I are currently attacking the non linear problem using numerical hydrodynamical methods on a supercomputer. The major limitation is the restriction to one-dimensional (spherically symmetric) motions.

HEARN: I am surprised by your question no. 2. Even a global instability needs triggering. An electronic oscillation produces an oscillation by the positive feedback of electron shot noise. For systems with rapid growth of the instability it will make little difference.

RYBICKI: In a system with an advective instability, the instability needs continual triggering. It is possible that the character of the resultant motions is relatively insensitive to the nature of the trigger, but the opposite may also be true. I am particularly interested in the possible role that non radial pulsations may play in setting the character of at least some of the unstable motions.

HEARN: The characteristic period of the instability is much more likely to be determined by the atmosphere between the point of generation and the other parts of the atmosphere. In the case of the supersonic wind instabilities there is a cut off wavelength, the Sobolev length.
Since in the non linear phase there is one shock per wavelength, which dissipates energy, the dominant period is likely to be that related to the Sobolev length.

RYBICKI: I agree that the scales of the instabilities may be initially set by the modes of maximal linear growth rates, but non linear effects (such as shock coalescence) might lead to very different scales in the non linear regime.

OWOCKI: Because the Sobolev length is so small, it does seem unlikely that the instability in winds could lead directly to macroscopic phenomena observable in the line profile, e.g. narrow components. It seems, however, that the driving by pulsations might introduce a lower frequency signature that could lead to such macroscopic phenomena.
A second comment: because the instability drifts outward, it does not seem likely that energy from this instability could feed back onto the atmosphere and thereby perhaps drive pulsations.

RYBICKI: Concerning your second point, it is likely that inclusion of scattering will provide a means of transmitting information inward in the wind. Whether this can feed back into the mechanism of non radial pulsations seems less likely.

CASSINELLI: Stan Owocki mentioned signal propogation speeds. I think the result of Abbott several years ago that the CAK critical point corresponds to the radiatively amplified sound waves was particularly interesting. Is that result still considered valid?

RYBICKI: This apparent contradiction with Abbott's result is discussed in Owocki and Rybicki (1986; in press).

WILLIAMS: How far out in the wind do the instabilities persist? Until the constant velocity v-infinity region?

RYBICKI: These instabilities weaken at large distances, where the radiation driving mechanism becomes weak. Structures that might have been formed during the instability phase may persist for a long time, although gas pressure again becomes able to even them out.
In fact I have wondered if such persistent density enhancements could be very important in causing high clumping factors, which could lower our

estimates of mass loss rates using radio and IR observations.

LAMERS: There is one problem posed by observations which does not seem to be explained by radiation driven instabilities; i.e. there are various cases in which narrow components which have been followed in time with a sufficient resolution to measure their 'acceleration' (dv/dt). In many of these, possibly most of them, the observed 'acceleration' is smaller than the expected one, based on our 'knowledge' of the velocity law. Could you comment on this?

RYBICKI: I see no reason why non linear structures in the wind could not behave in this way. In some of our preliminary numerical hydrodynamic calculations we have seen such things in fact. Remember that the 'velocity law' may not have a unique meaning in the presence of instabilities, and at times the material may be accelerated to a greater or smaller velocity.

CASTOR: In the dynamical calculations by Owocki, Rybicki and myself we do see some wave structures that propagate inward in mass (outward in radius, but more slowly than the wind). These are rarefactions, not shocks, a puzzling and therefore tentative result. This is the kind of feature that could produce a low apparent acceleration, such as the narrow components show.

HENRICHS: There are three cases known in the literature for producing material in the stellar wind that moves slower than the terminal velocity of the wind.
1) Material with a higher density than the ambient wind and which is driven by saturated lines will undergo a slower acceleration compared to the 'normal' stellar wind.
2) Material released from a point above the photosphere (Underhill & Fahey, 1984) but which is subject to the same acceleration as the 'normal' wind from the stellar surface, will also reach a lower terminal velocity provided that the initial kick is small.
3) Mullan (1984) proposed Corotating Interacting Regions (CIR's) in a stellar wind, arising from interacting fast and slow streams in the wind, analoguous to the solar case.
Here I cannot discuss the details of these models, but it seems that there are many ways to produce material with a 'terminal' velocity lower than the terminal velocity of the wind.

HENRICHS: Maybe I should clarify a little more on the timescales involved with the discrete components. It seems to me that we observe two typical time scales of different origin.One of them is the short one I just commented on. The other one is of order of years in Be stars, and represents the overall characteristics of the features, i.e. the components might disappear completely for several years and appear again for a while and so on. If they are present we see them vary on the short timescale. It is my impression that for the hotter and more luminous stars this second (longer) timescale becomes shorter and shorter so that actually discrete component phases overlap, giving the impression of a 'steady' feature. This would of course very much complicate a disentangling of the two effects, but in no way rules out that instabilities as you discussed might not apply.

VREUX: How close to the star do you observe narrow components with the properties you describe (km s^{-1} as unit). Do you have a sizable number of observations in the lowest part of the velocity curve?

HENRICHS: We have several cases (γ Cas, ξ Per and many other O7 stars) where we observe additional variable absorptions at velocity as low as 0.4 times the 'terminal' velocity of the wind. In the case of ξ Per it is very clear that this rather low velocity absorption gradually develops in a high-velocity narrow (discrete) component. Prinja & Howarth (Ap.J. Sup. 1986 in press) give several examples.
Assuming an arbitrary velocity law one concludes that these features will be formed within an hour or so if they would originate from the surface, i.e. within a few stellar radii.

HEARN: I think it is necessary to say continually that in theoretical development it is necessary to walk before you run. Experience has always shown that to keep up with observations you have to run very fast indeed. For all the mathematical sophistication of the work of Owocki & Rybicki, it is still in one dimension. The table of Martens shows that 2 dimensional Rayleigh Taylor type instabilities are possible and they seem to have growth rates that are just as rapid as the 1 dimensional instabilities. Even the large computer program of Castor at Livermore is only working in one dimension.

OWOCKI: This is very true. I think it is important to emphasize the need to proceed carefully. In any theoretical work one can always think of more physics to put in. For example, even if we relax obvious approximations and treat, for example, two or even three dimensions, there still remain other complications, e.g. magnetic fields, rotation etc. One is then forced to a kind of 'kitchen sink' approach, where it becomes increasingly dubious whether one can really understand, or believe, one's results. I prefer a much more pedantic approach, in which one builds blocks of physical understanding. I would ask observers to be patient, and we will work hard to keep up.

RYBICKI: The Rayleigh-Taylor instability is likely to have a very short horizontal scale, so it is unlikely to be a good candidate for explaining the discrete components, which require coherence over the stellar disk.

V.D. HUCHT: Have you ever observed stationary narrow features? If so, could these perhaps originate in the outer decelerating part of the wind?
In three WC9 stars (single!) we observe, in a spectrum infested with emission lines, narrow absorption lines (FeIII UV38) which appear to be stationary in profile and Doppler shift. (see: Van der Hucht, K.A., Conti, P.S. and Willis, A.J.: 1982 in C. de Loore & A.J. Willis, IAU

Symposium #99).

HENRICHS: I presume that the 'shells' where you are referring to are formed at a much larger distance from the star which make them look more stationary.

Attentive audience during the discussions. First row: Joe Cassinelli, Paul Wesselius and Huib Henrichs. Second row: Bernhard Wolf, Roberto Viotti and Bert de Loore. John Castor (standing) is waiting for his turn.

A serious conversation at lunch:
Roberta Humphreys, George Rybicki and Stan Owocki

"The Instability Table":
Cassinelli, Owocki, Castor and Rybicki

STELLAR INSTABILITIES IN THE UPPER PART OF THE HERTZSPRUNG-RUSSELL DIAGRAM

Cornelis de Jager
Astronomical Observatory and
Laboratory for Space Research
Utrecht

1. INTRODUCTION

The pleasure of attending such a remarkable scientific meeting is great indeed. I am deeply obliged to the members of the Scientific and Local Organizing Committees for having organized the Workshop on 'Instabilities in Luminous Early Type Stars' in such an efficient, humanly pleasant, and very well considered way. I am particularly thankful to the many eminent scientists from abroad and from this country who participated so intensively in this Workshop and thus contributed to its success. I think the Workshop greatly helped in clarifying ideas on the questions why there are no stars brighter or more massive than certain upper limits, and what happens to stars close to these limits.

A summary of a Workshop like this one has necessarily a subjective character, even after having obtained some comments on a preprint of this paper from invited reviewers. At a few places I may have introduced viewpoints that are not generally accepted.

2. OBSERVATIONAL REVIEW

The main characteristic of stars in the upper part of the Hertzsprung-Russell diagram were given by Humphreys. There is an upper limit of stellar existence, above which no star seems able to exist. This limit, which is approximately the same for our Galaxy and a few nearby galactic systems is now generally called the Humphreys-Davidson Limit. It lies at log $(L/L_{\odot}) \approx 6.5$ for early O-type, slopes down to 5.7 at log $T_{eff} = 3.75$ and remains <u>nearly</u> horizontal for cooler stars. Refinements may be possible. There is an open space - or a dip? - in the area of A and F type stars (due to rapid evolution?) and it is not yet certain that there are no red super- or hyper-giants above the presently assumed HD limit; in other words: it may not be excluded that the limit will eventually show to slightly bend upward in its red end.

The region below the limit is characterized by the occurrence of the Luminous Blue Variables (LBV's), a name for the various kinds of objects hitherto classified either as Eta Carinae variables, P Cygni

H. J. G. L. M. Lamers and C. W. H. de Loore (eds.), Instabilities in Luminous Early Type Stars, 191–198.

variables, S Doradus variables, Hubble-Sandage variables or Alpha Cygni variables. Although it is realized that the five groups named above do not have strictly identical properties their differences are sufficiently small to justify their being grouped together in one new class, the LBV's. These stars are irregularly variable, in luminosity and in spectrum, with for each object several time scales per object, the longest one being of the order of years. Some of them (P Cyg, Eta Car) show more or less violent mass-ejections of which the largest ones occur on time scales of centuries. It is remarkable that there are also stars likely close to the HD limit that do not show the LBV characteristics. Why is that? Reversely, some LBV's have luminosities considerably below those of the HD limit.

Apart from the fairly violent characteristics of the LBV's, virtually all stars in the whole upper part of the Hertzsprung-Russell diagram are slightly pulsationally variable. Their atmospheric instability is in addition shown by their mass loss (increasing with luminosity) and by the occurrence and occasional formation of dust shells around cooler stars. We mention those around VX Sgr and NML Cygni. The variable dust shell of IRC 10420 (F8 Ia^+) is already a classic case. The formation of such shells of gas and dust may cause (semi-)permanent changes of spectral type. Maser emission is a characteristic of cool super- and hypergiants.

While the diversity of high-luminosity stars seems bewildering, we should never forget that we are actually observing stars with the same origins, but in different evolutionary phases. The only relevant parameters are in fact their ZAMS mass M_Z, and the time t_Z elapsed since their arrival at the ZAMS. The following evolutionary scenarios are assumed:

for $M_Z > 60$: O → Of → LBV → WR(?)

50 - 60 : O → Blue Supergiant → { LBV → WR ; OH/IR → LBV → WR. }

Scenarios for less massive stars have also been developed. Essential in these scenarios is that for $M_Z \gtrsim 60$ ($\log (L/L_\odot) \gtrsim 5.7$) the evolutionary tracks return blueward after having reached the HD limit, apparently because of excessive mass loss near and beyond the limit. Less luminous stars, with consequently smaller ZAMS luminosity can move redwards all the way and return only thereafter. Variable A in M33 may be a case on the return track. It is obvious that for an understanding of this evolutionary behaviour the study of medium-type (G and K) hypergiants becomes necessary.

An interesting question is why the rate of mass loss increases when a star approaches the HD limit. In the temperature range around 10^4 K the HD limit lies well below the Eddington limit calculated on the basis of the star's continuous absorption. But additional absorption by the many strong lines in the ultraviolet multiply the absorption coefficient by a considerable factor, thus strongly reducing the Eddington limit luminosity and bringing it close to the HD limit, particularly in the supersonic part of the stellar winds, which are therefore strongly accelerated. The energy needed for driving these winds at the photospheric level may come from dissipation of the kinetic energy of the violent turbulent motions that are observed in these atmospheres.

There are more problems that remain open. The role of OH/IR stars in the evolutionary scenarios has to be further clarified. The assumption that the WR stars represent a final non-degenerate stage of evolution of massive stars is another topic for study. Let us note, finally, that in some cases evolutionary changes are so rapid that they occur in the human time scale.

3. THE MOST MASSIVE STARS

This topic was reviewed by Appenzeller. In massive stars, where the nuclear reaction rate is strongly temperature-dependent, pulsations of the stellar cores may be amplified (the epsilon mechanism) but more outward, in the surface layers they are damped by radiative losses through a kappa mechanism. This seems not to be so for very massive stars; for these, first-order investigations show that there is a net gain of pulsational energy yield over the losses making $\int pdVdt > 0$, and thus such stars may become vibrationally unstable. In the HR diagram this instability occurs in a small region at and slightly redwards of the top of the ZAMS. The lower luminosity of this region at the ZAMS sets an upper limit to stellar masses: stars with higher ZAMS masses cannot originate or persist. The precise influence of the surface layers on stellar instability deserves further investigions. For the time being it is only possible to conclude that stars are:

surely stable for $M \lesssim 130\ M_\odot$,
surely unstable for $M \gtrsim 10^3\ M_\odot$,

with a terra incognita in between.

The difficulty in these calculations is that the influence of the outer layers on stellar stability is hard to include: such a study demands knowledge of their composition and extension. It makes difference if these layers are stationary or expanding. An increase of the core/envelope density ratio stabilizes the stars. Therefore stars become more stable when moving away from the main sequence.

In relation to the above results it is of interest to remark that Eta Car, to which a mass of ~ 120 $M_\odot$ had been attributed on the basis of its location in the HR diagram now appears to be a quadruple trapezium system occupying an area of 500 AU diameter. The most luminous component provides about 70% of the total light (corresponding with 85 $M_\odot$?). The object R136a in Doradus (2000 $M_\odot$) consists of at least seven components.

For stars at the HeZAMS the expected upper mass limit for vibrational stability is $\approx 10\ M_\odot$. Therefore, if Wolf-Rayet stars have the high effective temperature sometimes ascribed to them, they must be close to the HeZAMS and, since they have a few tens of solar masses, they should be vibrationally unstable!

4. VARIABILITY OF BLUE STARS

This topic was reviewed by Baade (for O and Of stars), Vreux (for WR stars), Lamers (for the Luminous Blue Variables), and Davidson (large outbursts in LBV's).

The O and Of stars are practically all variable, both in luminosity and in radial velocity, slightly in colour. The amplitudes of the brightness and v_R-variations tend to increase towards evolved stars and with L, as does the rate of mass loss $\dot{M}$. This may suggest a relation between $\dot{M}$, L and the amplitude of the variations, but that is far from proven yet. Although the variations are not strictly periodic, a glance at the light- or v_R-curves suggests the existence of a quasi-period P_q (i.e. the main component in a Fourier analysis of the variations). It is important that Period-Luminosity laws can be established; they are equally valid for m(t) and $v_R(t)$. These laws vary with the spectral type. No O-B star is known to pulsate in a radial mode (excluding the Beta Cephei stars), but among the medium and later spectral types there are certainly non-radial pulsators (cf. further Section 5). There are cases of weak correlation between photospheric variations and those in the low winds, but in most stars there is none; the correlation is still weaker and more often absent for the distant stellar winds. Observed variations in the winds show that these are due to the emissions of blobs of gas with masses between 10^{-9} and 10^{-12} $\dot{M}$, and with various velocities.

About half of the Wolf-Rayet stars show some variability; there is evidence for a period of less than one day in the case of WR136. For the binaries among the WR stars the light variation is partly due to binary effects (occultation, etc.) but there remain effects of another nature, perhaps intrinsic and due to stellar pulsations. The late WR stars have the highest degree of 'random' brightness fluctuations. Cases have been observed in which the amplitude of the variations abruptly increased in a dramatic way while the period remained constant. Two models for the brightness variation are likely: the binary model: a Wolf-Rayet primary with a compact secondary, or the single model in which the variations are due to non-radial pulsations. The latter case applies particularly to the cooler WR stars. These stars also show variation of polarization in contrast to the hotter stars. In the single-star model the pulsations are thought to arise in a burning shell. Predicted periods for such models are of the order of a few hours and the amplitudes may be depth-dependent. Core pulsations are predicted to have periods of 15-30 min, but variations of these periods have not yet been observed. The strong winds of the WR stars may conceal such variations. That the winds are directional and variable follows from their observed variation of polarization. The variability is also apparent from the observed occasional emission of 'puffs', semi-regularly spaced in time, with velocity increasing in time till a (constant) velocity ceiling is reached. Such a phenomenon was also observed in P Cyg, one of the Luminous Blue Variables. Episodical formation of dust shells associated with a large increase of brightness (2 mag) has been observed in WR 140, a double-lined spectroscopic binary with a period of 7.9 years.

Luminous Blue Variables are of spectral types O, B, A (and F). They vary in spectrum, as well as in m_v, but m_{bol} remains more or less constant. Hence at brightness maximum T_{eff} is lowest. This property makes their tracks in the theoretical Hertzsprung-Russell diagram nearly horizontal. There is a true spectrum of brightness variations, with largest amplitudes (one or two magnitudes) corresponding to the longest periods (years); true outbursts are still larger and have a time scale of centuries.

At maximum R71 (LMC) had P Cygni profiles in all Balmer lines which means: a strong stellar wind. At minimum the P Cygni profiles have disappeared but the spectrum shows emission and forbidden lines, which means: a teneous circumstellar gas envelope. The large-scale variations of R71 have time scales of a few years, on which 'micro-variations' are superimposed. These microvariations occur also in other LBV's and have $\Delta m \approx 0.2$ mag and time scales of about a month. The cool dust shell of R71 extends to 8000 R_{star}, and has a mass of approx. 3×10^{-4} $M_{\odot}$. Assuming (conventionally) a hundred times more gas, then, with $\dot{M} = -5 \times 10^{-5}$ M yr^{-1}, this shell must have been formed in about 600 years.

As in all stars close to the HD limit the limiting wind velocity v_{∞} is smaller than for normal supergiants or giants.

Episodical mass loss, an important property of LBV's, has also been well studied in P Cyg. Remarkable is the ejection of thin shells at intervals of about 60 d, observable as variations in $H\alpha$ and in the photospheric continuum, and of thick shells at intervals of about 200 d, observable in the UV resonance lines, but not photometrically. In several LBV's mass ejection seems related to the occasional formation of transient dust shells.

Large outbursts occur next to the episodical mass loss described above, but have so far been observed only in Eta Car and P Cyg. Like the smaller outbursts the large ones seem associated with the mere expulsion of gas, rather than being a (partial) stellar explosion: the bolometric magnitude of Eta Car seems to have changed hardly during and after the large outburst of 1835/40. Besides that outburst there was perhaps a smaller one near or before 1751; smaller outbursts occur regularly.

Eta Car is certainly an evolved star, having nuclear reaction products (He, N, O) in its photosphere. Observations of the gas and dust cloud around that star show blobs of gas moving with various velocities; these may be attributed to various outbursts. The mass of the homunculus, the nebula just around the star is a few solar masses, and assuming that this mass has been ejected gradually since the big outburst of 1835-1840, the average mass loss is estimated a few times 10^{-2} $M_{\odot}$ yr^{-1}. The 'quiet time' mass loss may be one or two orders smaller, a value of 10^{-3} to 10^{-4} $M_{\odot}$ yr^{-1} is assumed. The expected life time of Eta Car is therefore of order 10^4 year. In passing we note the type V supernova 1961v for which the pre-SN mass is estimated at 2×10^3 $M_{\odot}$ (Eddington luminosity of the pre-SN). A supernova or an extreme LBV??

Should these outbursts be explained by a kind of valve-mechanism? The epsilon-mechanism surely does not work because of the high ratio between the density at the stellar center and the average density.

5. PULSATIONS OF MASSIVE STARS

This topic was reviewed by Burki and Osaki. For the Beta Cephei stars various modes have now been detected, partly by studies of light- and v_{rad}-curves, partly also by investigating line profile variations. For 12 Lac one radial and three non-radial modes have so far been found.

For the 53 Persei stars (B3-B7, III-IV; P = 1-9d) the mode of pulsation is not well known. Also for the bright supergiants and hypergiants over the whole spectral range (O through M) the situation is not yet very clear yet. Empirical relations have been established between the quasi-period P_q and the luminosity L. About 40% of a group of well observed medium and late type supergiants stars pulsate in the fundamental mode and 60% have $P/P_{fund} > 1$, which is attributed to non-radial pulsations in the g-mode.

For the formal analysis of non-linear oscillations of stars Legendre polynomials are used to describe the deformations of the stellar surface and their time behaviour. The modes of oscillation are in that symbolism described by the integers n, l, and m. It remains to be investigated if the non-radial pulsations of super- and hyper-giants can really be described by Legendre polynomials in the strict mathematical sense. In view of the enormous size of these objects the characteristic time for the propagation of a disturbance ($\approx v_s/R$) is much larger than the quasi-period P_q. Therefore the pulsations may be stochastic in character rather than non-radial in the mathematical sense. 'Chaotic' motions can generate various modes of pulsation.

For explaining the physics of the pulsations the three conventional types of waves are introduced: the p-waves (in which pressure is the restoring force), the g waves (gravitation) and the f-waves (surface waves driven by gravitation). In the supergiants the atmospheric instability is greatly enhanced by the fact that the thermal and dynamic time scales are nearly equal. Their observed non-radial pulsations are probably g-modes. It is a matter of dispute whether these modes are initiated by convective motions above a H burning shell. In order to find out if waves generated in the interior parts of stars can reach the surface, and to what extent they have been attenuated when arriving there, the propagation diagrams (one for each mode of oscillation) are most useful as a diagnostic tool. In constructing these diagrams knowledge of the cut-off frequences (Brunt-Väisälä frequence for gravity modes, Lamb frequency for acoustic modes) is essential.

The importance of non-radial pulsations for stellar structure is that - like convection - they can transport angular momentum through the star. Is this process perhaps related to the observed episodical mass loss in Be stars?

It is of interest to know the energies involved in the various modes. Henrichs made an attempt to describe the energy diagram for the pulsations in a 12 $M_\odot$ ZAMS star (B0 V) with v_{rot} = 50 km s^{-1}. The energies per mode appear to range from 10^{44} to 6 x 10^{44} erg, a small number as compared to the potential or thermal energy contents of that star ($\approx 10^{51}$ erg).

6. PHOTOSPHERIC AND ATMOSPHERIC DISTURBANCES

Reviews on these topics were given by Zahn (convection and rotation), Castor (shock-driven instabilities) and Rybicki (radiation driven instabilities). Convection in luminous stars is heavily turbulent, because the Prandtl-number is small ($\approx 10^{-6}$), while the Rayleigh number is large. Early type stars do not have (sub-)photospheric convection zones even not in the region where they should exist in view of the Schwarzschild criterion, because convection cannot develop there in view of the strong radiative losses, but instabilities of a stochastic (turbulent) character can develop. This may be the reason why fairly strong turbulence is observed in photospheres of stars close to the Humphreys-Davidson limit. In some stars the turbulence is even supersonic.

Stars cannot rotate as a rigid body as von Zeipel's theorem states. But departures from uniform rotation lead to instability. Differential rotation is related to meridional circulation and this, in turn, may induce smaller-scale turbulence. Open questions are what energy is involved in these motions, and what is the character of this turbulence. For cooler stars the influence of magnetic fields should be considered; it can modify the acoustic energy output of stars by orders of magnitude. Photospheric shock waves may occur in photospheres of stars where the turbulent velocity amplitude approaches or surpasses the velocity of sound. Can these shock waves lead to mass loss? If we restrict ourselves to considering the momentum transfer, shocks can only lead to mass loss if the cooling time of the shocked gas is much larger than the time interval between successive shock waves. If, for a given star, the pulsation period is long, then the rate of mass loss is small. For red giants, with their low surface gravity and low intrinsic cooling efficiency, shock-driven mass loss can be important.

In addition, dissipation of turbulent energy in shock waves may be of importance. In any way the energy contained in the turbulent motions is sufficient to drive the stellar winds, but the way this energy transfers into wind energy has still to be investigated. A possibility still to be explored is that of the occurrence of radiation-driven shocks, and their influence on photospheric instability and stellar mass loss.

Radiation-driven disturbances can occur in the outer ('coronal' or stellar wind) parts of stars. In order to originate they must be triggered in some way and thereupon amplified by radiation. The strongest instabilities are expected to arise in the supersonic flow, and their physics seems to be well understood, contrary to the situation in the subsonic part of the winds where the numerical-mathematical treatment, as well as the physics is more complicated. So far, most research has been restricted to the linear approximation. The development of a non-linear theory, which is actually necessary for any instability treatment is still needed. 'Brute computer force' might help, perhaps. The observed superionisation in stellar winds, and the observed X-ray emission may also be attributed to radiation-driven instabilities.

One of the remarkable properties of the stellar winds is the occasional occurrence of discrete components. Most of them are transients but some are semi-stationary. The moving components tend to move through the winds, mostly somewhat faster, but slower components do exist too. A true miracle is the stationary component in the wind of P Cyg which has been there for at least 8 years! In the three brightest WC9 stars stationary narrow absorption components have been observed at 0.6 v_∞; these components must originate in the outer, decelerating part of the stellar wind sphere.

ACKNOWLEDGEMENTS

Many thanks for comments on a preprint of this review are due to drs. Baade, Castor, Nieuwenhuijzen, Van der Hucht and Vreux.

Cees de Jager giving his excellent review of the workshop

Section II

Poster Papers

Discussions at the end of a session: Cees de Jager with Jean-Paul Zahn, and Rens Waters (who recorded the discussions) with Pieter Mulder.

Françoise Praderie and Stan Owocki

MASSIVE STARS IN NEARBY GALAXIES

P.Massey (1), J.Hutchings (2), L.Bianchi (3)
(1) Kitt Peak National Observatory, U.S.A.
(2) Dominion Astrophysical Observatory, Canada
(3) Osservatorio Astronomico di Torino, Italy

ABSTRACT. IUE spectra and optical data (MMT blue spectra and CCD photometry) have been obtained for OB supergiants in M31 and M33. UV and visible data yield consistent values for the effective temperature, bolometric magnitudes, and extinction for the stars, which we classify as late O to early B. The UV resonance lines have very low outflow velocities compared to galactic stars, and no P Cygni profiles, suggesting that there may be significant differences in mass loss mechanisms among Local Group galaxies, as a consequence of different metallicities. The UV interstellar extinction in both galaxies appears like the LMC and not like the Galaxy.

1. INTRODUCTION.

Stellar winds from hot luminous stars are known to be radiatively driven. They are accelerated to supersonic outflow velocities mainly through lines of C,N,O though lines from other metals (especially Ne through Ca) may play an almost equal role.

Therefore differences can be expected in stellar wind velocity laws and mass loss rates for hot stars in galaxies with different metallicity than our own Galaxy.

Such studies have been carried out so far only for stellar winds in the MCs, showing that (i) O,B supergiants in the MCs have weaker resonance lines, lower values of V_∞ and global differences in their optical spectra respect to galactic ones (Hutchings 1982) (ii) m.s. and giant O,B stars have lower V_∞ but mass loss rates comparable to galactic stars (Garmani and Conti 1985) (iii)the extinction laws in the UV range are different from the galactic law (cf. e.g. Hutchings 1982).

Recent wind models by Kudritzki et al. (1986) for hot stars in the MCs show that significantly weaker winds, lower wind velocities <u>and</u> smaller mass loss rates are expected by lowering the metallicity. The results are in general agree-

H. J. G. L. M. Lamers and C. W. H. de Loore (eds.), Instabilities in Luminous Early Type Stars, 201–204.

ment with the terminal velocity measurements in the MCs.

We are extending this study of stellar winds to the hottest, most luminous stars in the Local Group galaxies. Some results for M31 and M33 are shown.

2. IDENTIFICATION OF THE MOST LUMINOUS STARS.

The first step of this study was the identification of the most luminous/ hottest stars in these galaxies. The information existing in the literature was in fact not sufficient to select the most massive stars: (a) in M33 Humphreys and Sandage (1980) catalogued blue and red stars and gave magnitudes and colors, but the errors on the colors e.g. were such to make not distinguishable an O-type from an early-A star. (b) for M31 no comparable information existed, except for Baade's field IV, which is however poor of massive stars.

Various optical data were collected: "grens" plates at the 3.6m CFHT in the 3500-5000Å range, at 1000Å/mm dispersion down to a limiting magnitude V=22, UBV CCD frames at the 4m KPNO for OB associations in M31, optical spectra at the MMTO with the photon counting dual array spectrograph.

In this way we were able to select the most massive/ most luminous stars, with little or no extinction, and at different distances from the galactic centre (to check for a possible metallicity gradient).

3. IUE OBSERVATIONS AND RESULTS.

The most luminous O,B supergiants in M31 and M33 are feasible with the IUE satellite. To date, we have partial wavelength coverage for 7 stars, and total coverage (1150-3200 Å) for 3 stars. Each low dispersion spectrum was obtained with uninterrupted 18hrs exposure.

In Figure 1 an exemple of an IUE SWP spectrum is shown for an object in M31. The target appears clearly double in the IUE image, and the two spectra were extracted separately. In the optical the two stars are not resolved, however the optical spectrum looks composite, as it shows Balmer lines, HeI 4143, 4471, OII4650, corresponding to a B0-2 supergiant, but also HeII4542, 4686 indicating a mid type O. The CCD photometry (V=17.84, U-B=-0.95, B-V=-0.09) obviously refers to the brightest object.

When comparing the UV continuum distributions with model atmospheres, it appeared that the spectra dereddened only for the galactic foreground extinction (E(B-V)=0.11) would have too low Teff values for their spectral types. While a further correction for interstellar reddening using

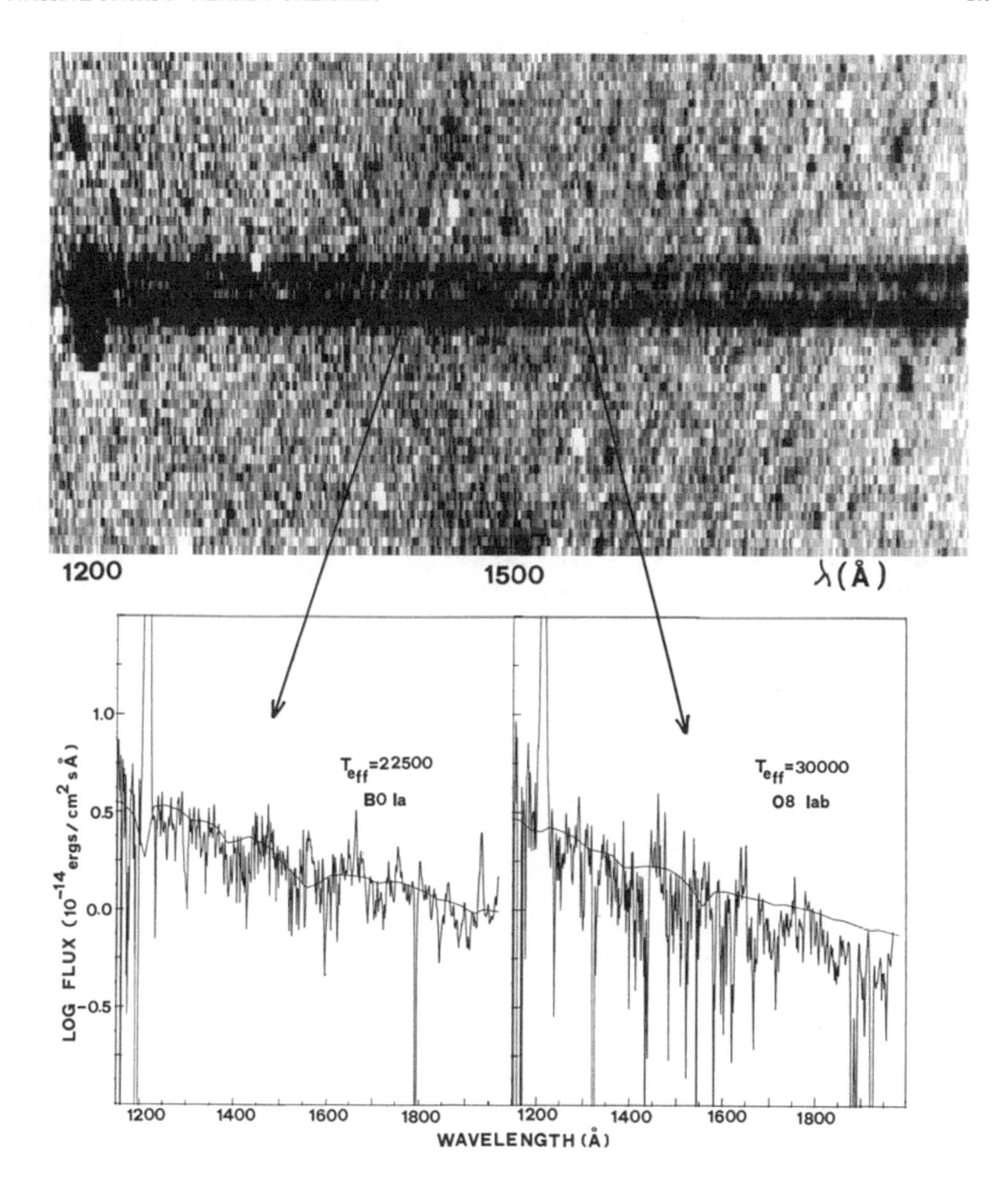

Figure 1. Top: a portion of the IUE SWP image of the object M31 NGC206-CFHT3. Two hot spectra are visible. Bottom: the two short wavelength spectra, dereddened for foreground galactic extinction with E(B-V)=0.11 plus additional LMC-type extinction with E(B-V)=0.05. Model atmospheres are shown for comparison.

a galactic-type law would not give reasonable results, it was possible to fit the UV spectra with model atmospheres consistent also with the optical data by assuming an LMC-type extinction in M31.

Extrapolation of the best fit models to the optical range gives colors and absolute magnitudes for the stars which are consistent with Teff and with the optical spectral type/ luminosity class if we use the recent statistical relations between stellar parameters derived by de Jager and Nieuwenhuijzen (1986) for galactic stars.

Similar results were found for a star in M33.

The analysis of the lines shows some differences with respect to galactic stars of similar type. The stars in M31 have wind velocities of the CIV and SiIV resonance doublets significantly lower than typical galactic values, the lines are weaker and no emission is detectable in any of them. The star in M33 seems to have lines of normal strength but again with very low terminal velocities. More details can be found in Massey et al. (1985). New observations of more stars seem to confirm the above results, but a detailed analysis is still under way and will be published elsewhere (Massey et al. 1986).

REFERENCES.

de Jager,C.,Nieuwenhuijzen,H.,1986,preprint
Garmany,C.,Conti,P.,1985,Ap.J., 293,407
Humphreys,R., Sandage,A.,1980,Ap.J. Suppl.,44,319
Hutchings, J.,1982, Ap.J., 255, 70
Kudritzki,R.,Pauldrach,A.,Puls,J.,1986,preprint
Massey,P.,Hutchings,J.,Bianchi,L.,1985,A.J.,90,2239
Massey,P.,Hutchings,J.,Bianchi,L.,1986,in preparation

Cornelis de Jager, Hans Nieuwenhuijzen,
and Karel A. van der Hucht
Laboratory for Space Research Utrecht
Beneluxlaan 21
3527 HS Utrecht

ABSTRACT. We have collected literature data on rates of mass-loss $\dot{M}$ for 264 O through M-type stars and a number of stars of other types. It appears possible to develop $\log(-\dot{M})$ into a series of Chebychev polynomia of the first kind in $\log T_{eff}$ and $\log(L/L_\odot)$ and their cross products. The average scatter of the data (one sigma) around this interpolation model is $\pm$ 0.46, while the intrinsic scatter of one single measurement is $\pm$ 0.37.

1. INTRODUCTION

At the Porto Heli symposium we described how the variation of the rate of mass-loss over the Hertzsprung-Russell diagram can be represented by a power series in $\log T_{eff}$ and $\log(L/L_\odot)$ and their cross-products (De Jager et al., 1986). We have now repeated that investigation. There are two important changes as compared to the previous investigation. First, the number of stars used for the Porto Heli paper was about 150, but the present material contains 264 stars of spectral types O through M, three O-type subdwarfs, 17 Be stars, 11 C stars, 4 S stars, 32 Wolf-Rayet stars and 15 nuclei of planetary nebulae. Secondly, the interpolation is now made by Chebychev polynomia of the first kind T_i and their cross-products T_iT_j; hence

$$- \log(-\dot{M}) = \sum_{n=o}^{N} \sum_{\substack{i=o \\ j=n-i}}^{i=n} a_{ij} \, T_i(\log T_{eff}) \, T_j(\log(L/L_\odot)) \quad (1)$$

2. MODEL ADAPTATION

A twenty-terms representation of the mass-losses of O through M-type stars appears sufficient (N=5). The one sigma value of the adaptation of the mass-loss data to the numerical model (eq. (1)) is $\pm$ 0.46, while the

H. J. G. L. M. Lamers and C. W. H. de Loore (eds.), Instabilities in Luminous Early Type Stars, 205–208.

average sigma value of a single measurement of log $(-\dot{M})$ is $\pm$ 0.37. This shows that the adaptation to eq. (1) is fairly good, but that in addition the rate of mass-loss slightly depends on other parameters than T_{eff} and L.

The adjacent Figure 1 shows the distribution of mass-loss over the Hertzsprung-Russell diagram. Dots indicate the positions of single stars and the lines connect points of equal rates of mass-loss. The labels give the negative logarithm of the rate of mass loss expressed in solar masses per year. It is of interest to note that the line of 10^{-4} $M_{\odot}yr^{-1}$ mass-loss fairly well follows the Humphreys-Davidson upper brightness limit. We also draw attention to the shape of the lines in the upper red part of the Hertzsprung-Russell diagram, where the curves of large mass-loss bend upward for decreasing T_{eff}-values. Finally, we draw attention to a few very well studied stars for which the mass-loss is known with a high accuracy:

Zeta Pup (O4 Ief) : log $(-\dot{M})$ = $-$ 5.325 $\pm$ 0.043
Eps Ori (B0 Iae) : $-$ 5.680 $\pm$ 0.057
VY CMa (M3-5 Iae) : $-$ 3.620 $\pm$ 0.047.

REFERENCE

C. de Jager, H. Nieuwenhuijzen and K.A. van der Hucht: 1986, in Proceedings Porto Heli Symposium.

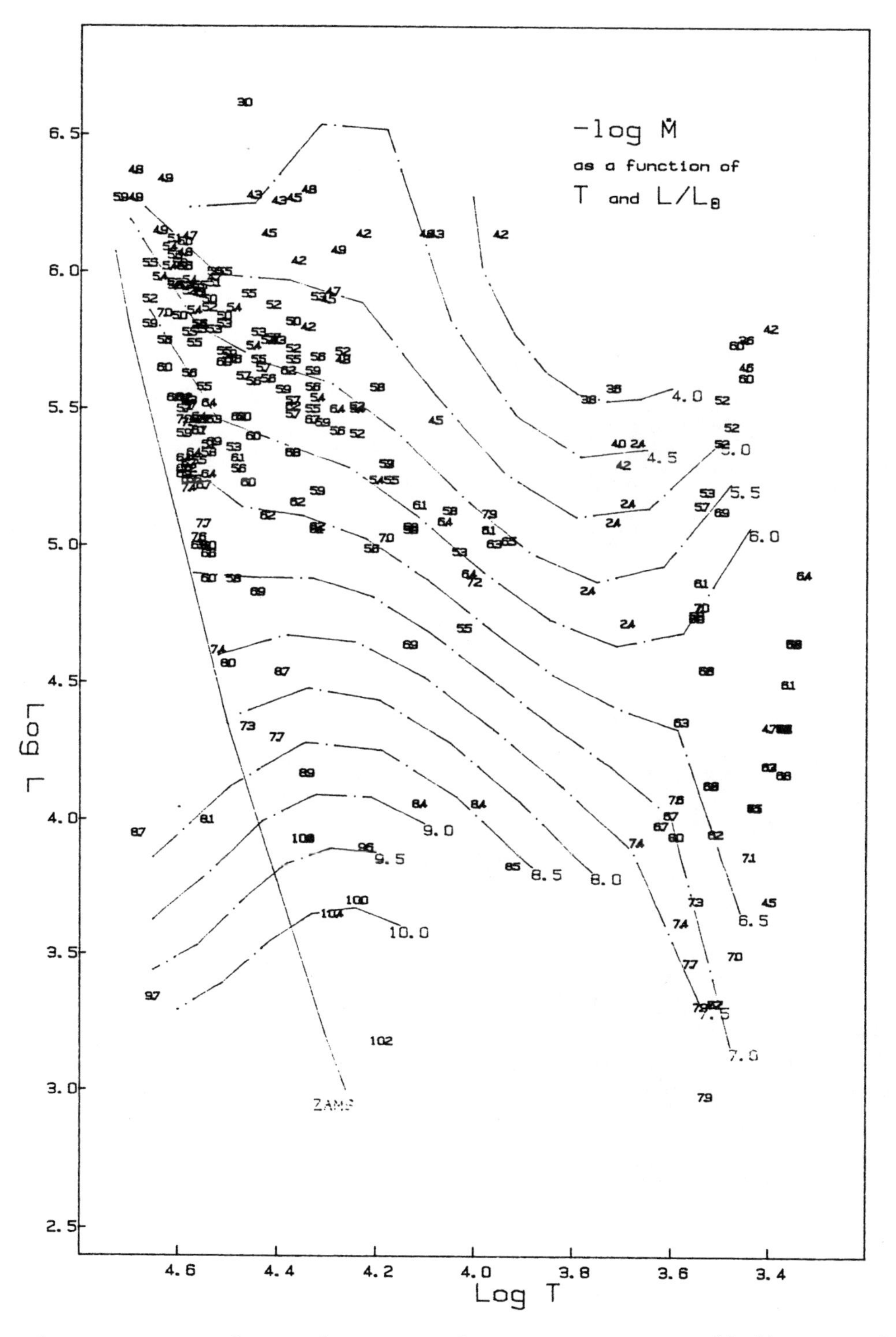

Figure 1. Rate of mass-loss over the Hertzsprung-Russell diagram. Lines are labeled with - log (- $\dot{M}$).

Cees de Jager in discussion with Luciana Bianchi

Mart de Groot teaches Mrs. Noels and her daughter how to play billiards in the Dutch way

THE UNSTABLE O6.5f?p STAR HD 148937 AND ITS INTERSTELLAR ENVIRONMENT

Claus Leitherer and Carlos Chavarria-K.
Landessternwarte Königstuhl
D-6900 Heidelberg
Germany

The massive, early-type star HD 148937 (spectral type O6.5f?p) is surrounded by a unique set of nebulosities. A spherically symmetric Strömgren sphere (Radius $\simeq$ 25 pc), an ellipsoidal filamentary nebulosity (semimajor axis $\simeq$ 5 pc) interpreted as a stellar-wind-blown shell, and a bipolar nebular complex (semimajor axis $\simeq$ 1 pc) with HD 148937 located in the apparent center of symmetry. Here we report on observations of these nebulosities (narrow-band CCD imaging, IDS spectrophotometry, high resolution ($\Delta V \simeq 7$ km sec^{-1}) spectroscopy) in an attempt to establish a consistent model of HD 148937 and its nebulosities.

The two bipolar nebulosities (known as NGC 6164/5) show striking resemblance. They are of similar extent and distance from HD 148937 implying a simultaneous origin due to an explosive event in HD 148937. A comparison of H_α- and H_β-images and IDS fluxes gives a rather homogeneous dust distribution all over the bipolar nebula with an extinction in agreement with E(B-V) of HD 148937. We conclude that the apparent morphology of NGC 6164/5 is not simulated by variable extinction but rather reflects the actual distribution of (excited) gas. Moreover, it is safe to assume that the distribution of H_α emission is identical with the total amount of gas in the inner nebulosities (i. e. NGC 6164/5 is density-bounded). The output of Lyman quanta of HD 148937 ($\log N_L \simeq 49.15$ sec^{-1}) even suffices to keep ionized the stellar-wind-blown shell and the Strömgren sphere. The ratio [O III]/H_β drastically decreases with increasing distance from HD 148937. This behavior is paralleled by a corresponding increase of [N II]/H_α indicating the transition from the high-ionization zone to the low-ionization zone. We stress the similarity to the structure of ejecta from novae (see Gallagher and Anderson 1976).

In establishing a geometrical model for the bipolar nebula we are led by its striking resemblance to ejecta from novae or symbiotic stars (e. g. Solf 1983). The high

H. J. G. L. M. Lamers and C. W. H. de Loore (eds.), Instabilities in Luminous Early Type Stars, 209–210.

rotational velocity of HD 148937 ($v \cdot \sin i$ = 200 km sec^{-1}, Conti and Ebbets 1977) strongly implies axial symmetry for the system and thus giving rise to the ejection of matter preferentially along the two axes of symmetry. The detailed kinematic structure is derived from our highly-resolved line profiles. We find a double-cone structure nearly perpendicular to the line of sight with a kinematic age of 10^3 - 10^4 yr. The ejection of the nebulosities may be due to instabilities in this star close to the Of-WR transition phase. The high rotational velocity may play a crucial role in these instabilities.

The influence of rotation on the stellar wind of HD 148937 and the resulting non-isotropic mass flow could also account for the oval-shaped structure of the wind-blown shell. We underline that the spectral appearance of this shell is typical of wind-blown shells from e. g. WR-stars and clearly excludes an origin as due to a SN explosion. The age of the bubble as determined from the luminosity of the wind is a few times 10^5 yr.

Since NGC 6164/5 is clearly ejected by the star and expands into space with virtually no contamination by ISM (the density in the stellar-wind cavity is ~10^{-2} cm^{-3}) it provides the unique possibility of studying the chemical abundances of an evolved Of star. Our physical analysis gives an electron density of $1 \cdot 10^4$ cm^{-3} for regions II, V and $5 \cdot 10^2$ cm^{-3} for III, IV. The electron temperature is 6700 K. In these respects the nebulosities closely resemble normal H II regions. We find no evidence for shock excitation in the lines. The abundance analysis is treated in the usual way following Peimbert and Costero (1969). We find an overabundance of nitrogen relative to hydrogen by a factor of 6 relative to the sun for NGC 6164/5. On the other hand, the outer border of the Strömgren sphere (NGC 6188) shows abundances typical of normal H II regions (cf. Orion). An overabundance of N by a factor of 6 is in good agreement with theoretical evolutionary modelling of stellar abundances by Maeder (1983). This provides strong support for the evolutionary state of Of stars as being intermediate between O- and WR-stars.

References:

Conti, P. S., Ebbets, D.: 1977, Astrophys. J. 213, 438
Gallagher, J. S., Anderson, C. M.: 1976, Astrophys. J. 203, 625
Maeder, A.: 1983, Astron. Astrophys. 120, 113
Peimbert, M., Costero, R.: 1969, Bol. Obs. Ton. Tac. 5, 3
Solf, J.: 1983, Astrophys. J. Lett. 266, L 113

TWO COMMENTS ON THE β CEPHEI VARIABLE 12 LACERTAE

M. Jerzykiewicz
Wroclaw University Observatory

1. THE DOMINANT PULSATION MODE

According to a popular notion, the dominant pulsation mode in the β Cephei variables is radial. There is, however, at least one exception to this rule. The exception is 12 Lacertae.

In 1956 this star was the subject of an international observing campaign, organized by de Jager (1963). Frequency analysis of the international campaign photometric observations led to the discovery that the light variation of 12 Lacertae consists of at least six short-period sinusoidal components (Barning 1963, Jerzykiewicz 1978). The four strongest components, which were also identified in the radial velocity data, have frequencies equal to 5.1790, 5.0667, 5.4901, and 5.3347 c/d. Note that the first frequency corresponds to the well-known primary period of $0^{d}.193089$ (Young 1915), while the second, to the secondary period of $0^{d}.197367$, discovered by de Jager (1953). Note, moreover, that three frequencies, the first, the fourth, and the third, form an equidistant triplet. This had led to the conclusion that these three frequencies result from a rotational splitting of a nonradial mode. There is a controversy in the literature whether the spherical harmonic order of the mode, ℓ, is equal to 3 (Jerzykiewicz 1978) or 2 (Smith 1980), but the fact that the primary component is nonradial is not disputed. The secondary component may be radial. However, it is much fainter in both light and radial velocity variations than the primary.

2. STABILITY OF THE COMPONENT AMPLITUDES

There are eight series of photoelectric observations of 12 Lacertae which can be used to investigate the long-term behavior of the amplitudes of the four sinusoidal components, mentioned in the previous section. Each series is confined to a single observing season. The three oldest ones include the 1937, 1939, and 1940 observations of Fath (1947), while the two most recent, the 1977 and 1979 observations of Jarzębowski et al. (1979) and Jerzykiewicz et al. (1984). The remaining three series consist of the 1951 observations of Nekrasova (1952), de

H. J. G. L. M. Lamers and C. W. H. de Loore (eds.), Instabilities in Luminous Early Type Stars, 211–212.

Jager's 1956 international campaign data (de Jager 1963), and the 1959 observations of Opolski and Ciurla (1961).

A least squares analysis of all these observations shows that the four component amplitudes did not change appreciably from 1937 to 1979. The mean values of the amplitudes, computed from the results of eight independent solutions (that is, one for each series of data) are equal to $0^m.0405$, $0^m.0150$, $0^m.0122$, and $0^m.0066$. The standard errors of these mean amplitudes amount to $0^m.0009$, $0^m.0006$, $0^m.0004$, and $0^m.0009$, respectively. They are of the same order of magnitude as the mean errors of the individual amplitudes, used to compute the mean values. The mean errors vary from about $0^m.0015$ for the older series of data, including the 1951 observations of Nekrasova, to $0^m.0006$ for the more recent ones. Thus, in the case of 12 Lacertae there is no evidence so far for long-term variations of the pulsation amplitudes such as those that were recently discovered in several other β Cephei stars.

REFERENCES

Barning, F.J.M.: 1963, B.A.N. **17**, 22.
Fath, E.A.: 1947, Publ. Goodsell Obs. No. 12.
Jager, C. de: 1963, B.A.N. **17**, 1.
Jarzębowski, T., Jerzykiewicz, M., Le Contel, J.-M., Musielok, B.: 1979, Acta astr. **29**, 517.
Jerzykiewicz, M.: 1978, Acta astr. **28**, 465.
Jerzykiewicz, M., Borkowski, K.J., Musielok, B.: 1984, Acta astr. **34**, 21.
Nekrasova, S.V.: 1952, Izv. Krym. astrofiz. Obs. **9**, 126.
Opolski, A., Ciurla, T.: 1961, Acta astr. **11**, 231.
Smith, M.A.: 1980, Astrophys. J. **240**, 149.
Young, R.K.: 1915, Publ. Dom. Obs. Ottawa **3**, 65.

Christoffel Waelkens and Mike Jerzykiewicz

NON RADIALLY PULSATING WOLF-RAYET STARS

A. Noels, R. Scuflaire
Institut d'Astrophysique
5, avenue de Cointe B-4200 Ougrée-Liège

A few non radial modes, some of which are trapped modes, can be amplified in H-burning shell models coming from the evolutionary sequence of a star of initial mass 100 $M_\odot$ losing mass through stellar wind. The periods range from half an hour to a few hours. The duration of this instability phase is rather short, of the order of a few e-folding times, which are of a few thousand years.

Some WR stars, particularly WN stars, show a variability in emission lines with periods in a narrow range of a few hours. Vreux (1985), Vreux et al. (1985) and Vreux (1986) have emphasized that variability, suggesting that it could be due to non radial oscillations.

It is well accepted that some WR stars can be formed from massive stars evolving with mass loss. Radial oscillations have been shown to be amplified in such models after the H-burning shell phase, if the structure of the star is close enough to a homogeneous helium star, more massive than the critical mass of such stars, of the order of 16 $M_\odot$ (Noels and Gabriel 1981, Noels and Gabriel 1984, Maeder 1985). The periods however seem too short, less than one hour, to explain the variability observed in WR stars. Longer periods can be found in non radial oscillations but the problem is to find favourable conditions to amplify them, i.e. to obtain a vibrational instability.

We have analysed models extracted from the evolution of a 100 $M_\odot$ star (Noels and Gabriel 1981) whose H-R diagram is given in figure 1. Some properties of the selected models are given in Noels and Scuflaire (1986).

In the vibrational stability analysis, the perturbation of any variable f is written in the form

$$\delta f(r,\theta,\varphi,t) = C\,\delta f(r)\,P_\ell^m(\theta,\varphi)\cos(m\varphi-\sigma t)\,e^{-\sigma' t}$$

H. J. G. L. M. Lamers and C. W. H. de Loore (eds.), Instabilities in Luminous Early Type Stars, 213–216.

P_ℓ^m is the associated Legendre polynomial of degree ℓ and order m, σ the adiabatic angular frequency and σ' the damping coefficient whose expression is

$$\sigma' = - \frac{\int \frac{\delta T}{T} \delta\varepsilon \, dm - \int \frac{\delta T}{T} \delta\left(\frac{1}{\rho} \operatorname{div} \vec{F}\right) dm}{2\sigma^2 \int |\delta r|^2 \, dm} = - \frac{E_N - E_F}{D}$$

A positive value of σ' means a damping of the oscillation while a negative value means an amplification and a vibrational instability. E_N comes from the nuclear terms and has always a destabilizing effect. E_F comes from the flux terms, its main contribution arises from the external layers and it has generally a damping effect.

<u>So, a necessary condition for a mode to be amplified is to have a large amplitude in a nuclear burning region and a rather small amplitude outside.</u>

In the case of non radial oscillations, the amplitude of $\frac{\delta P}{P}$ goes to zero at the center, so core burning models can be discarded at once. Figure 2 shows the first g^+ modes for l = 1 in the case of model 2 which is a H-core burning model. We can see that the amplitudes are large outside the nuclear burning region and all the non radial modes are damped.

We find more favourable conditions in H-shell burning models, and two different kinds of situation arise.

1. <u>Low l, low order modes</u>

Figure 3 shows that for model 3, the amplitude of the g_1^+ mode for l = 1 is very large in the H-burning shell, and rather small in the external layers. The amplification term E_N dominates the damping term E_F and this mode can reach a finite amplitude. The g_2^+ and g_3^+ modes have a node in side the H-burning shell, which lowers E_N and these modes are damped (Noels and Sculfaire, 1986). Model 4 is also vibrationally unstable towards the g_2^+ mode for which the second extremum is just inside the nuclear burning shell. Due to the sharp increase in the central condensation, $\frac{\rho_c}{\rho}$, the amplitude near the surface becomes too large in the following models and stability is restored at model 5. The periods obtained here are of the order of 4 hours which comes closer to the observed value. The amplification time, $\frac{1}{|\sigma'|}$, is of the order of a few thousands years, about 10 times shorter than the whole duration of the unstable phase, so this instability is rather mild.

2. Moderately high l, trapped modes

Trapped modes are modes which have an oscillatory behaviour in the r variable, in a narrow trapping zone in the star and which are evanescent, with a decreasing amplitude, outside. The local condition for a mode to have an oscillatory behaviour is that its angular frequency σ must be greater or smaller than both the acoustic cut-off frequency σ_a and the gravity cut-off frequency σ_g, given by

$$\sigma_a = \frac{\sqrt{\ell(\ell+1)}\,c}{r} \qquad \sigma_g = \sqrt{-\left(\frac{d\ln\rho}{dr} - \frac{1}{\Gamma_1}\frac{d\ln P}{dr}\right)g}$$

where c is the local speed of sound.
Figure 4 shows σ_a^2 for l = 5 and l = 10 and σ_g^2 in model 4. The very high peak of σ_g near the H-burning shell allows modes, for moderately high l, to be trapped in that region if their angular frequency is smaller than the height of the peak of σ_g. For l = 5, the g_6^+, g_5^+, g_4^+, g_3^+, f and p_6 modes are trapped and for l = 10, the g_6^+, g_5^+, g_3^+, g_1^+, p_2 and p_7 modes are trapped (Scuflaire and Noels 1986). Figure 5 illustrates the difference of behaviour between two consecutive p modes. The p_7 mode is trapped in the H-burning shell while the p_6 mode has a negligible amplitude in the nuclear burning shell. Amplification occurs for the p_7 mode while damping is found for the p_6 mode.
The situation is similar but less favourable in model 3 for which however the low order g^+ modes are marginally stable. All the trapped modes are either vibrationally unstable or marginally stable. Their periods range from half an hour to a few hours and their amplification time is of the order of a few thousand years, but here again the duration of the unstable phase is rather short, of the order of only a few e-folding times.

Some non radial modes can be amplified in H-burning shell models, with periods in the range of the observed periods. Such unstable models can only represent WN stars, and maybe even only late type WN stars, as they must still have enough hydrogen in the external layers but this is still in agreement with the observations, showing such variations mostly in WN stars.

Maeder A. 1985, Astron. Astrophys. 147, 300.
Noels A., Gabriel M. 1981, Astron. Astrophys. 101, 215.
Noels A., Gabriel M. 1984, Proc. 25th Liège Int. Astrophys. Coll., 59.
Noels A., Scuflaire R. 1986, Astron. Astrophys.
Scuflaire R., Noels A. 1986, submitted to Astron. Astrophys.
Vreux J.M. 1985, P.A.S.P. 97, 274.
Vreux J.M. 1986, this workshop.
Vreux J.M., Andrillat Y., Gosset E. 1985, Astron. Astrophys. 149, 337.

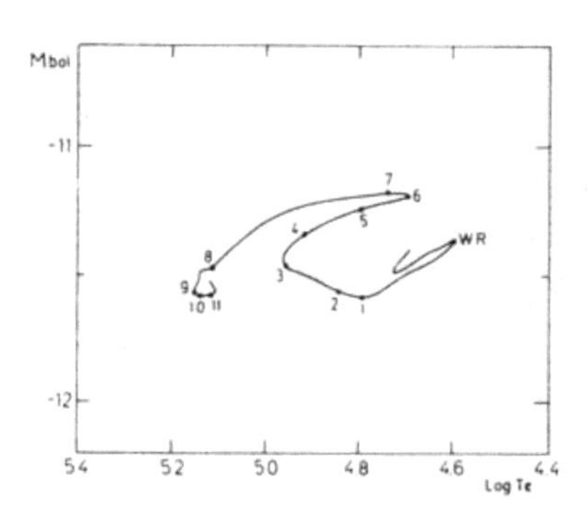

Figure 1

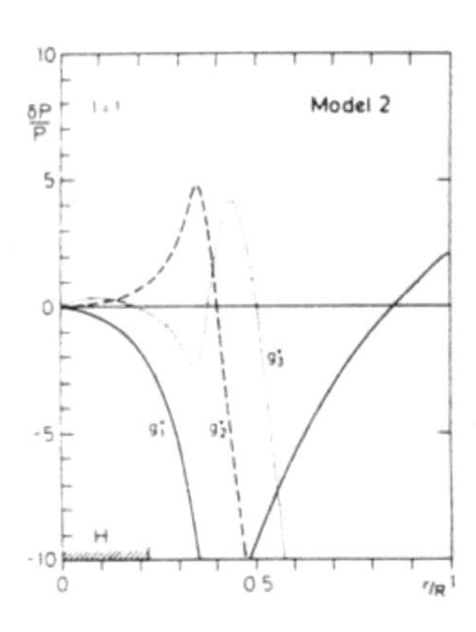

Figure 2

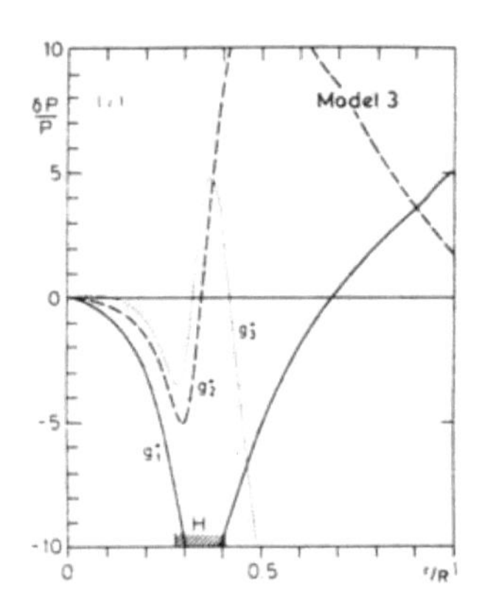

Figure 3

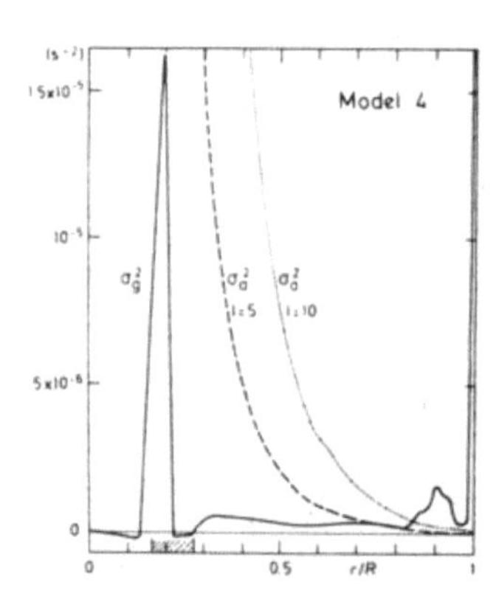

Figure 4

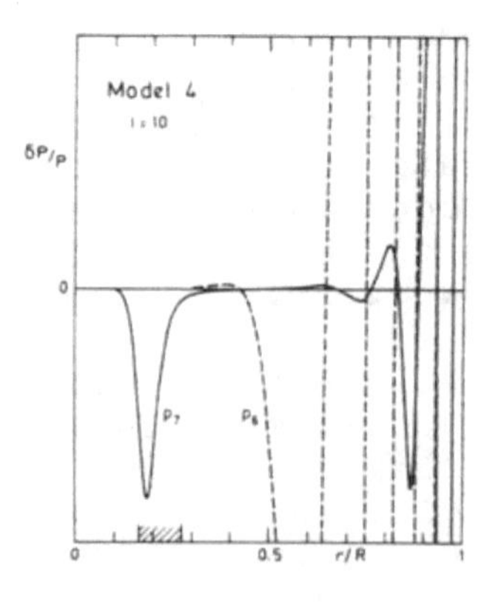

Figure 5

SOBOLEV TYPE LINE PROFILE IN CASE OF NON RADIAL WIND DENSITY PERTURBATIONS

R. Scuflaire, J.M. Vreux
Institut d'Astrophysique
Université de Liège
Avenue de Cointe, 5
B-4200 Ougrée (Belgium)

ABSTRACT. We have investigated the modifications induced on the P Cygni line profiles of an outwards accelerating wind by density fluctuations modulated by non radial pulsations. The results obtained in a first approach of the problem compare favourably with some observed time dependent profiles of ultraviolet lines.

1. INTRODUCTION

In a recent paper (Vreux, 1985), one of us has suggested that the variability observed in some Wolf-Rayet stars might be due to non radial pulsations. In this context, we have decided to investigate the signature on line profiles of perturbations of the wind density modulated by non radial pulsations. We have tackled this problem in the frame of a linear theory.

2. COMPUTATION OF THE PERTURBED PROFILE

We start with a classical Sobolev type profile produced by a spherically symmetric wind accelerating outwards. We have chosen, as our stationary model, a model described by Castor and Lamers (1979) with parameters corresponding to the case $\mathcal{T} = 1$ of their figure 8B. As a working hypothesis we assume that the stellar oscillation affects the wind only through a modulation of the density at the base of the wind and has no effect on the velocity law. The density perturbation at the base of the wind is assumed to be given by the theory of non radial pulsations of stars, i.e.

$$\delta\rho/\rho = \varepsilon \sqrt{4\pi}\, Y_{\ell m} (\theta,\phi)\, e^{-i\sigma t}$$

where ε is a small parameter, $Y_{\ell m} (\theta,\phi)$ a spherical harmonics and σ the angular frequency of the pulsation. This perturbation of density moves through the envelope at the velocity of the wind. This affects the resulting line profile through the perturbation of the optical depth.

H. J. G. L. M. Lamers and C. W. H. de Loore (eds.), Instabilities in Luminous Early Type Stars, 217–220.

One obtains the perturbation of the flux, relative to the continuum in the following form

$$\delta F(\lambda)/F_c = A(F_1(\lambda) \cos \sigma t + F_2(\lambda) \sin \sigma t)$$

where A depends on ℓ, m and θ_0, the angle between the axis of rotation of the star and the line of sight. F_1 and F_2 depends on λ, σ and ℓ only. The details of the computations and the exact form of A, F_1 and F_2 are given elsewhere (Scuflaire, Vreux, 1986).

3. RESULTS AND DISCUSSION

It is convenient to introduce the dimensionless parameter $\omega = R\,\sigma/v_\infty$ which measures the frequency of the stellar pulsation in a natural unit associated with the wind. Figures 1 and 2 give perturbed line profiles at different phases Φ of the pulsation for $\ell = 2$ and $\omega = 0.5$ and 2.

As presently our computations are made with only one velocity law and one opacity law, we have not tried to reproduce a given observed profile. Nevertheless a comparison between the perturbed profiles of figure 1 at phases 0.25 and 0.75, for example, leads to the description of a profile variation which is surprisingly similar to the one observed by Willis et al. (1986) in the P Cygni profile of N IV λ 1718 in the spectrum of WR6.

At higher values of ω, as displayed in figure 2, the perturbations sometimes give the impression of moving absorption features superimposed to a more or less normal P Cygni profile. This is reminiscent of the signature of the "puffs" described by Gry et al. (1984) and of the narrow components described by Henrichs (1984).

REFERENCES

Castor J.I., Lamers H.J.G.L.M., 1979, Astrophys. J. Suppl. Ser., 39,481.
Gry C., Lamers H.J.G.L.M., Vidal-Madjar A., 1984, Astron. Astrophys., 137,29.
Henrichs H., 1984, ESA-SP-218, Proc. 4th European IUE Conf., 43.
Scuflaire R., Vreux J.M., 1986, Astron. Astrophys., submitted for publication.
Vreux J.M., 1985, PASP, 97,274.
Willis A.J., Howarth I.D., Conti P.S., Garmany C.D., 1986, in Luminous Stars and Associations in Galaxies, eds C. de Loore, A. Willis and Laskarides.

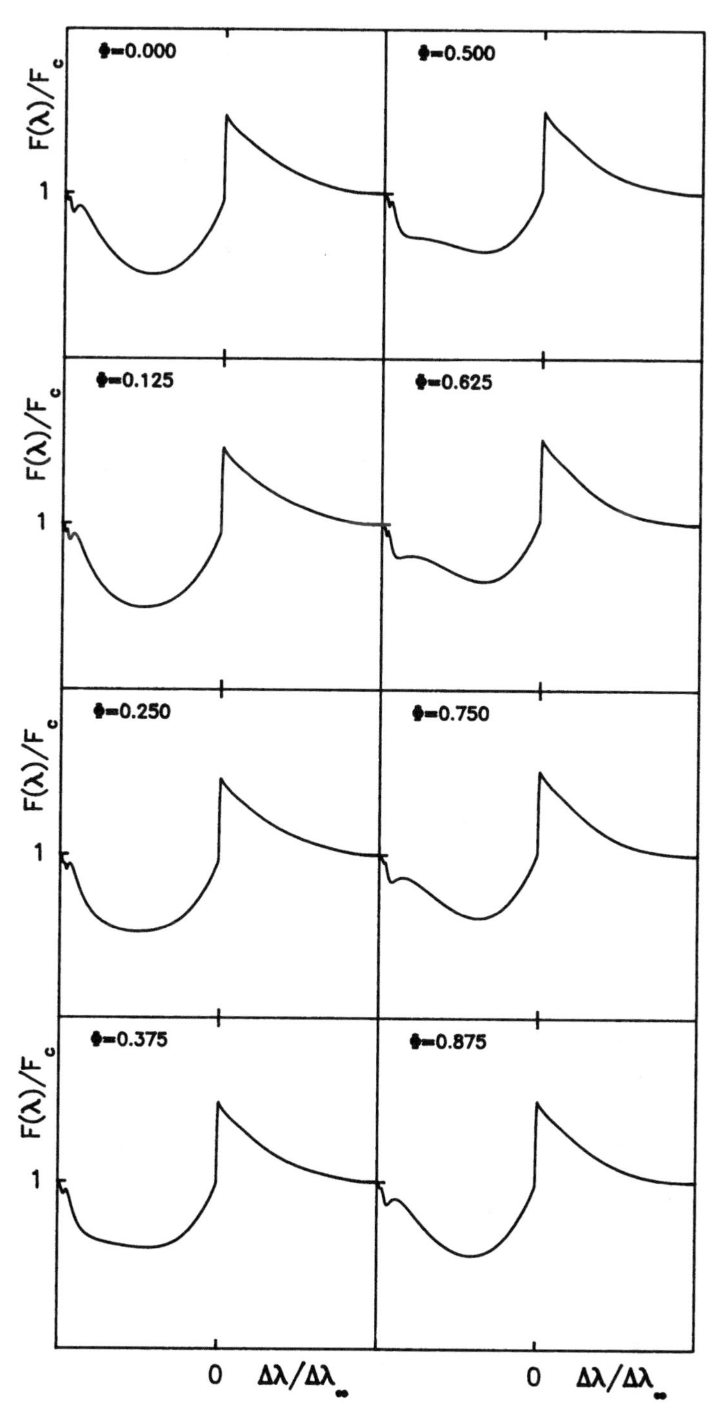

Figure 1. The line profile $F(\lambda)/F_c$ for $\ell = 2$ and $\omega = 0.5$

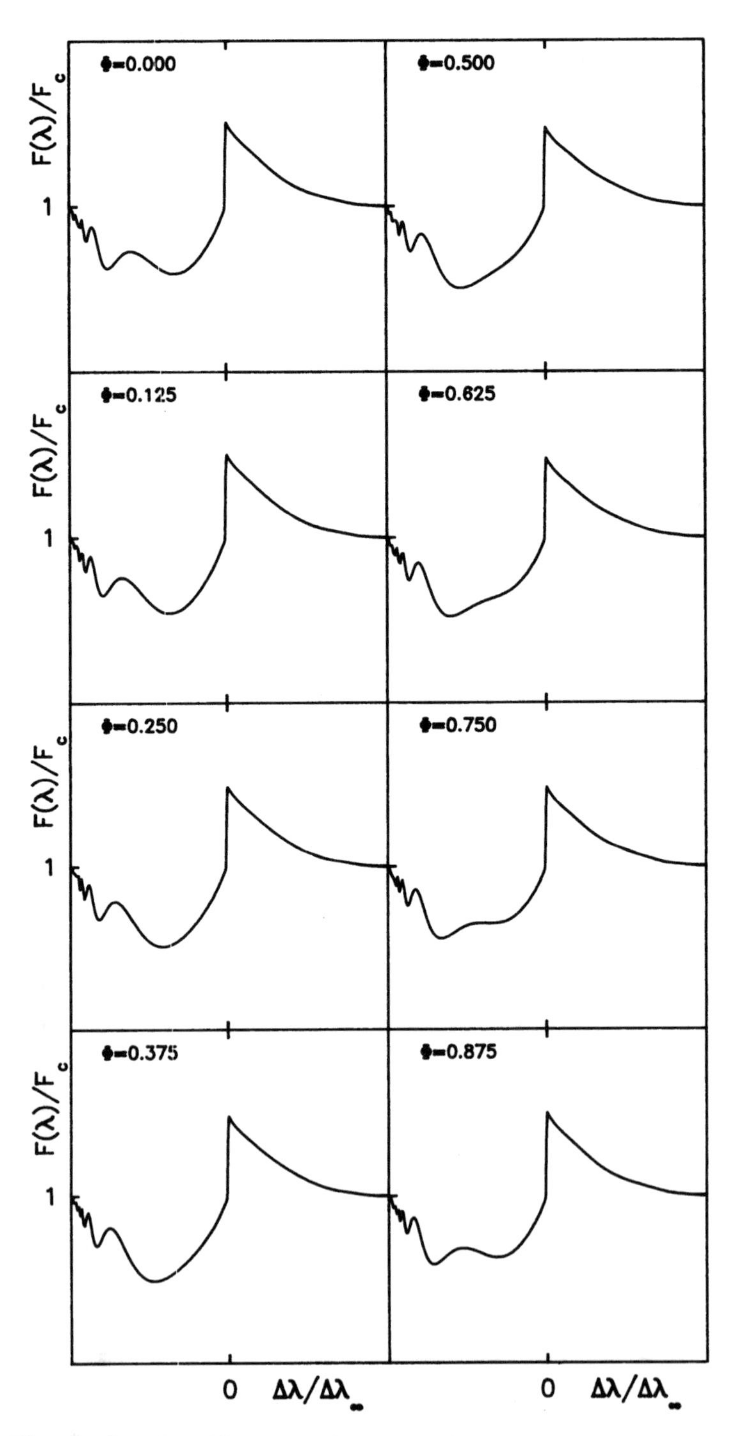

Figure 2. The line profile $F(\lambda)/F_c$ for $\ell = 2$ and $\omega = 2$

EPISODIC DISTORTION AND DUST FORMATION IN THE WIND OF WR 140

P.M. Williams,[1] K.A. van der Hucht,[2]
H. van der Woerd,[2] W.M.Wamsteker,[3] T.R.Geballe,[4,5]
C.D.Garmany[6] & A.M.T.Pollock[7,8]

1. Royal Observatory, Blackford Hill, Edinburgh, Scotland.
2. SRON Laboratory for Space Research, Utrecht, Beneluxlaan 21, Utrecht, The Netherlands.
3. ESA-IUE Tracking Station, Villafranca del Castillo, Madrid, Spain
4. United Kingdom Infrared Telescope, 665 Komohana Street, Hilo, Hawaii 96720, U.S.A.
5. Kapteyn Astronomical Institute, Rijksuniversiteit te Groningen, The Netherlands.
6. JILA, University of Colorado, Box 440, Boulder, CO 80309, U.S.A
7. Dept Space Research, University of Birmingham, Birmingham. B15 2TT, England.
8. ESA-EXOSAT Observatory, ESOC, Robert Bosch Strasse 5, 6100 Darmstadt, F.R.G.

ABSTRACT. In 1985 April, the WC7+abs star WR140 = HD 193793 was observed to have brightened by over two mag. in the infrared owing to the formation of a new dust shell. The growth and evolution of the shell was monitored by infrared observations using UKIRT during the remainder of 1985. Examination of infrared photometry of this star since 1979 and previously published data indicate that the dust formation occurs at intervals of 7.9 years. Phasing the published radial velocities of the absorption line component with this period confirms that it is a member of an eccentric (e = 0.7-0.8) binary system having periastron passage shortly before dust formation. Further evidence of the distortion of the stellar wind from the WC7 component is provided by changes in the P Cygni profile of the C IV lines observed with IUE while the dust was forming. The X-ray spectrum also changed between 1984 and 1985 in becoming significantly "harder" while the non-thermal radio source was extinguished. An interpretation is sought in terms of a model wherein the non-thermal radio and X-ray emission is excited by collision of the WC and O star winds and modulated by the orbital motion. A suitably compressed region of the wind must be advected far enough from the stars to form dust.

1. INFRARED LIGHT CURVE AND ORBITAL MODULATION.

In 1977, HD193793 (= WR140) was found to be abnormally bright in the infrared indicating formation of a dust shell (Williams *et al.* 1977, 1978 and Hackwell, Gehrz & Grasdalen 1979 [= HGG]). The shell was observed to cool over a timescale of months suggesting that it was expanding and that shell formation

H. J. G. L. M. Lamers and C. W. H. de Loore (eds.), Instabilities in Luminous Early Type Stars, 221–226.

might be a recurrent phenomenon (Williams & Antonopoulou 1979 and HGG). For this reason, and to study the fading and cooling of the 1977 shell, WR140 was monitored in the infrared from UKIRT. Persistence was rewarded in 1985 April when WR140 was observed to have brightened by almost 2 mag in L' over an interval of only 6 weeks! Observations between 1 and 20μm were made several times during 1985. The infrared spectra were those of heated amorphous carbon dust grains condensing in and moving with the stellar wind.

The 1985 maximum is well defined by the observations. By matching it to the 1977-78 data, we derive an interval of 7.9 yrs (2885 days) between dust formation episodes. The 1970-75 decline discussed by Hackwell *et al.* (1976) is entirely consistent with this interval and reinforces it as a period. The infrared data are re-plotted in Fig. 1 phased with a 7.9 yr period, zero phase representing the L' maximum in 1985.40. We conclude here from the light curve that the variations of WR140 are periodic with P = 2885 $\pm$ 25 days.

Given that WR140 shows periodic episodes of grain formation, it is necessary to seek the "clock" triggering these events and we re-examine its status as a binary. Although the composite nature of the spectrum led McDonald (1947) to adopt the types WC6 + O6 with the O6 component being 1.2 mag brighter than the WC6, she was unable to detect orbital motion. Lamontagne, Moffat & Seggewiss (1984) derived a 3-year period from the emission line data but Conti *et al.* (1984) showed this to be spurious and found no evidence for binary motion.

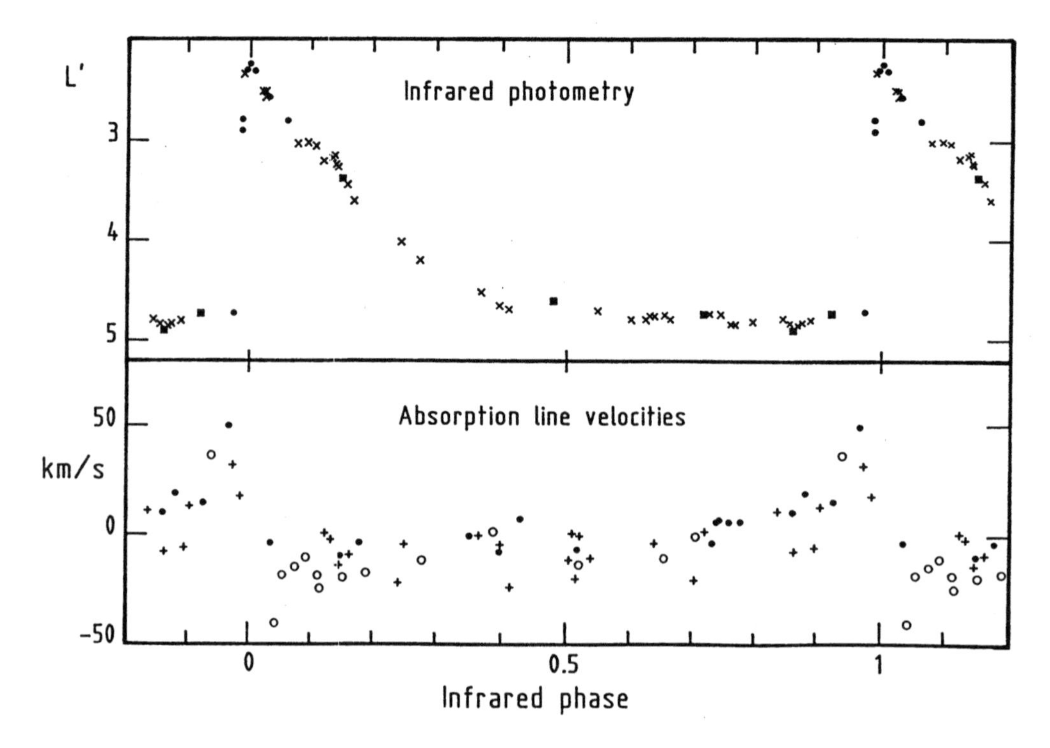

Fig. 1. Infrared photometry and absorption line radial velocities phased to 7.9 yr. Different symbols mark different cycles (top) or different data sets (below).

We therefore plotted the absorption line velocities from both Lamontagne *et al.* and Conti *et al.* - covering over 60 years - against the period established by the infrared data (Fig 1). It is immediately apparent that all the significantly more positive velocities occur at the same phase ($\phi \sim 0.95$) although coming from different cycles. The absorption line velocity data demonstrate that the O-type star is a member of a high eccentricity ($e \sim 0.7$-0.8) binary system having periastron passage shortly before infrared maximum.

We conclude that orbital motion provides the timing of the outbursts and continue our discussion on the basis that WR140 is a binary of high eccentricity and 7.9 yr period. We will not attempt further analysis of the velocities here but point to the urgent need for more velocity data especially in the year leading up to the next infrared maximum expected in 1993.3.

2. ULTRAVIOLET OBSERVATIONS OF THE STELLAR WIND

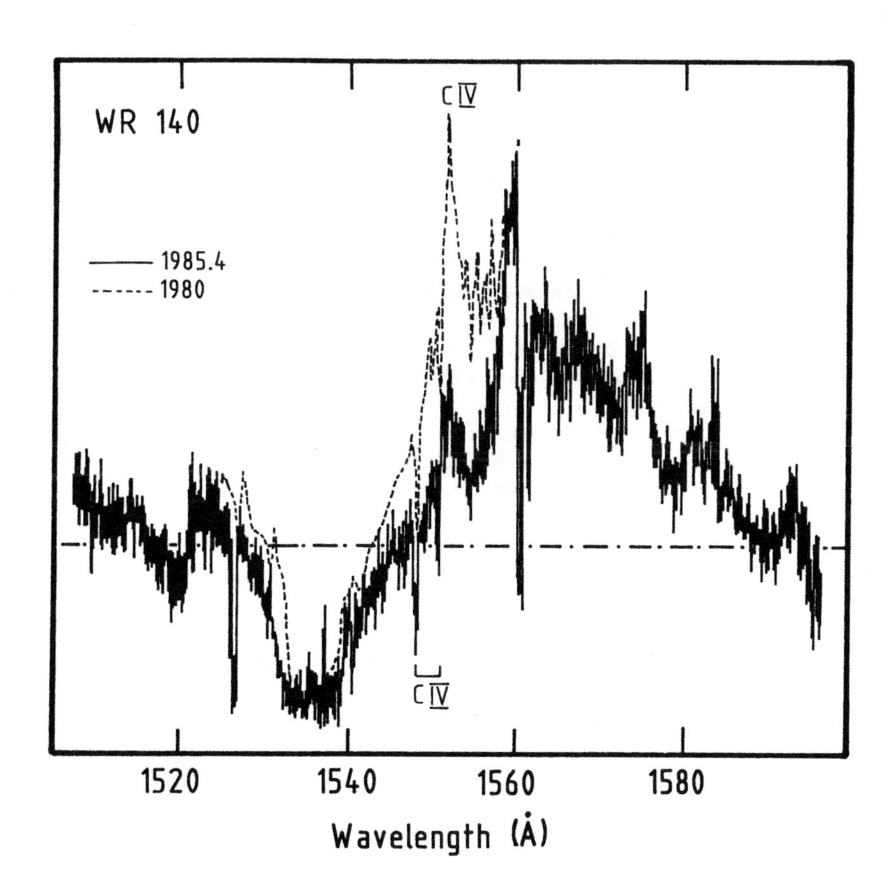

Fig 2. The IUE high resolution spectrum near infrared maximum compared with that in 1980 (broken line). Note the change in intensity of the C IV resonance lines and the broadening of their P Cygni absorption components.

High and low resolution ultraviolet spectra of WR140 as a target of opportunity were taken with the IUE on 1985 April 27, while dust formation was still increasing. The low resolution spectra taken through the large aperture allowed measurement of the continuum flux in the ultraviolet. Comparison of this with fluxes measured from previous IUE spectra (1979 and 1980) or our spectrum taken during infrared decline (1985 November) showed no significant change. We conclude that the dust emission episode was not caused or accompanied by any change in the total luminosity of the system.

The profiles of the CIV resonance lines ($\lambda\lambda$ 1548, 1552) on the high resolution spectra, however, were significantly different during dust formation than either before or since. This is illustrated in Fig. 2. The 1980 spectrum, taken from IUE Archives and discussed by Fitzpatrick *et al.* (1982), shows the P Cygni profile of the CIV doublet having a sharp violet cut-off consistent with the 3000 km/s terminal velocity. During dust formation, the violet edge is less sharp and the absorption broader indicating disturbance of the previously steady expanding wind. There is still an expanding shell, but it is apparent that the velocity structure in the line of sight is more complex. The intensity of the CIV emission peak is also significantly lower, while other lines remain of comparable strength. These spectra will be studied in more detail elsewhere.

3. THE NON-THERMAL SOURCE AND X-RAY OBSERVATIONS

In 1975.8, Florkowski & Gottesman (1977) observed WR140 to have a radio flux spectrum between 3.7 and 11 cm of index -0.19, quite different from that observed from γ Velorum or expected from free-free radiation from a steadily expanding shell. Similar non-thermal emission has been observed from other hot stars and it has become recognised that electrons are being accelerated to relativistic energies in their winds. In 1984.4, Pollock (1985) observed a strong X-ray source associated with WR140, making it the brightest X-ray WR star known and confirming the presence of relativistic electrons in its wind. Further radio observations of this source will be reported and discussed elsewhere; here we note that Becker & White (1985) found strong non-thermal emission in 1983.8 but a much weaker source in 1985.1 having a radio spectral index of ~0.8, close to the "thermal" value. Combined with our data for this epoch, we also find a spectral index of 0.8 between 3.8 μm and 6 cm, indicating extinction of the non-thermal radio source in 1985.1.

As soon as the infrared brightening was reported, WR140 was reobserved with EXOSAT using the LE telescope on 1985.3 and both the LE and ME instruments on 1985.5 and 1985.8. The spectra are strikingly different to those observed before outburst. The low energy cut off in 1985 is significantly greater than before whereas the ME flux is actually higher. The 1984 spectrum can be modelled by a power law of index ~ -1.4 suffering absorption equivalent to that caused by a column density of ~ 4 x 10^{21} hydrogen atoms cm^{-2} and compatible with the interstellar extinction to WR140. During the infrared outburst, the spectrum indicates an increase in absorption by a factor of ~5. The additional extinction clearly occurs in the circumstellar shell and is due to the wind, not the dust. This demonstrates that the X-ray source has moved further into the circumstellar wind during periastron.

4. A COLLIDING WIND MODEL

X-ray emission has been observed from the WR binaries γVel, θMus, EZ CMa and V444 Cyg. The time-variable fluxes observed from the last two (Moffat *et al.* 1982) suggest that the X-rays probably originate between the two components, presumably by the collision of stellar winds. Because the WR winds are 1-2 orders of magnitude denser than the O-star winds, we expect the shocked regions wherein the non-thermal radio to X-ray (and γ ray) radiation originate to be near the O-type components. Why, if this is the case, does WR140 show non-thermal radio emission (at least some of the time) while γVel and θMus do not? And why is its X-ray flux greater? The answers lie in the relative scales of the systems. The period of WR140 is 36 times that of γVel and 320 that of θMus. For reasonable stellar masses, the semi-major axis of its orbit is 2.6 x 10^{14} cm; 13 times that of γVel and 50 that of θMus. Consequently, the O-star in the WR140 system and the region of shocked material responsible for the non-thermal emission are much further out in the dense Wolf-Rayet wind than in the other systems. At radio wavelengths, Wright & Barlow (1975) define a characteristic radius in the wind analagous to a radio photosphere. In each of γVel, θMus and WR140, this has a value near 5 x 10^{14} cm. The 5 GHz optical depth here is 0.244. Any non-thermal radio emission arising between the components of γVel or θMus occurs deep in the Wolf-Rayet stellar winds, at most 10^{13} cm from γVel and even closer to θMus, well below the radio photospheres. We expect such emission to be totally absorbed in the winds and to see only the thermal radio emission arising further out. Only in the case of WR140 during apastron, when the separation of the components and hence the distance of the region responsible for the non-thermal emission from the WC component is around 4.6 x 10^{14} cm, is there a possibility of the non-thermal radio source being observed. The high eccentricity of the orbit modulates both the intrinsic intensity of the non-thermal source and its extinction. The non-thermal radiation observed by Florkowski & Gottesman (1977) and Becker & White (1985) was seen at comparable phases (ϕ = 0.78 and 0.80). By $\phi \sim 0.96$ (January 1985) the source was too deep in the wind to be detectable and only the weak, thermal source was measured. The modulation of the radio flux around the cycle will enable us to track the movement of the shocked region in the Wolf Rayet wind.

A parallel process must be responsible for modulating the LE cut-off to the X-ray spectrum attributable to increased extinction in the wind. The higher ME intensity near infrared maximum and periastron passage is a consequence of the increase of Wolf-Rayet wind density experienced by the O star at this time.

Only the triggering of the dust formation needs to be explained; we have to postulate that part of the stellar wind is sufficiently compressed by the shock that it remains dense enough to form dust when it is sufficiently far from the stars to be cool enough. Apparently, this compression is sufficient near periastron passage only. Noting that the first dust formation occurred some 2.5 x 10^{15} cm from the stars, we infer from the wind velocity that perastron passage and formation of a compact region must have occurred some 90 days before infrared maximum, or at ϕ (infrared) ~0.97. This appears to be the case (Fig. 1).

ACKNOWLEDGEMENTS

We thank the ESA-IUE and EXOSAT establishments for timely and continuous response to our requests for target-of-opportunity status. Some of the infrared monitoring was conducted during UKIRT Service time but most observations were made during time generously made available to PMW by colleagues and visiting astronomers.

REFERENCES

Becker, R.H. & White, R.L., 1985. *Astrophys.J.*, **297**, 649.
Conti, P.S., Roussel-Duprè, D., Massey, P. & Rensing, M., 1984. *Astrophys.J.*, **282**, 693.
Fitzpatrick, E.L., Savage, B.D. & Sitko, M.L., 1982. *Astrophys.J.*, **256**, 578.
Florkowski, D.R. & Gottesman, S.T., 1977. *M.N.R.A.S.*, **179**, 105.
Hackwell, J.A., Gehrz, R.D. & Grasdalen, G.L., 1979. *Astrophys.J.*, **234**, 133.
Hackwell, J.A., Gehrz, R.D., Smith, J.R. & Strecker, D.W., 1976. *Astrophys.J.*, **210**, 137.
Lamontagne, R., Moffat, A.F.J. & Seggewiss, W., 1984. *Astrophys.J.*, **272**, 258.
McDonald, J.K., 1947. *Pub. Dominion Astrophysical Obs.* **7**, 311.
Moffat, A.F.J., Firmani, C., McLean, I.S. & Seggewiss, W., 1982. In: *Wolf-Rayet Stars: Observations, Physics, Evolution (I.A.U. Symp. 99)*, eds: C.W.H. de Loore & A.J. Willis, p. 577.
Pollock, A.M.T., 1985. *Space Science Reviews*, **40**, 63.
Williams, P.M. & Antonopoulou, E., 1979. *M.N.R.A.S.*, **187**, 183.
Williams, P.M., Beattie, D.H., Stewart, J.M. & Lee, T.J., 1977. *I.A.U. Circular 3107*.
Williams, P.M., Beattie, D.H., Lee, T.J., Stewart, J.M. & Antonopoulou, E., 1978. *M.N.R.A.S.*, **185**, 467.
Williams, P.M. & Smith, M.G., 1985. *I.A.U. Circular 4056*.
Wright, A.E. & Barlow, M.J., 1975. *M.N.R.A.S.*, **170**, 41.

PHOTOMETRIC VARIABILITY OF WOLF-RAYET STARS

A.F.J. Moffat[1], M.M. Shara[2] and R. Lamontagne[1]

[1] Département de Physique, Université de Montréal
[2] Space Telescope Science Institute, Baltimore

A potentially important factor in the initiation and acceleration of the winds in WR stars may be related to instabilities in the core or in the wind.

To explore this possibility, we are in the process of making extensive surveys of complete samples of WR stars using broadband, optical photometry during contiguous-night observing runs spread over several weeks at a time. Some results which have emerged so far are:

(1) Somewhat over half of WR stars show continuum light variations of amplitude $\geq$ 0.02 mag, only slightly more frequent than in a randomly selected sample of non-WR stars. Some WR are constant to within the photometric errors (typically 0.005 mag).

(2) Light variations of well-known WR binaries can be accounted for by phase-dependent occultation or transparency effects, with additional variability probably of random nature appearing with increasing amplitude towards the cooler WR subclasses.

(3) WN8 stars appear to show the highest level of (probably) random noise variability among WR stars.

(4) We can make no definitive claim yet for the source of variability in the remaining stars, among (a) random wind fluctuations, (b) pulsation, (c) rotation or (d) binary modulation involving a close, low-mass companion.

Several examples demonstrate the basis for conclusions (1)-(3). These are drawn from stars observed during a 14-night run in Israel in 1984 June (Moffat and Shara 1986, A.J., in press) and an 18-night run in Chile in 1985 March/April.

We also show our best guess as to the overall variability of WR stars (after allowing for duplicity) compared to normal H-burning stars.

H. J. G. L. M. Lamers and C. W. H. de Loore (eds.), Instabilities in Luminous Early Type Stars, 227–230.

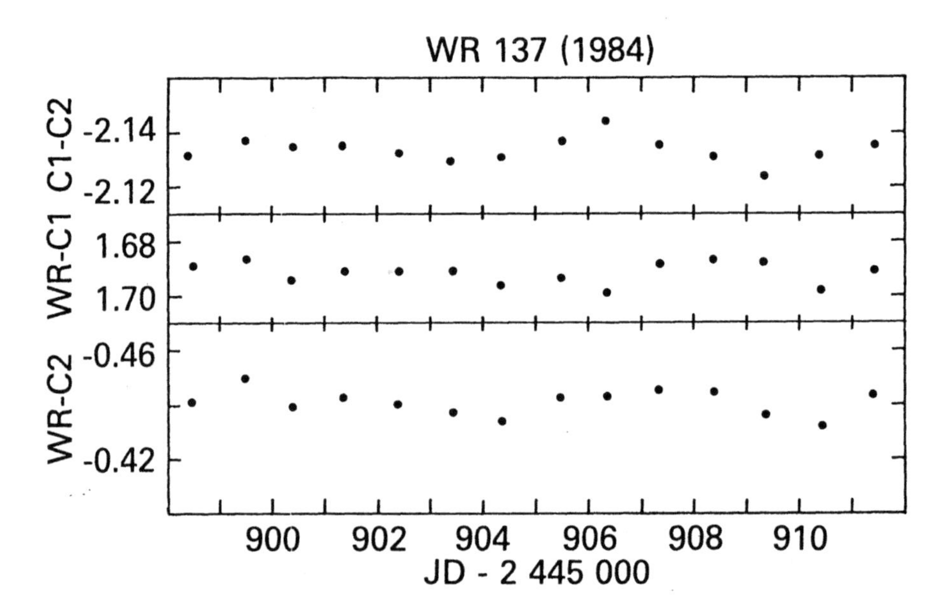

FIG. 1: Broadband B photometry of the WC7 + abs star HD 192641 = WR 137 versus time. With σ (WR - C) = 0.005 mag for either comparison star and σ (C1 - C2) = 0.005 mag, we conclude that this WR star is constant in contiuum light within the instrumental errors, at least over a time scale of 2 weeks. The WC8 star HD 192103 = WR 135 shows similar behaviour. Among the whole, complete magnitude-limited sample of 18 northern WR stars down to b = 11.5 mag, these two WR stars stand out as being the most likely non-variables during the 1984 observing run.

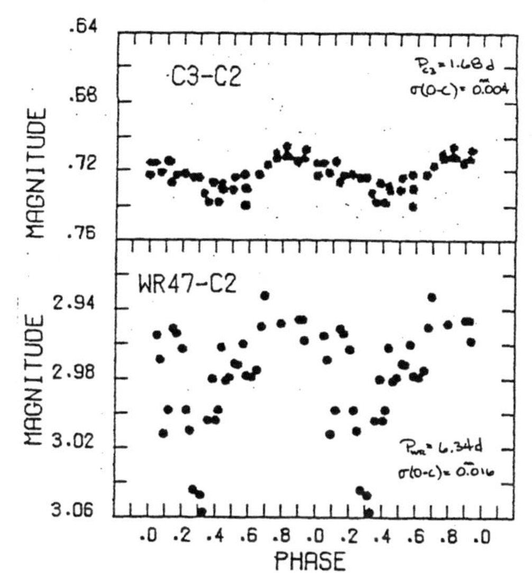

FIG. 2: Broadband V photometry of the WN6+O5 star HDE 311884 = WR47 versus orbital phase. The data show moderately deep (0.1 mag), narrow dips (eclipses?) every orbital cycle, although the published velocity orbit (Niemela, Conti and Massey 1980, Ap.J., 241, 1050) is not accurate enough to predict zero phase precisely. The light curve does not repeat itself exactly from one cycle to the next as indicatd by the increased scatter σ (O-C) about a simple curve through the data, compared to the comparison stars. This is compatible with the trend that late type WR stars (especially late WN) in general tend to show a significant component of (random?) variations. Note that C3 is variable with P = 1.68 d, which is used to calculate the phase for C3 - C2.

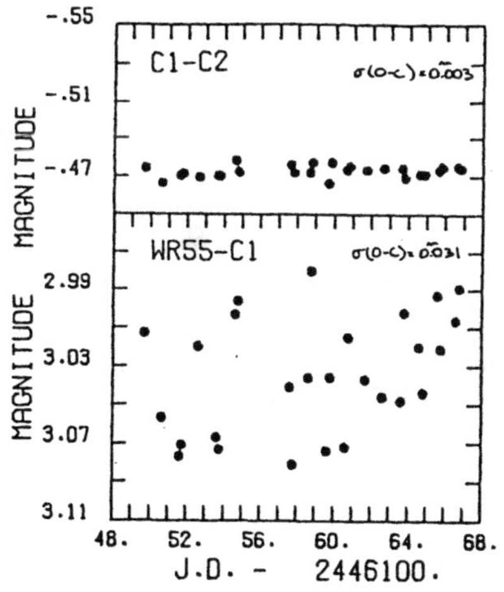

FIG. 3: Light curve of a sample of WN8 star (WR 55 = HD 117688) among a complete sample of 5 southern WN8 stars monitored on 1985 with v 13 mag. All 5 WN8 stars (the other 4 are WR16 = HD 86161, WR40 = HD 96548, WR66 = HD 134877 and WR 82) show different but generally high degrees of variability. The nature of these variations can only be settled by further intensive monitoring.

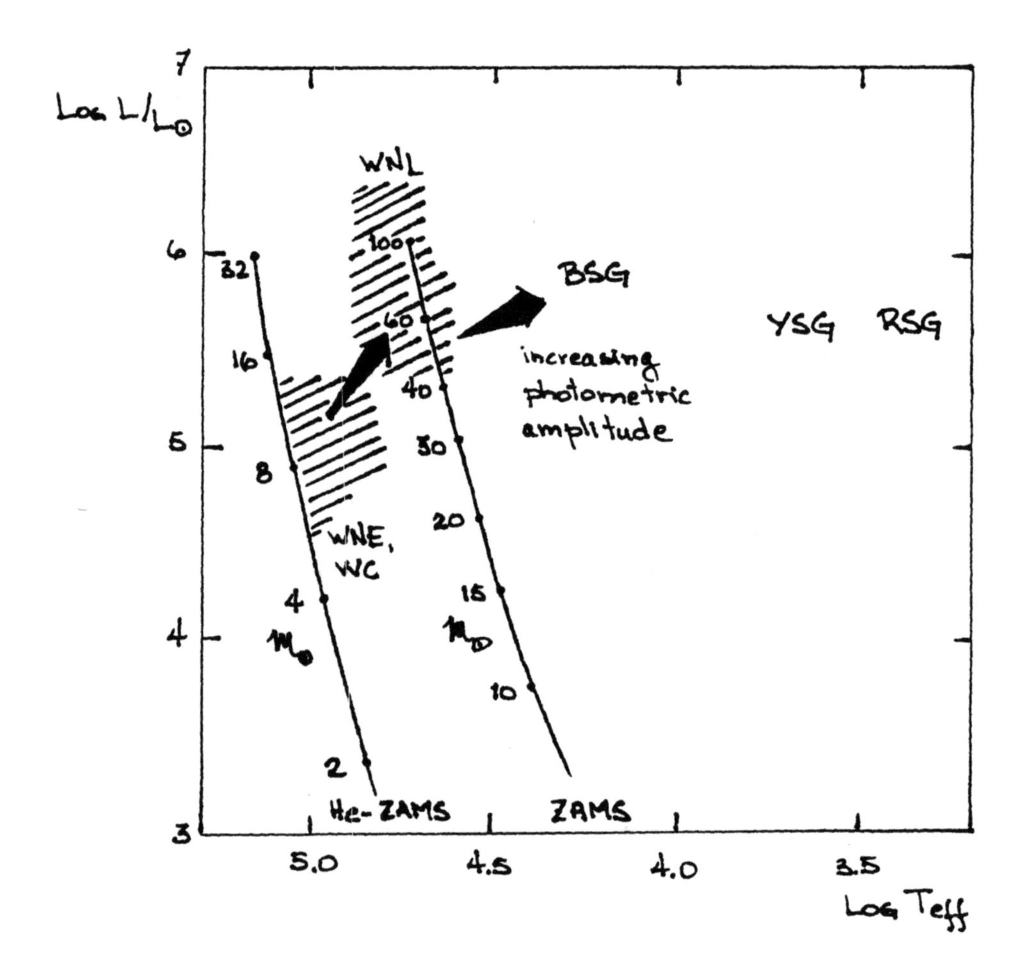

FIG. 4: Schematic positioning of various types of luminous stars in the theoretical H-R diagram. Amplitudes of photometric variability tend to increase up and to the right of the ZAMS both for normal stars and for WR stars, as indicated by the arrows. The positions of WNL versus WNE, WC stars is based on the few but important analyses of 4 eclipsing WR binary systems (cf. Lipunova and Cherepashchuk 1982, Sov. Astron. 26, 569.).

IMPROVED MASS LOSS RATES FOR WC7-9 STARS: THEIR EFFECT ON THE WOLF-RAYET STELLAR WIND MOMENTUM PROBLEM

J.P. Cassinelli[1] and K.A. van der Hucht[2]
[1]Sterrewacht Sonnenborgh, Utrecht, Nederland, and
Washburn Observatory, University of Wisconsin, USA
[2]SRON Ruimteonderzoek Utrecht, Utrecht, Nederland

ABSTRACT. Applying recent surface abundance determinations for Wolf-Rayet stars of notably the WC sequence, the effect of model atmosphere energy distributions on the ionization and mass loss rates is investigated. We conclude that these stars lose more mass than previously considered, which makes it even more difficult to explain WR stellar winds by radiation pressure alone.

1. MASS LOSS RATES

The most accurate estimates of the mass loss rate of Wolf-Rayet stars are made from measurements of their free-free radio continuum. In deriving mass loss rates for WC stars, Abbott et al. (1986) have assumed that their atmospheres consist of pure ionized helium. However, evolutionary calculations of e.g. Maeder (1984) and Prantzos et al. (1986) show large abundances of carbon and oxygen. This has an effect on the radio flux interpretation, because the free-free flux is converted to a mass loss rate by applying a value for the ionic mass per free electron. The ionization of helium, carbon, and oxygen in the radio continuum forming region is also important. We derive the ionization from a partly empirical approach. The optical spectra of the WC9 stars show that CII and CIII are both present and are of roughly equal strength. The lines are produced near the optical photosphere of the stars near the base of the wind.

We have computed model atmospheres applying the abundances of WC stars as calculated by Prantzos et al., and have adjusted the effective temperature so that at two stellar radii equal abundances of CII and CIII are derived. This occurs at 19000 K. We assume that same radiation field is simply geometrically attenuated out to the radio continuum forming region. There we derive the ionization stages of all the elements and convert the observed radio flux to a mass loss rate, using the well known equation of Barlow & Wright (1975). For the other WC types we have applied the effective temperatures given by Willis (1982).

H. J. G. L. M. Lamers and C. W. H. de Loore (eds.), Instabilities in Luminous Early Type Stars, 231–236.

Tabel 1. Revised mass loss rates for WC stars.

WR	Type	d^a (kpc)	$v_\infty{}^b$ (km/s)	$S_\nu{}^b$ (mJy)	$T_{eff}{}^c$ (K)	$R_*{}^d$ ($R_\odot$)	$R_r{}^e$ (R_*)	C_5 (10^{-6})	$\dot{M}$ (10^{-5} $M_\odot$/yr)	$\dot{M}/\dot{M}^b$
111	WC5	1.58	3550	0.33	40000	8.9	550	1.50	2.5	1.6
79	WC7+O6	2.00	3300	1.0	30000	11.8	925	3.05	15.3	3.1
86	WC7+a	2.18	2400	0.5	30000	11.8	710	2.94	7.5	3.0
93	WC7+a	1.74	3100	0.9	30000	11.8	760	2.96	10.7	3.4
137	WC7+a	1.82	2700	0.37	30000	11.8	510	2.90	5.1	2.6
140	WC7+O4-5	1.26	3200	1.27[f]	30000	11.8	655	2.90	8.8	2.8
11	WC8+O9I	0.48	1600[g]	29.	26000	13.2	1060	3.22	11.6	1.5
113	WC8+O8V	2.00	2900	$\leqslant$ .4	26000	13.2	520	3.06	$\leqslant$ 7.0	2.2
135	WC8	2.09	1900	0.60	26000	13.2	660	3.12	6.7	2.7
81	WC9	1.78	1300	0.3	19000	14.6	360	4.37	3.0	2.3

Notes: μ = 7.7 throughout; $T_e(R_r)$ = 10000 K for all cases; a) Hidayat et al., 1986; b) Abbott et al., 1986; c) Willis, 1982; for WC9 determined by van der Hucht et al., 1986; d) Willis, 1981, extrapolated for WC9; e) radius of 6 cm radio photosphere; f) at radio minimum on 21 May 1985, Bieging, priv. comm.; g) Barlow & Roche, priv. comm.

Because of the higher carbon and oxygen abundances and the greater ionic mass per free electron with respect to the case of pure and fully ionized helium assumed by Abbott et al. (1986), we derive for the WC7-9 stars mass loss rates which are a factor 2 to 3 larger (see Table 1).

The same method was also applied to WN stars.

Particulars of this work are published in van der Hucht et al. (1986).

2. MASS LOSS RATES AND MASSES

Abbott et al. (1986) have been searching for a correlation between mass loss rates and masses for WR binaries with RV solutions, using the compilation of Massey (1981). In the meantime however, new polarimetric determinations of orbital inclinations have been published for the WC binaries HD152270 = WR79 (Luna, 1982) and γ^2 Vel = WR11 (Luna, 1985), which result in lower masses than those quoted by Abbott et al. For WR79 Luna (1982) finds i = 35±8°, resulting in M(WC7) = 11±5 $M_\odot$ and for WR11 Luna (1985) finds i = 75±5°, resulting in M(WC8) = 19±3 $M_\odot$. Notably the lower mass of the WC7 star in WR79 upsets the correlation suggested by Abbott et al.

This implies that for the five objects listed in Table 2, on which Abbott et al. based their $\dot{M}$-M correlation (and of which the $\dot{M}$ values of WR113 and WR133 are only upper limits, because the radio fluxes used are upper limits), there is actually <u>no</u> reliable correlation between $\dot{M}$ and M (see Fig. 1), and thus also no $\dot{M}$–L correlation.

The points of interest which emerge from Table 2 and Fig.1 are that in binaries WC stars tend to have larger mass loss rates than WN stars, and that among the WC stars the one with the highest mass loss rate, i.e. WR79, has the lowest mass. This may indicate that WR79 is a relatively old WC binary. It also has among the WC binaries the smallest $a\sin^3 i$ value (34 $R_\odot$, cf. Massey, 1981), which indicates that mass transfer may have been fairly effective in this particular case.

Table 2. WR mass loss rates, masses and luminosities.

WR	HD/Name	Type	log $\dot{M}(M_\odot/yr)^a$	$M/M_\odot{}^b$	$\log(L/L_\odot)$	η
133	190918	WN4.5+O9.5Ib	$\leq$-4.95 ±.16	8 – 13	5.2	$\leq$ 6
139	V444 Cyg	WN5+O6	-4.86 ±.16	11 ± 1	5.3	9
79	152270	WC7+O5-8	-3.81 ±.11	11 ± 5	5.2	147
11	γ^2 Vel	WC8+O9I	-3.93 ±.12	19 ± 3	5.6	23
113	CV Ser	WC8+O8V	$\leq$-4.16 ±.16	11 – 14	5.3	50

Notes: a) $\dot{M}$ values from van der Hucht et al., 1986); formal error from errors in d, v_∞ and S_ν from Abbott et al., 1986.
b) M values: WR11: Niemela & Sahade, 1980; Luna, 1985. WR79: Seggewiss, 1974; Luna, 1982. WR113: Massey & Niemela, 1981. WR133: Fraquelli & Horn, 1981, in preparation. WR139: Ganesh et al., 1967; Rudy & Kemp, 1978.

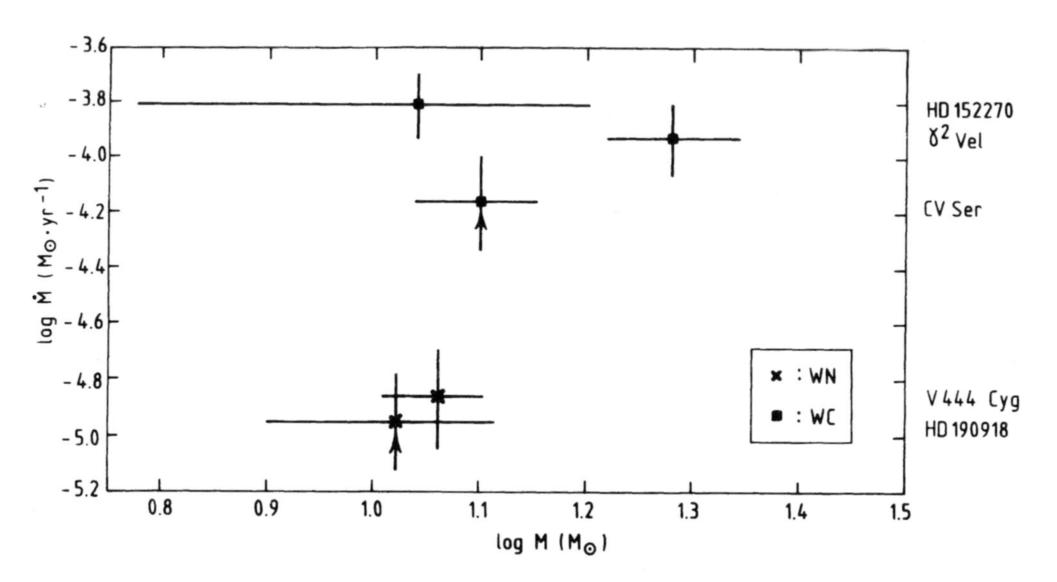

Fig. 1. Mass loss rate vs. mass for five WR stars in binary systems.

3. THE MOMENTUM PROBLEM

The major difficulty presented by Wolf-Rayet stars is the momentum problem. The mass loss rates exceed the single scattering upper limit (Cassinelli & Castor, 1973) that can be driven by the transfer of radiative momentum L/c to the mass flow momentum $\dot{M}v_\infty$.

It is convenient to decribe the momentum problem using the ratio

$$\eta = (\dot{M}v_\infty)/(L/c).$$

Barlow et al. (1981) already found that for WR stars η ranges between 4 and 53. Their value relied on best effort estimates of the luminosities of the stars, as observed from the UV to the IR. Willis (1982) studied the UV resonance lines and found WR temperatures, typically 40000 K for early type WR stars (WN4, WC5) and 26000 K for late type WR stars (WC8).

Recently, it has been argued that WR stars may have much larger effective temperatures than assumed by Willis et al., and that because of this the luminosities should be larger and the value of η could be reduced to about 3, small enough that radiation driven wind theory could explain the WR mass loss if multiple scattering is accounted for. Pauldrach et al. (1985) for example state that their radiation driven wind model with T = 90000 K can explain the mass loss rate of the WN5 star in V444 Cyg. They note in passing that the luminosity used is not consistent with interior theory, being a factor of two too large.

Here we focuss on the five stars in Table 2. The interior models of Prantzos et al. (1986) provide as mass - luminosity relation:

$$\log(L/L_\odot) = 3.60 + 1.57\log(M/M_\odot).$$

The resulting L values are given in Table 2.

Note that the values of η are far above the maximum of $\approx$ 3 that allows for an explanation using radiation driven wind theory. Note also that we have not assumed a value for the effective temperature of the stars, but have only used the theoretical L-M relation and thus have avoided the problem encountered by Pauldrach et al.

To get around the momentem problem one cannot simply appeal to a larger luminosity, because the values that we have used for the WC7-8 stars are already near the Eddington Limit: let

$$\Gamma = 0.2L/4\pi cGM,$$

then we have for WN4.5-5 stars $\Gamma \cong .24$ and for WC7-8 stars $\Gamma \cong .27$. So even by increasing L to its maximal value L_{Ed} gives only $\eta \cong 13$, which is still clearly not achievable by radiation driven wind theory.

The results suggest that an alternative mechanism is required to explain the very large mass loss rates of the WR stars, or perhaps that the outflows are not spherically symmetric and that $\dot{M}$ and v_∞ are determined by different regions of the wind.

REFERENCES

Abbott, D.C., Bieging, J.H., Churchwell, E., and Torres, A.V.: 1986, Astrophys. J. **303**, 239.
Barlow, M.J., Smith, L.J., and Willis, A.J.: 1981, Monthly Notices Royal Astron. Soc. **196**, 101.
Cassinelli, J.P. and Castor, J.I.: 1973, Astrophys. J. **179**, 189.
Ganesh, V.S., Bappu, M.K.V., and Natajaran, V.: 1967, Kodaikonal Obs. Bull. **A184**.
van der Hucht, K.A., Cassinelli, J.P., and Williams, P.M.: 1986, Astron. Astrophys., in press.
Luna, H.G.: 1982, Publ. Astron. Soc. Pacific **94**, 695.
Luna, H.G.: 1985, Rev. Mexicana Astron. Astrofis. **10**, 267.
Massey, P.: 1981, Astrophys. J. **246**, 153.
Massey, P. and Niemela, V.S.: 1981, Astrophys. J. **245**, 195.
Niemela, V.S. and Sahade J.: 1980, Astrophys. J. **238**, **244**.
Pauldrach, A., Puls, J., Hummer, D.G., and Kudritzki, R.P.: 1985, Astron. Astrophys. (Letters) **148**, L1.
Prantzos, N., Doom, C., Arnould, M., and de Loore, C.: 1986, Astrophys. J. **304**, 695.
Rudy, R.J. and Kemp, J.C.: 1978, Astrphys. J. **221**, 200.
Seggewiss, W.: 1974, Astron. Astrophys. **86**, 670.
Willis, A.J.: 1982, in: C. de Loore & A.J. Willis (eds.), *Wolf-Rayet Stars: Observations, Physics, Evolution*, Proc. IAU Symp. No.**99** (Dordrecht: Reidel), p.87.
Wright, A.E. and Barlow, M.J.: 1975, Monthly Notices Royal Astron. Soc. **170**, 41.

Discussion at dinner: Moffat, Williams,
van der Hucht, Schulte-Ladbeck

Difficult problems?
Joe Cassinelli, Allan Willis and Peter Conti

VARIABILITY OF WOLF-RAYET STARS IN LINEAR POLARIZATION

A.F.J. Moffat, P. Bastien, L. Drissen, N. St-Louis and
C. Robert
Département de Physique
Université de Montréal
C.P. 6128, Succ. "A"
Montréal, P.Q. H3C 3J7

The main reasons for observing variations in linear polarization among WR stars are:
(1) to look for inhomogeneities and asymmetries in the winds and
(2) to determine some basic parameters of WR binary systems.

We have been acquiring extensive, repeated observations in linear polarization for a large sample of WR stars, including the eight bright Cygnus stars, the 14 brightest WR stars in the south and several selected close binaries in the north (e.g. HD 197406: Drissen et al. 1986, Ap.J., 304, 188; CQ Cep: Drissen et al. 1986, Ap.J., 306, in press).

Each data point was obtained with a precision of $\sigma(p) \simeq 0.01$ to 0.03%, depending on the brightness of the star, in a broadband (FWHM = 1800 A) filter centred at 4700 A. Thus, the observations refer mainly to continuum light. Previously observed stars show nearly identical variations in different spectral bands, (cf. McLean 1980, Ap.J. (Letters), 236, L149; Luna 1982, P.A.S.P., 94, 695; Rudy and Kemp 1978, Ap.J., 221, 200) so there appears to be little urgent need for multiband observations, at least in the optical part of the spectrum.

The principal results emerging so far are:

(1) Most WR stars show some kind of polarization variability; nevertheless, some show no variability above the instrumental scatter. The latter tend to be the hotter WR subtypes without known or suspected companions, suggesting that such stars are inclined to have more homogeneous winds.

(2) Well-known binaries typically show a double-wave modulation per orbital cycle, with total amplitude in polarization ranging from 0.8% for the shortest binary CQ Cep (P=1.64d) down to 0.2% for the 78.5-day system γ^2 Vel. From these variations, one can obtain the orbital inclination and hence the net masses of the components by combining with spectroscopic data, which yield only $m \sin^3 i$.

(3) As in the case of photometric variations, WN8 stars tend to exhibit the greatest random variability in polarization, among all WR subtypes.

H. J. G. L. M. Lamers and C. W. H. de Loore (eds.), Instabilities in Luminous Early Type Stars, 237–240.

We illustrate these findings for three stars observed recently during a 40-night polarimetric run in Chile.

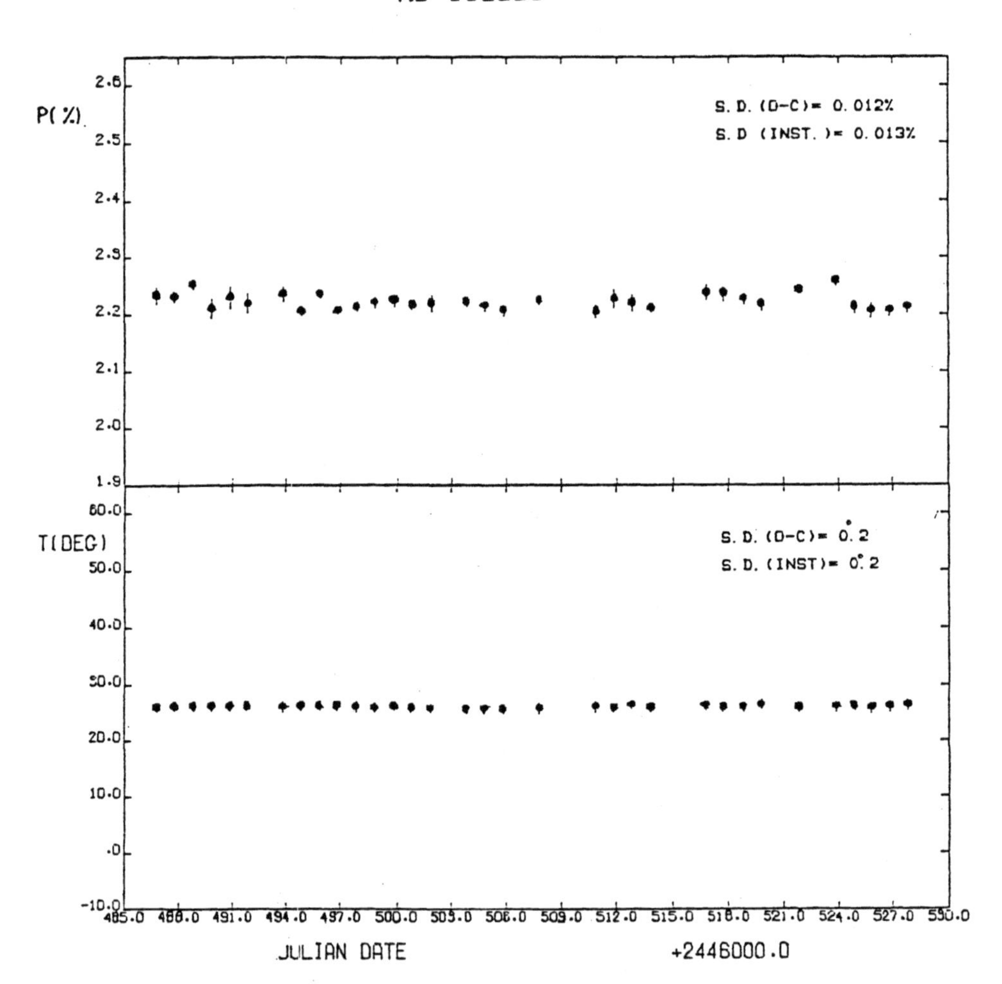

FIGURE 1

Polarization amplitude P and polarization angle Θ versus time for the single WC7 star of constant light and radial velocity (Davis, Moffat and Niemela 1981, Ap.J., 244, 528), HD 156385 = WR 90. No variations are evident beyond the instrumental scatter. A similar situation prevails for the single WC5 star HD 165763 = WR 111.

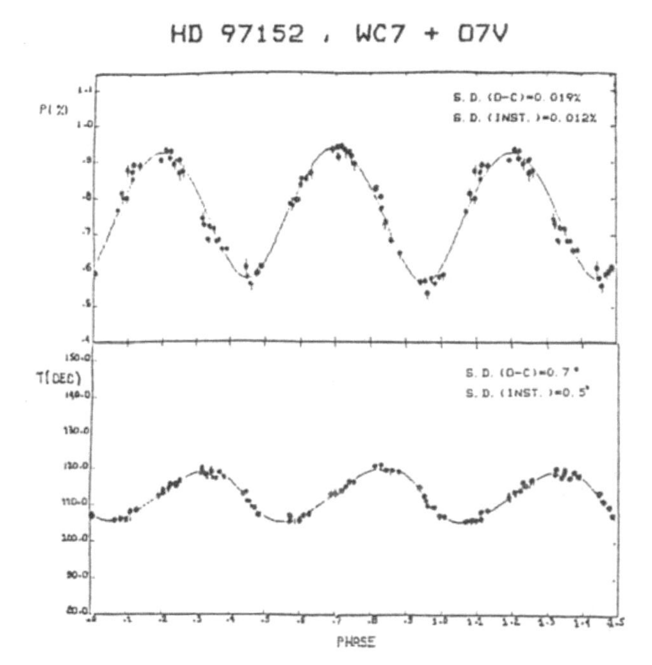

Figure 2a:

P and Θ versus phase for the double-line WC7+O7V binary HD 97152 = WR 42. The phase is based on the 7.886 -day velocity orbit of Davis, Moffat and Niemela (1981). The data here cover 5 complete orbital cycles. The curve is from a Fourier fit to the Stokes parameters Q and U, with sine and cosine terms in λ (=2πΦ) and 2 λ. The complete domination of the terms in 2λ implies that we are seeing the simplest kind of binary modulation according to the model of Brown, McLean and Emslie (1978, Astr. Ap., 68, 415) i.e. a circular orbit with corotating non-occulting scatterers located symmetrically about the orbital plane. The O-C scatter is only marginally larger than the instrumental scatter.

Figure 2b:

Stokes parameters for linear polarization Q=Pcos2 Θ versus U=Psin2 Θ for HD 97152 from the fit as noted in figure 2a. The shape of the mean curve (broken) yields an orbital inclination i=43.5° ± 5°. Hence the absolute masses are:

$M_{WC7}=(3.6\pm0.3)/\sin^3 i=11\pm 3M_\odot$

and

$M_{O7V}=(6.1\pm0.5)/\sin^3 i=19\pm 5M_\odot$

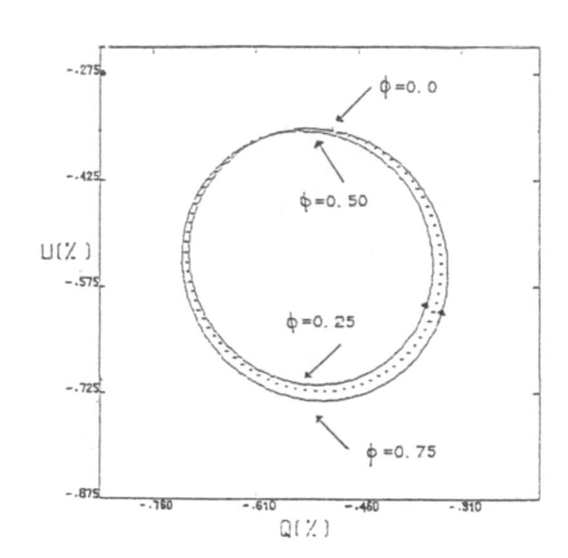

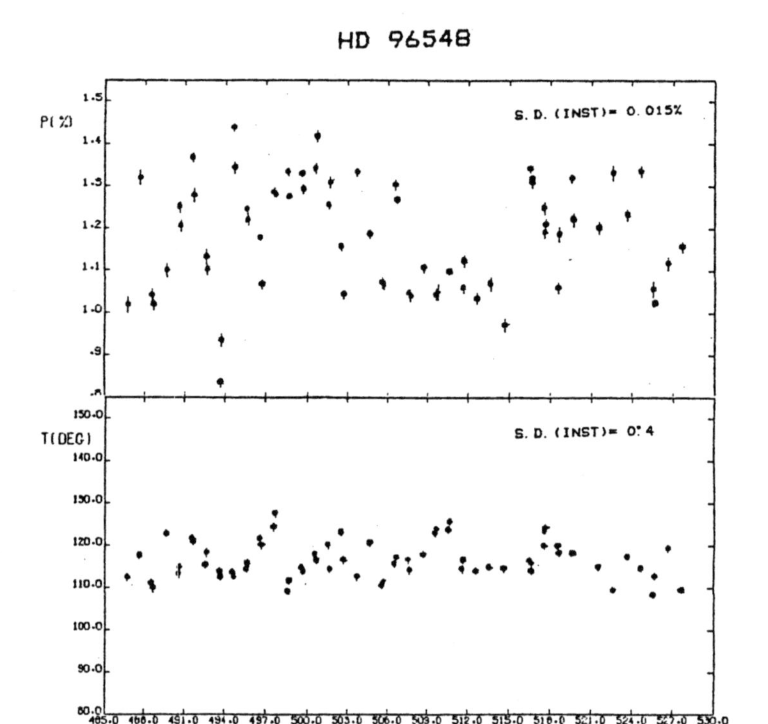

Figure 3a:

P and Θ versus time for the single-line WN8 star WR 40 (=HD 96548). Period searches (0.2 - 80 d) using various techniques fail to reveal any convincing period.

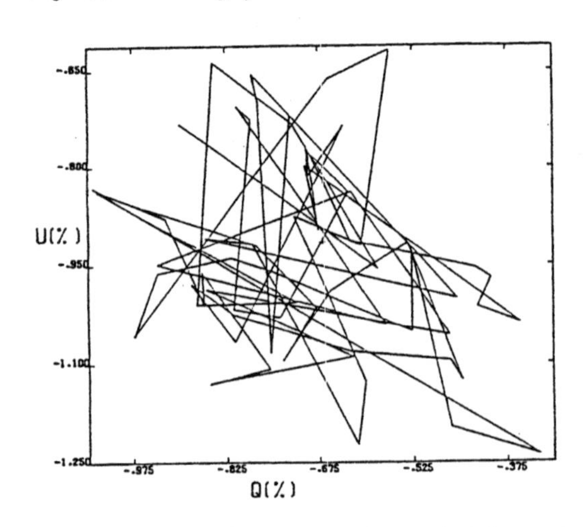

Figure 3b:

Q versus U for HD 96548 with points adjacent in time connected by a straight line. Parallel observations in a red filter (λ=7500A, FWHM= 2700A) show identical variations. The random nature of the variations is striking when compared to fig 2b. As in the case of the relatively slow-wind supergiant P Cyg (Hayes 1985, Ap.J., 289, 726), we interpret the present variations in HD 96548 to be the result of blobs of plasma being accelerated randomly in time and position in the relatively slow (i.e. WN8-type) WR wind.

AN EXTENDED NEBULOSITY SURROUNDING THE S DOR VARIABLE R 127

I. Appenzeller, B. Wolf
Landessternwarte, Königstuhl
D-6900 Heidelberg
Federal Republic of Germany

O. Stahl
European Southern Observatory
Karl-Schwarzschild-Str. 2
8046 Garching bei München
Federal Republic of Germany

ABSTRACT. New high resolution spectrograms of R 127 show the presence of an extended (~4" ≙ 1 pc) expanding (v = ±28 km s^{-1}) gaseous nebula around this high-mass loss S Dor variable.

The LMC star R 127 was discovered to be an S Dor variable by Stahl et al. (1983). Before its S Dor outburst, it had been classified as an Of supergiant by Walborn (1977). Since the beginning of the S Dor outburst the visual brightness of R 127 has further increased (see Figure 1). R 127 is now (visually) brighter than S Dor and has become the (visually) second-brightest star of the LMC.

In order to study R 127 in this extraordinary bright phase we obtained in March 1986 new high-resolution spectrograms in the blue and red spectral range using the CASPEC echelle spectrograph of the European Southern Observatory, La Silla, Chile. The slit orientation was E-W and a 2" x 6" slit was used. The two most interesting results of these new observations are:

(1) Even at high resolution and very good S/N the blue spectrum of R 127 now appears undistinguishable from that of the prototype S Dor at maximum light (see Figure 2).

(2) The red spectrogram, which contains the [N II] 6548 and 6583 and the [S II] 6717 and 6731 lines shows the presence of a well resolved extended gaseous nebula around R 127 (see Figures 3 and 4). The nebula (which is also detected at the Balmer lines) shows blueshifted and redshifted emission (projected) on the position of the stellar continuum, and no wavelength-shift at the maximum (East-West)

H. J. G. L. M. Lamers and C. W. H. de Loore (eds.), Instabilities in Luminous Early Type Stars, 241–244.

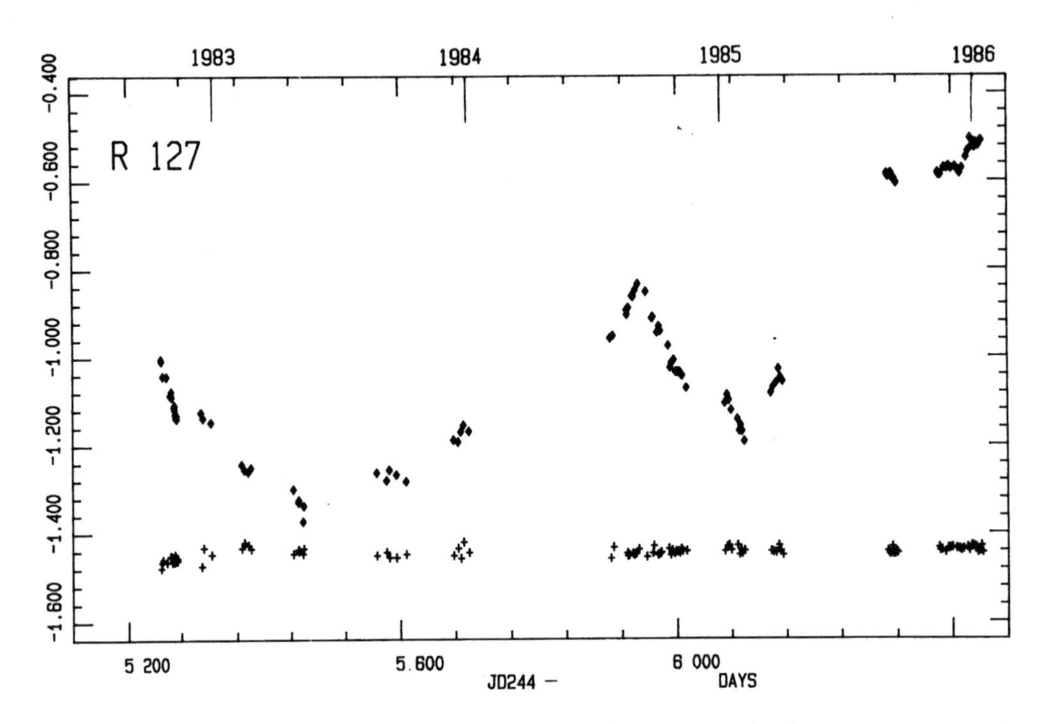

Figure 1. Visual (Strömgren y) differential lightcurve (◊) of R 127. The y-magnitude of C_1 is 8.86. (+) denotes differences between comparison and check star.

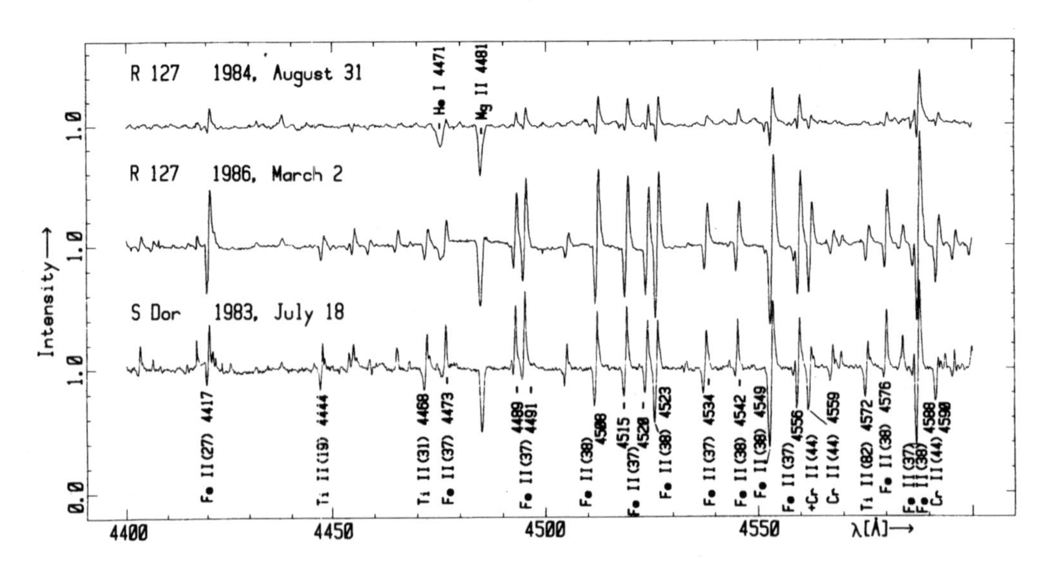

Figure 2. The spectral evolution of R 127 and comparison with S Dor.

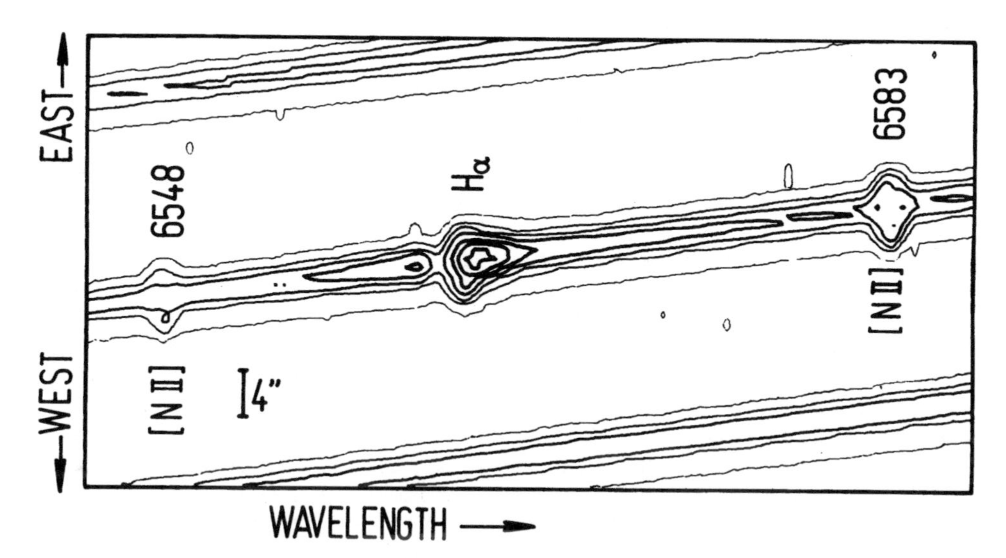

Figure 3. Contour plot of a section of the "red" echellogram of R 127, showing the angular extent of the [N II] and H_α lines. The intensity difference between two contours corresponds to a factor of two. The lowest contour corresponds to about 1 % of the maximum intensity.

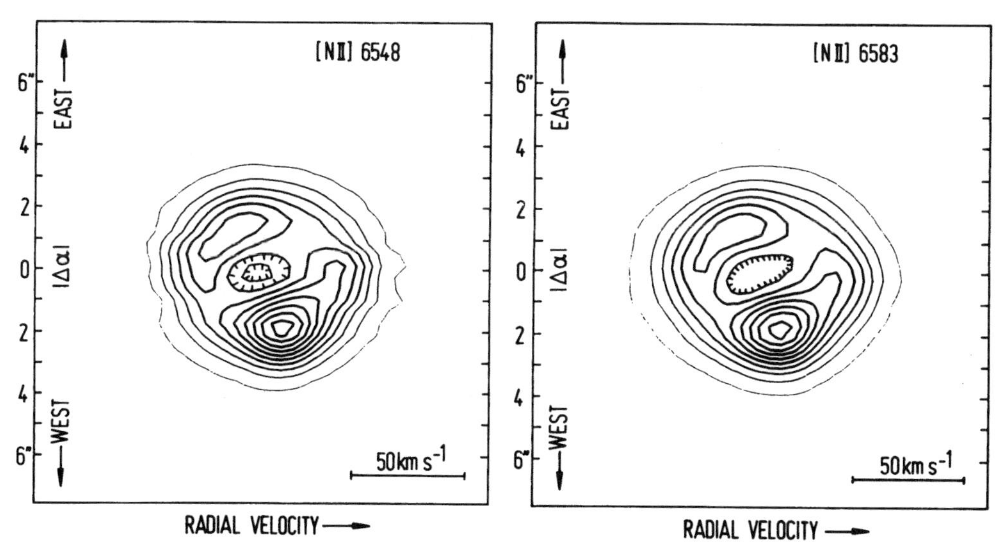

Figure 4. Contour plots of the velocity distribution of the [N II] emission along the projected spectrograph slit east and west of the star. The lowest intensity contour corresponds to 5 % of the maximum emission. Between the contours the intensity increases by 10 %. These plots have been derived from Figure 3 by subtracting the stellar continuum and rebinning to obtain equal ordinate values for points of identical angular distance from the star.

distance from the star. Hence, the nebulosity appears to be an expanding shell, reminiscent of the nebula around the galactic extreme supergiant AG Car. The angular diameter (or East-West extension) of the nebula around R 127 is of the order 4", corresponding to ≈ 1 pc at the distance of the LMC. Hence, the nebula around R 127 is about twice as large as the ring nebula of AG Car. The expansion velocity of the R 127 nebula is found to be 28 km s^{-1} from our spectrograms. Hence, assuming a constant expansion velocity we derive a kinematic age of the R 127 nebula of $\approx 2 \cdot 10^4$ years. This corresponds closely to the expected lifetime of the S Dor evolutionary phase. As noted above, in first approximation the velocity distribution along the slit indicates an isotropic expansion. However, there are obviously deviations from total spherical symmetry or/and obscuration effects: The nebular emission to the west of the star contains more redshifted material, while the (weaker) emission to the east of the star appears to be on the average slightly blueshifted (relative to the systemic velocity). This may indicate that the expansion velocity field contains a bidirectional component.

REFERENCES:

Stahl, O., Wolf, B., Klare, G., Cassatella, A., Krautter, J., Persi, P., Ferrari-Toniolo, M.: 1983, Astron. Astrophys. 127, 49

Walborn, N. R.: 1977, Astrophys. J. 215, 53

THE LMC-S DOR VARIABLE R 71: AN IRAS POINT SOURCE

B. Wolf, F.-J. Zickgraf
Landessternwarte Königstuhl
D-6900 Heidelberg
Germany

ABSTRACT. The S Dor variable R 71 of the LMC was found to be an IRAS point source. Its flux densities at 12, 25, and 60 μ are explained to be due to the radiation of an extended (d ≈ 8000 stellar radii) dust envelope of very low temperature (T_{dust} ≈ 140 K). The total amount of matter (i. e. gas and dust) of the envelope is estimated to M ≈ 3 · 10^{-2} $M_{\odot}$. About this mass has been ejected within the kinematic age (t_{kin} ≈ 500 yrs) of the extended dust envelope if the previously derived mass loss rate ($\dot{M}$ ≈ 5 · 10^{-5} $M_{\odot} yr^{-1}$) of R 71 is adopted.

R 71 is one of the best investigated S Dor-variables of the LMC (Wolf et al. 1981 and literature quoted therein). It had an outburst during the seventies and is now in a phase of minimum light (see Fig. 1).

Recent spectroscopic observations of high resolution and high S/N-ratio revealed double peaked [N II]-lines (separation of the peaks V = 39 km s^{-1}) indicating the presence of an extended circumstellar shell (Stahl and Wolf, 1986) with a low expansion velocity.

Glass (1984) using ground-based IR observations found a 10 μ-excess of 6 mag compared to the L-magnitude. But note that R 71 has never been detected in the M-band. Hence R 71 must be surrounded by a very cool dust shell. Therefore we searched the IRAS-point source catalogue (IRAS Explanatory Supplement, 1985) and found an entry (IRAS 05027-7124) for the position of R 71. Since this star is completely isolated and in an unobscured region of the LMC the IRAS-flux densities 1.02, 11.90 and 3.83 Jy at 12, 25 and 60 μ, respectively (see Fig. 2) can only be explained to be due to dust radiation from R 71.

From a black-body fit to the far infrared flux values we derived a dust-temperature of T_{dust} ≈ 140 K only. With this temperature we obtained approximate values for the dis-

H. J. G. L. M. Lamers and C. W. H. de Loore (eds.), Instabilities in Luminous Early Type Stars, 245–248.

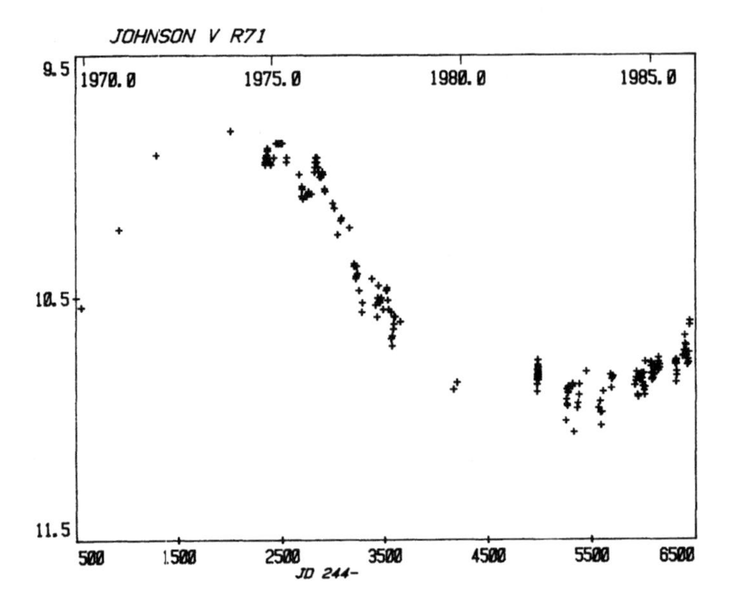

Fig. 1. Light curve of the LMC-S Dor variables R 71. The observations prior to 1978 were taken from van Genderen (1979) and the sources quoted by van Genderen. Observations after 1982 are from the "Long-term photometry of variables" group, initiated by C. Sterken.

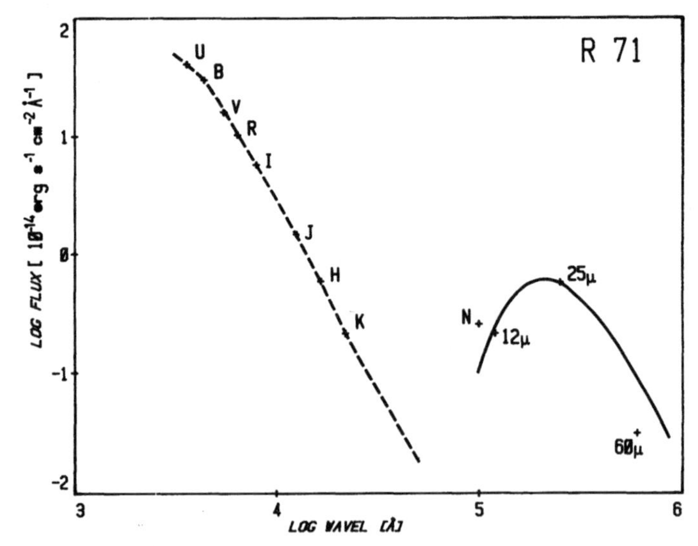

Fig. 2. Broad-band fluxes of R 71 during minimum phase from ground-based observations (UBVRIJHK and N) and IRAS-measurements at 12, 25 and 60 microns. Note the presence of a very strong excess in the far infrared which is ascribed to a very cool (T_{dust} ≈ 140 K) dust shell around R 71. The solid line is the black-body curve for T = 140 K.

tance d of the dust from the star. If the IR radiation is emitted by an optically thick, spherically symmetric envelope its size can be derived by

$$d = D \sqrt{F_{\lambda dust}/\pi B_{\lambda}(T_{dust})} = 7900\ R_* \qquad (1)$$

(D = 50 kpc; $F_{\lambda dust}$ = observed flux at $\lambda = 25\mu$; $R_* \approx 80\ R_\odot$, see Wolf et al.,1981, van Genderen,1979)

Assuming the dust grains as black bodies in thermal equilibrium with the stellar radiation field the radius of the dust shell is given by

$$d = \frac{1}{2} R_* \ (T_{eff}/T_{dust})^2 = 8300\ R_* \qquad (2)$$

(T_{eff} = 18000 K from the spectral type during minimum light).

With the observed expansion velocity of 20 km s^{-1} (see above) a kinematic age of t_{kin} = 400 yrs is derived for the dust forming region.

Using the relation of Gehrz et al. (1980)

$$M_{dust} = \frac{1}{2} \ \frac{L_{IR} \cdot \varrho}{\delta \ T^4_{dust} \ (\Omega_e/a)} \qquad (3)$$

and assuming the grain material to be olivine (graphite is less likely since its formation requires C/O > 1 which is not expected in the case of an S Dor variable (cf. Maeder, 1983)) with the Planck mean emission cross section of Gilman (1974), we derive $M_{dust} = 3 \cdot 10^{-4}\ M_\odot$. With the canonical gas-to-dust ratio of 100 a total mass of the envelope of $3 \cdot 10^{-2}\ M_\odot$ is derived. With the mass loss rate $\dot{M} = 5 \cdot 10^{-5}\ M_\odot yr^{-1}$ given by Wolf et al. (1981) this amount of matter could be ejected within 600 yrs, which is in good agreement with the kinematic age.

ACKNOWLEDGEMENTS: We are grateful to the observers of Dr. C. Sterken's "Long-Term Photometry of Variables" group for providing photometric data for the lightcurve of R 71. This work was supported by the Deutsche Forschungsgemeinschaft (Wolf 269, 2-1 and SFB 132).

REFERENCES:

Gehrz, R. D., Hackwell, J. A., Grasdalen, G. L., Ney, E. P., Neugebauer, G., Sellgren, K.: 1980, Astrophys. J. 239, 570
Gilman, R. C.: 1974, Astrophys. J. Suppl. 28, 397
Glass, I. S.: 1984, Mon. Not. R. astr. Soc. 209, 759
IRAS Explanatory Supplement, eds. C. A. Beichman, G. Neugebauer, H. J. Habing, P. E. Clegg, T. J. Chester, 1985, JPL D-1855
Stahl, O., Wolf, B.: 1986, Astron. Astrophys. (in press) = ESO Preprint no. 414
van Genderen, A. M.: 1979, Astron. Astrophys. Suppl. 38, 381
Wolf, B., Appenzeller, I., Stahl, O.: 1981, Astron. Astrophys. 103, 94

Listening to a question from the end of the room.
Wolf, van Genderen and Lamers

ECLIPSE SPECTRUM OF THE LMC P CYG STAR R 81

O. Stahl
E.S.O., Karl-Schwarzschild-Str. 2, D-8046 Garching
B. Wolf and F.-J. Zickgraf
Landessternwarte Königstuhl, D-6900 Heidelberg

R 81 ($B2Ia^+$) of the LMC was found to be a close counterpart of the galactic star P Cygni by Wolf et al. (1981). Subsequent photometric observations mainly within the "long term photometry of variables" organized by Sterken have revealed several brightness minima ($\Delta m = 0.5$) with a period of 74.6 days indicating eclipses of the primary star (see Fig. 1). (A period of 149.2 days can be excluded from the colour variations during the eclipses.)

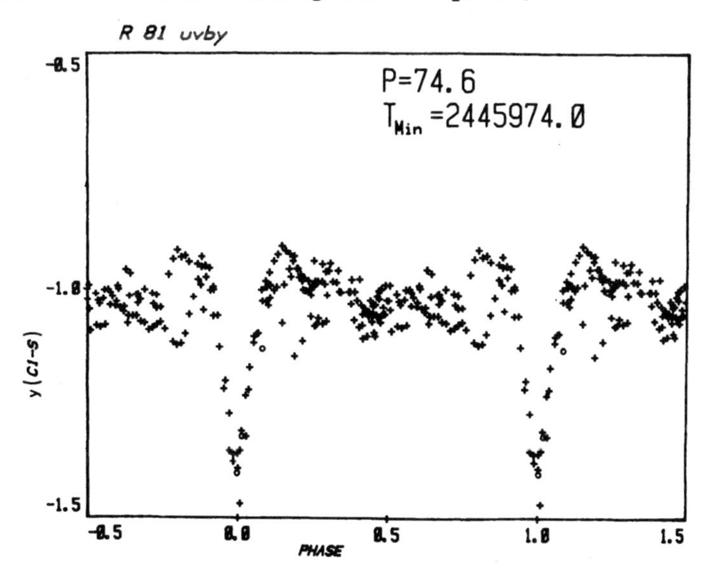

Fig. 1: Lightcurve in the Strömgren y-band of R 81 showing the eclipse with a period of 74.6 days from observations taken during more than 3 years. The scatter which is much larger than the observational error ($\sigma = 0.005$) is due to the intrinsic variability. Such variations are usually observed for luminous early-type stars. In addition Appenzeller (1972) found a rise in brightness of R 81. These measurements are also included (O) and show that the star was occasionally observed during the rise from the minimum. These measurements carried out 15 years ago define the period particularly well.

H. J. G. L. M. Lamers and C. W. H. de Loore (eds.), Instabilities in Luminous Early Type Stars, 249–252.

During the most recent eclipse in March 1986 a CASPEC spectrogram was taken in the blue spectral range. The most striking features in the spectrum of this early B-type star are FeII-absorption lines showing two components. The velocities of these components are v_{a1} = 257 km s^{-1} and v_{a2} = 70 km s^{-1}. These components were not found in spectra of the visual range taken previously (Stahl et al., 1985). However, on our IUE high dispersion LWR spectrogram taken earlier at phase ϕ = 0.15 we found that this spectral region is dominated by strong FeII absorption lines (as expected for this close counterpart of the galactic star P Cygni (cf. Cassatella et al., 1979). These UV lines are also split into three components with radial velocities of +100, +145, and +230 km s^{-1}, respectively. The fastest and the slowest components could correspond to the two components in the visual range. The remaining difference in velocity could be due to time variations caused by accelerated shells as e.g. suggested for P Cygni by Lamers et al. (1985).

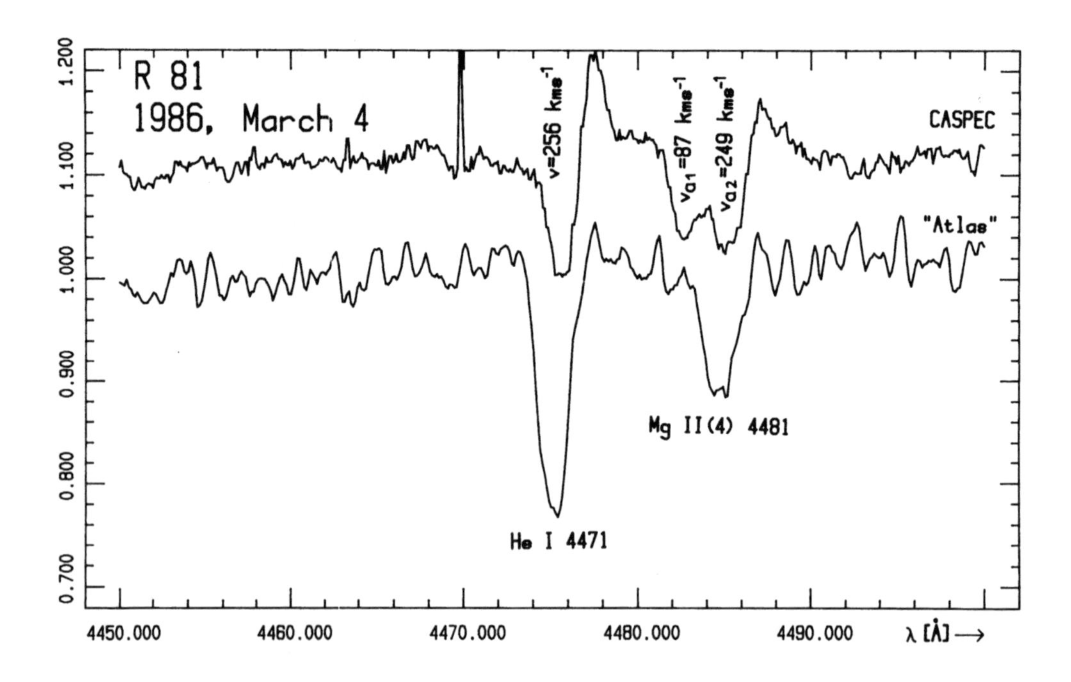

Fig. 2: Comparison of the HeI 4471 and MgII 4481 profiles observed during eclipse with those from the "Atlas". This comparison and the FeII lines indicate the existence of a phase-dependent wind.

In the optical spectrum we also observed the stronger component in the CrII(30) multiplet (v_{a1} = 61 km s^{-1}). MgII λ4481 is also split into two components of radial velocities (v_{a1} = 249, v_{a2} = 87) and has in addition a P Cygni emission. The profile is rather different from that observed previously (see Fig. 2). Surprisingly the HeI lines are weaker than usual and show a marked P Cygni profile (see Fig. 2). A few apparently undisturbed photospheric lines (SiII and SiIII, see Fig. 3) were identified from which we derived the system velocity of v_{sys} = 271 km s^{-1}. From that it is almost certain that the blueshifted component of the singly ionized metals is formed in the stellar wind of R 81. Likewise, it is most probable that also the red component of the FeII and MgII lines is formed in the wind. Firstly, the secondary star must be rather cool and visually faint since at least no pronounced secondary minimum is observed. Secondly, the measured velocity differs by 15 to 20 km s^{-1} from the system velocity but agrees on the other hand with the velocity measured e.g. from the P Cygni absorption components of the HeI lines. Finally a multi-component structure of the FeII lines is also observed in the UV range, both in the case of P Cygni (Cassatella et al.) and of R 81 (see above).

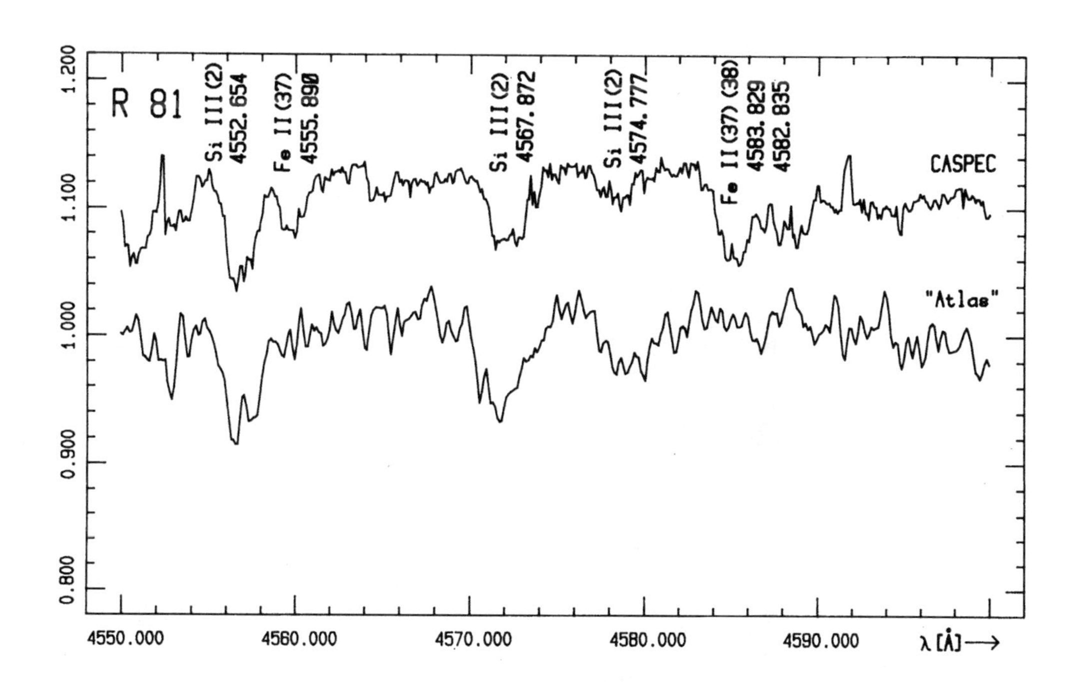

Fig. 3: Same comparison as above for the range 4450 < λ < 4600 containing photospheric SiIII lines. From the SiIII lines of the eclipse spectrum we derive the system velocity of v_{sys} = 271 km s^{-1}. In addition, FeII envelope lines (only present in the eclipse spectrum) are denoted.

The FeII, MgII, and HeI lines (also the strength of the Balmer lines) provide strong evidence that the stellar wind is phase-dependent and seems to be particularly strong during the eclipse.

R 81 is a particularly interesting P Cygni star since its

absolute luminosity is reliably known (due to its membership to the LMC) and it offers the possibility to derive the mass of a P Cygni star. For this purpose phase-dependent spectroscopic observations of high resolution and high S/N ratio are badly needed.

References:

Appenzeller, I.: 1972, Publ. Astron. Soc. Japan 24, 483.

Cassatella, A., Beeckmans, F., Benvenuti, P., Clavel, J., Heck, A., Lamers, H.J.G.L.M., Macchetto, F., Penston, M., Selvelli, P., Stickland, D.: 1979, Astron. Astrophys. 79, 223.

Lamers, H.J.G.L.M., Korwaar, P., Cassatella, A.: 1985, Astron. Astrophys. (in press).

Stahl, O., Wolf, B., de Groot, M., Leitherer, C.: 1985, Astron. Astrophys. Suppl. Ser 61, 237.

Wolf, B., Stahl, O., de Groot, M.J.H., Sterken, C.: 1981, Astron. Astrophys. 99, 351.

Reunion of colleagues from Heidelberg: Leitherer (back), Stahl, Schulte-Ladbeck and Wolf

The light- and colour variation of Eta Carinae for the years 1983-1986 in the VBLUW system*

A.M. van Genderen[1] and P.S. Thé[2]

[1] Leiden Observatory, Postbus 9513, 2300 RA LEIDEN, The Netherlands
[2] Astronomical Institute 'Anton Pannekoek', Roetersstraat 15, 1018 WB AMSTERDAM, The Netherlands

Abstract. - New VBLUW photometry of the peculiar variable star Eta Carinae, was made in the interval 1983 to 1986, to extend the detailed light- and colour curves for the epoch 1974 to 1983. A new light maximum has been observed in 1986, with a height similar to the maximum of 1981/1982.

1. Introduction

In a paper by van Genderen and Thé (1984) the historical and the recent light variation of the peculiar variable star Eta Carinae has been extensively discussed. The modern light history is based on VBLUW photometry (Walraven system), made from 1974 up to 1983.

The object, consisting of a central star and a surrounding nebula (the "homunculus"), showed small amplitude light- and colour variations ($0^m.1$ to $0^m.2$) with time scales of 1 to 3 yr. The characters of the reddening (foreground as well as in the homunculus), and photometric parameters of the central star and the homunculus, were estimated.

The trend of the Balmer jump derived from the photometry, gives the impression that the temperature of the central star is fluctuating. These fluctuations are of the order of 1000 K. On account of this Balmer jump study and considering the trend of the correlation between light- and colour curves, we presumed that the central star had expelled a new shell in 1981/1982. We have noticed a sudden reddening of the colours, while the visual light was still rising. (This is the so called S Doradus effect). Spectroscopic work of Zanella et al. (1984) confirmed the expulsion of a new shell at that epoch.

Another peculiar feature which Eta Carinae showed during the decade 1974 to 1983 was a steady rise in the ultraviolet light. This rise was $\sim 0^m.1$ more than for V.

* Based on observations collected at the ESO, La Silla, Chile.

H. J. G. L. M. Lamers and C. W. H. de Loore (eds.), Instabilities in Luminous Early Type Stars, 253–256.

2. The new photometry made during the epoch 1983 to 1986

Because of the importance to know what Eta Carinae will be doing in the near future, we decided to continue the observations a few times per year. They were made in the VBLUW system by several observers with the 90-cm Dutch Telescope at our request at the ESO, La Silla, Chile, between February 1983 and April 1986.

Figure 1 shows, apart from the long term observations of 1974 to 1983, also the new ones. It is evident that after the outburst of 1981/1982, the visual light declined by $0^m.12$, while V-B became bluer by $0^m.03$. Thus, just as in previous years, both changes are in antiphase. The rise in the ultraviolet was still going on, especially through the U pass-band (just shortward of the Balmer jump).

The latest observations of February through April 1986 however, show a new light maximum with an amplitude of $\sim 0^m.1$ and a reddening of B-L (the L pass-band contains the Balmer limit) and B-U by $\sim 0^m.02$ and $\sim 0^m.01$, respectively. It would be of great importance to check spectroscopically, whether a new shell has been expelled.

Acknowledgements.
We are very grateful to those who made the additional observations: E.R. Deul, J.W. de Bruyn, P.T.M. Feldbrugge, F. van Roermund, W. Tijdhof, H. Bovenchen and W.J.G. Steemers.

References

van Genderen, A.M., Thé, P.S.: 1984, Space Sc. Rev. 39, 317.
Zanella, R., Wolf, B., Stahl, O.: Astron. Astrophys.. 137, 79.

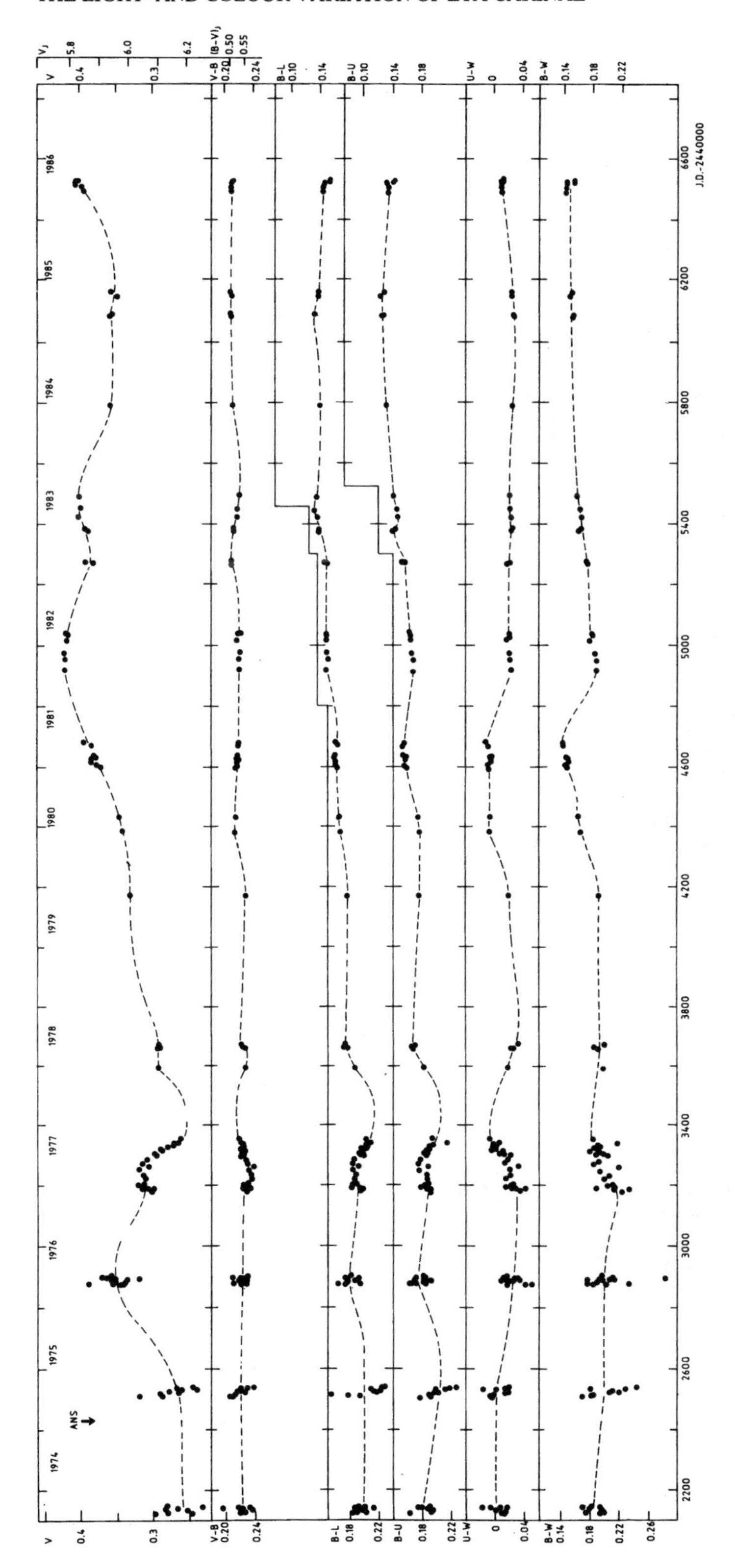

Caption for Figure 1

The light- and colour curves of Eta Carinae for the epoch 1974-1986 made in the VBLUW system of Walraven (in log int. scale). The V and B-V scales of the UBV system (in mag scale and with subscript J) are shown at the right.

Top:
Looking at books:
Claus Leitherer and
Edith Müller

Bottom:
Roberto Viotti
Arnoud van Genderen
Pik-Sin Thé

DO SUPERLUMINOUS STARS REALLY EXPLODE?

Roberto Viotti
Istituto Astrofisica Spaziale, CNR, 00044 Frascati, Italy

ABSTRACT

Many categories of astrophysical objects are characterized by sudden and deep light variations during their life, which are usually called "outbursts". In many cases the physical processes that are responsible of these outbursts are unknown, mostly because the phenomenon has not been observed at all the wavelength ranges and during the whole light evolution. We shall discuss three rather well documented cases of superluminous stars: Eta Carinae, P Cygni and AG Carinae, which are well representative of the uppermost part of the H-R diagram. It is concluded that the observed historical variations of these objects <u>occurred at almost constant bolometric magnitude</u>, i.e. the use of the term outburst is not appropriate. The implications for other cases, including the Hubble-Sandage variables, are discussed.

Very luminous stars are very few in number, but their interest is very large as: (i) they may represent a pahse, or different phases of the evolution of massive stars, after having left the main sequence, and (ii) for the complex physical processes taking place in the external boundary of their structure and in the expanding envelopes (Viotti 1985).

Variability is one major characteristic of these objects. Their light curve frequently presents luminosity brightenings or fadings of 1-2 mag or more, which are normally referred to as NOVA-LIKE EXPLOSIONS. The origin of these events and the associated physical processes is still an open problem, whose solution requires coordinated multifrequency observations covering the whole time evolution of the event. And this is a problem since these events are unpredictable. Another characteristic is the presence of shells or <u>compact nebulae</u> originating by matter ejected from the central star. In many cases there is also evidence of <u>circumstellar</u>

H. J. G. L. M. Lamers and C. W. H. de Loore (eds.), Instabilities in Luminous Early Type Stars, 257–262.

dust, a feature in contrast with the relatively high excitation temperature of the spectral lines. Is this circumstellar matter produced by huge explosions, or by continuum ejection of matter? Is the dust condensing in the stellar wind? Are the outbursts associated to a phase of the stellar evolution of massive stars, or of interactive binaries? An answer to these questions can be given only from the detailed study of a few well studied objects, such as Eta Car, P Cyg and AG Car discussed below.

ETA CARINAE

This sixth magnitude star was 150 years ago one of the brightest stars in the sky. Since 1856 its visual luminosity gradually decreased in about 14 years to the seventh magnitude. Previously, during its bright phase of 1830-1850 Eta Car underwent sudden brightenings. In one of those (1843) the star became even brighter than Canopus. For this behaviour Eta Car was (uncorrectly) classified by Payne-Gaposhkin (1957) as a very slow nova. Presently, the star is very bright in the infrared. Andriesse et al (1978) found that the bolometric magnitude ($0^m.0$) derived from the infrared energy distribution is close to the estimated bol mag during the bright phase of last century. This suggests that Eta Car is presently surrounded by a dusty envelope formed during the large luminosity fading, which is absorbing most of the optical and UV radiation of the central object(s). During the maximum phase most of the radiation was emitted in the visual (i.e. the B.C. was close to zero). The total radiation power at that phase is $\sim 10^7$ L_o, that is equal to the sum of the present mechanical power (stellar wind) and the radiative (mostly IR) power of Eta Car (see Andriesse et al, 1978 for more details). The lower luminosity of the star during the beginning of 1800 ($V \cong 2$) can be explained by a hotter temperature of the atmospheric envelope of the star with BC= -2, as in the cases of AG Car and P Cyg discussed below.

After the fading phase, there were some "events", such as the small nova-like outburst of 1889 when the spectrum was similar to that of novae near maximum with a (probably) cool continuum, and prominent P Cygni profiles of H and FeII, but with a rather low P Cygni velocity of -180 km s^{-1}.

In the following years, the continuum weakened, and a rich emission line spectrum appeared with low (FeII) and high ionization emission lines (HeI, FeIII, [NeIII]) similar to the present one.

During three epochs, in 1948, 1961 and 1981, temporary decreases of the intensity of the high temperature lines were detected, which are probably associated to transient phenomena in the atmosphere of Eta Car.

The light curve of this puzzling object may be therefore schematically described as follows:

beg. 1800 V=+2; hot atmosphere with BC=-2; M_{bol}= -13.0
1830-1850 V=-1/+1 variable at const M_{bol}=-13; BC=-2 to 0.
around 1850 increase of mass loss rate with $L_{mech} \cong 1.5\ L_{rad}$
1856-1870 dust formation; visual fading at const M_{bol}
1889-1900 nova-like outburst; cooler atmosphere (F_{eq}); ejection of shell
1948, 1961, 1981 events: cooler atmosphere; ejection of shells

AG CARINAE

This is a P Cygni star in the southern sky which has displayed in recent years large luminosity variations from V=6 to V=8, associated to large spectral changes, the star being "hotter" at minimum. From the analysis of the UV, optical and IR data during 1981-1983, Viotti et al. (1984) concluded that during this period AG Car remained at almost constant bolometric magnitude ($\sim$-8.3), in spite of the large visual luminosity variations. That is these variations are only apparent, and explained by temperature (and mean radius) variations of the atmosphere. Similar conclusion was reached by the Heidelberg group in their study of the luminosity and spectral variations of the Magellanic Cloud stars. The equivalent spectral type is variable from A0eq to B0-2eq (Caputo and Viotti 1970). More recently Stahl (1986) discovered Of features in the visual spectrum of AG Car, when the star was in a deep minimum (V=8). Schematically:

1949-50 eq. sp.: A0; star probably at max; BC$\simeq$0.
1953-59 eq. sp.: B0÷2; hotter atmosph.; V=8; BC$\cong$-2.
1981-83 A-type at max (V=6), B-type at min.(V=7/8); M_{bol}=const =$\sim$-8.3
1985 Of-type; V=8; BC$\cong$-3

This behaviour of AG Car can be easily explained by structure changes of the expanding atmosphere (on time scales of months and probably days), occurring at constant bolometric magnitude. They are not necessarily associated to large explosions and huge ejection of matter.
Actually, AG Car is surrounded by a ring nebula. Spectral observations in the optical and UV suggest that most of the light from the nebula is radiation from the central star scattered by dust grains in the nebula. We think that this nebula was not produced by an expolsion, but is the residual of the stellar wind during a previous red supergiant phase of AG Car.

P CYGNI

This star is rather stable near V$\sim$5 since 1780, but previously P Cyg was found variable with light maxima in 1600-06 and 1639-59 when the star reached the third magnitude and was "reddish". Deep minima were recorded during 1606-56 and 1959-83 with occasional fadings below visibility (see de Groot 1969). If the star at max was redder than today, then most of its light was probably emitted at optical wavelengths and its BC should have been close to zero. Presently according to Lamers et al.(1983) BC should be close to -1.6. Thus the difference in BC between the bright phase and now is within the errors equal to the different in the visual magnitudes. The obvious conclusion is that P Cyg during the 17th century maxima was probably at the same bolometric luminosity than now, and that no major "outbursts" occurred at that time. Finally, to explain the large fadings during 1600 we might think of a large increase of the envelope temperature at constant M_{bol}, as has recently happened in AG Car. This however cannot decrease the luminosity below visibility. We therefore need another process, most likely the dust formation like in Eta Car. The following brightening is probably the result of the distruption of the dusty envelope. To schematize the light history of P Cyg:

beg. 1600 brightening to V=+3; disappearence of dust envelope?
1600-1606 V=+3, reddish; cooler atmosphere; BC≙ 0; larger mass loss rate
1606-1656 var. down to V≲6; dust formation; ejection of shells?
1656-1659 V=+3, reddish; cooler atmosphere; BC= 0; high mass loss rate
1659-1683 var. down to V≲6; dust formation; ejection of shells
1683-1780 brightening to V 5; disappearence of residual dust and/or cooling from a previous Of/WR phase
1781-1786 V=5.2 to 5.6; shell event
1780-now V=+4.9; BC= -1.6; hot (Beq) atmosphere; small shell events

DISCUSSION

The presence of an intense emission line spectrum in these stars is an indication of the presence of an extended stellar atmosphere. P Cygni profiles show the presence of strong mass outflows. The structure of these atmospheres is probably "critical", in the sense that small changes of the matter outflow could produce large opacity variations, resulting in different energy distribution, line excitation/ionization.
Probably, an occasional consistent but not large increase of the mass loss rate might produce a drastic decrease of the temperature, even below the grain condensation temperature (typically 1500-1700 K). Formation of dust in the outer parts of the atmosphere results in a large, but apparent fading of the stellar luminosity, like in the R CrB stars and in novae. We believe that this process is also working in the Hubble-Sandage variables, which display large luminosity fadings on rather short time scales. Again this variability is probably only apparent, and could be attributed to luminosity redistribution at constant M_{bol} and to dust formation and distruption. Multifrequency observations of these LBV's during different light stages, in particular in the IR, are urgently needed to confirm this interpretation.

REFERENCES

Andriesse, C.D., Donn, B.D., Viotti, R.: 1978, Mon. Not. R. astr. Soc. 185, 771.

Caputo, F., Viotti, R.: 1970, Astronomy Astrophys. 7, 266.

de Groot, M.: 1969, Bull. astr. Soc. Netherlands 20, 225.

Lamers, H.J.G.L.M., de Groot, M., Cassatella, A.: 1983, Astronomy Astrophys. 128, 299.

Viotti, R.: 1985, The Messenger 39, 30.

Viotti, R., Altamore, A., Barylak, M., Cassatella, A., Gilmozzi, R., Rossi, C.: 1984, NASA Conference Publication 2349, p.231.

Hammerschlag-Hensberge, Bianchi, Henrichs and Burki

Angelo Cassatella and Mart de Groot discussing P Cygni stars

RADIATION PRESSURE IN ACOUSTIC WAVE CALCULATIONS OF EARLY-TYPE STARS

Bernhard E. Wolf and Peter Ulmschneider
Institut für theoretische Astrophysik,
Im Neuenheimer Feld 561
D-6900 Heidelberg, Federal Republic of Germany

ABSTRACT. A new method to treat radiation pressure in spectral lines has been developed which avoids the Sobolev-approximation. This method has been used for time-dependent acoustic wave calculations in early-type stars.

1. INTRODUCTION

For early-type stars spectral lines constitute the most important contribution to the radiation pressure force. In the treatments of Castor (1974) as well as Castor et al. (1975) the Sobolev-approximation has been used which however breaks down when acoustic waves in regions of small wind velocity are considered. For these cases a new method has been developed to compute the time-dependent radiation pressure force.

2. METHOD TO COMPUTE THE RADIATION PRESSURE

To take into account the effects of an intense radiation field in the time-dependent hydrodynamic equations (for a review of time-dependent atmospheric wave calculations in stellar atmospheres see Ulmschneider and Muchmore 1986) the energy equation does not need to be modified and reads (Wolf 1985, 1986b)

$$\Phi_R = -4\pi\kappa_R'(J-B) \quad , \tag{1}$$

where Φ_R is the net radiative cooling rate, J the frequency integrated mean radiative intensity, B the Planck function and κ_R' the Rosseland opacity without scattering. In the Euler equation a radiative force term

$$\rho g_R = \frac{1}{c}\int_0^\infty \kappa_\nu F_\nu d\nu \simeq \frac{1}{c}(\kappa_R+\kappa_L)\ F \quad , \tag{2}$$

has to be added where ρ is the density, F_ν is the monochromatic and F

H. J. G. L. M. Lamers and C. W. H. de Loore (eds.), Instabilities in Luminous Early Type Stars, 263–266.

the frequency integrated radiative flux. Here κ_ν is the monochromatic opacity with scattering, κ_R the Rosseland opacity (continuum contribution) and κ_L is a flux weighted mean opacity (line contribution). To make the problem tractable Wolf (1985) assumed LTE, a two beam approximation with no ingoing radiation field and that the line contribution arises at heights, where the continuum optical depth has fallen below unity and the lines can be considered generated by pure absorption in a fixed incident continuum radiation field. We then find

$$\kappa_L = \sum_{i,k} \chi_L(x) \int_o^\infty \varphi_\nu(x) \frac{F_\nu}{F} \exp(-\tau_\nu^L(x))\, d\nu \quad , \tag{3}$$

where F_ν and F are photospheric fluxes. The contribution of element i and ionization stage k, assuming an atom with two levels L,U (Mihalas 1978, p.336) is

$$\chi_L = \frac{\pi e^2}{m_e c} f_{LU}\, n_a\, X_i\, \frac{n_{ik}}{n_i} \frac{1-\exp(-h\nu_{LU}/kT)}{1+g_U\exp(-h\nu_{LU}/kT)/g_L} \quad . \tag{4}$$

Here n_a is the number density of heavy particles, X_i the fractional abundance by number, f_{LU} the oscillator strength, and n_i, n_{ik} are number densities computed in LTE. The flux F_ν is attenuated over the optical depth τ_ν^L from the photospheric height x_p of the star to height x along a ray inclined by angle cosine $3^{-1/2}$:

$$\tau_\nu^L(x) = \int_{x_p}^{x} \chi_L(x')\varphi_\nu(x')\, 3^{1/2} dx' \quad , \tag{5}$$

As the flux F_ν relevant for the radiative force at height x (using the Barbier-Eddington approximation) emerges from a distance $\Delta\tau$=2/3 closer to the star, the integration in Eq. (5) is carried out only to height $x_{2/3}$ (c.f. Wolf 1985). Following Castor et al. (1975) we approximate the ratio F_ν/F by the ratio B_ν/B of the Planck functions which can be taken out of the integral of Eq. (3). Assuming a rectangular line profile for which the central depth ϕ_L is chosen to match the relevant Voigt profile we have a line width $2\Delta\nu_L=1/\phi_L$. Fig. 1 shows the line profile band traced out by an acoustic wave motion in the (x,ν) plane. The fluid element at height x sees the radiative flux F_ν emerging from height $x_{2/3}$. The flux F_ν is present only if the frequency is outside the shaddow zone $\nu_R \leq \nu \leq \nu_B$ because it is assumed that any flux inside the shaddow zone has been absorbed. The limits ν_B and ν_R are given by

$$\begin{aligned} (\nu_B-\nu_{LU})/\Delta\nu_D &= \Delta\nu_L/\Delta\nu_D + 3^{-1/2}\, u_{MAX}/v_{th} \quad , \\ (\nu_R-\nu_{LU})/\Delta\nu_D &= -\Delta\nu_L/\Delta\nu_D + 3^{-1/2}\, u_{MIN}/v_{th} \quad , \end{aligned} \tag{6}$$

where ν_{LU} is the frequency of the line center, $\Delta\nu_D$ the Doppler width and v_{th} the thermal speed of the line forming ion. For every x $u_{MAX}(x)$ is th

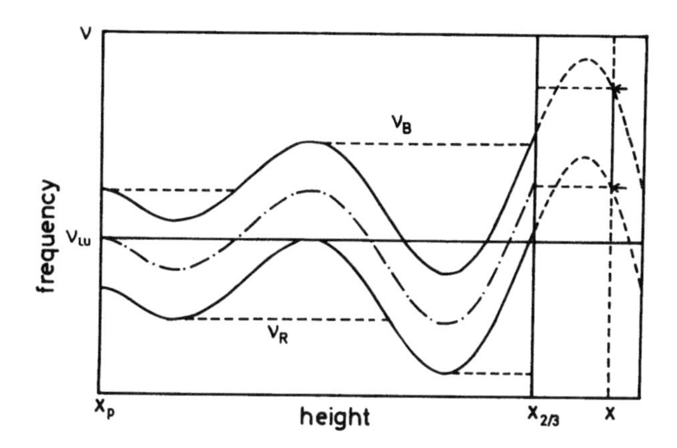

Figure 1. Height-frequency diagram for an acoustic wave

maximum value of the velocity of the wave in the height interval x_p to x, and $u_{MIN}(x)$ the corresponding minimum of the velocity in the same height interval. Assuming that $\tau_\nu^L(x)$ of Eq. (5) can be approximated by ∞ if $u_{MIN} \leq u(x) \leq u_{MAX}$ and by 0 else, then from Eq. (3) using

$$
f(u,x) = \begin{cases} 2\Delta\nu_L/\Delta\nu_D & , \text{ if } u > u_{MAX}+u_{th} \text{ or } u < u_{MIN}-u_{th} \\ 0 & , \text{ if } u_{MIN} \leq u \leq u_{MAX} \\ (u-u_{MAX})/v_{th} & , \text{ if } u_{MAX} \leq u \leq u_{MAX}+u_{th} \\ (u_{MIN}-u)/v_{th} & , \text{ if } u_{MIN}-u_{th} \leq u \leq u_{MIN} \end{cases} , \quad (7)
$$

$$
u_{th} = 12^{1/2}\, v_{th}\, \Delta\nu_L/\Delta\nu_D \quad , \quad (8)
$$

we obtain

$$
\kappa_L = \sum_{i,k} \chi_L(x)\, B_\nu/B\, \phi_L\, \Delta\nu_D\, f(u,x) \quad . \quad (9)
$$

For the line contributions there are three different effects: strong resonance lines with little flux in the line core can be Doppler shifted into the continuum radiation field, optically thin lines contribute because there are so many of them, new ions which see an undiluted radiation field can be created by ionization in the hot compression region behind the shocks. The thin line contribution can be evaluated by using the first case in Eq. (7), by which $\phi_L \Delta\nu_D f(u,x) = 1$ in Eq. (9). This contribution thus becomes independent of u and is a function of temperature T and pressure p only. The same is the case for the new ions which see an undiluted flux. Using some mean thermal speed v_{th} in Eq.'s (7) and (9), the velocity dependent contribution of the resonance lines can likewise be evaluated as function of T and p. We thus find

$$\rho g_R = F\{\kappa_R(T,p)+(1-w)\kappa_1(T,p)f(u,x)+w\kappa_2(T,p)+(1-w)\sum_{i=3}^{10}\kappa_i(T,p)\}/c \quad , \quad (10)$$

where κ_R is the Rosseland opacity with scattering and κ_1 to κ_{10} are line opacities. w is a weighting function which decreases with height. The Rosseland opacity describes the continuum contribution. κ_1 is the contribution from the resonance lines, κ_2 from the optically thin lines and κ_3 to κ_{10} from the strong lines of new ions. f(u,x) describes the Doppler shift of resonance lines into the undiluted continuum. The contribution from the thin lines and new ions does not depend on velocity. Consider a grid point at some height x with temperature T and pressure p, while at the next grid point closer to the star one has T' and p'. There are eight possibilities of T, p being greater, equal or less than T', p'. If between T', p' and T, p a new ion appears its opacity contribution is listed in the tables κ_3 to κ_{10}. The advantage of this approach is that the opacities κ_1 to κ_{10} can be pretabulated and summed over a large number of lines. Wolf(1985, 1986b) took up to 650 lines and showed that the important contributions came from the elements H, He, C, N, O, Na, and Fe.

3. RESULTS AND CONCLUSIONS

The present method avoids the Sobolev approximation, and thus is well suited for wave calculations. A large number of wave computations for stars of spectral types O3V to A0V, B0III to B5III and O5I to B5I were made (Wolf 1985, 1986d) using initial acoustic fluxes corresponding to a Mach number 0.01 and periods of 0.4 times the photospheric cut-off period. The calculations were started at a continuum optical depth of between 3 and 10^{-3}. It was found that there is a radiative damping zone up to an optical depth of typically 10^{-5} where the waves are essentially isothermal with a relative temperature fluctuation $\Delta T/T$ of around 10^{-7}. This also applied to the shocks formed in this zone. After passing the upper limit of the damping zone the waves by radiative acceleration rapidly grew to strong shocks with postshock temperatures of the order 3 10^6 K. The generated stellar wind could not be followed because of the limited width of the atmospheric slab which extended to continuum optical depths of around 10^{-9}. The magnitude of the radiative amplification in the various stars was discussed in detail. Radiation pressure in lines was found to be the dominant contribution.

REFERENCES

Castor, J.I.: 1974, Mon. Not. R. Astr. Soc. 169, 279
Castor, J.I., Abbott, D., Klein, R.: 1975, Astrophys. J. 195, 157
Mihalas, D.: 1978, Stellar Atmospheres, 2nd ed., Freeman, San Francisco
Ulmschneider, P., Muchmore, D.: 1986, Proceedings: Small Magnetic Flux Concentrations in the Solar Photosphere, Abhandlungen der Akademie der Wissenschaften, Göttingen
Wolf, B.E.: 1985, Ph.D. thesis, Univ. of Heidelberg
Wolf, B.E.: 1986a-d, Astron. Astrophys. submitted

THE INFLUENCE OF PHOTOSPHERIC TURBULENCE ON STELLAR MASS LOSS

Cornelis de Jager and Hans Nieuwenhuijzen
Astronomical Observatory and
Laboratory for Space Research Utrecht
Zonnenburg 2 and Beneluxlaan 21
3512 NL Utrecht and 3527 HS Utrecht

ABSTRACT. We show that the stellar rate of mass loss is positively correlated with the average microturbulent photospheric velocity, and that the energy contained in the microturbulent motions is of the same order of magnitude as the wind energies.

1. PHOTOSPHERIC MICROTURBULENCE

Since some time we are involved in determining the line of sight photospheric microturbulent velocity components ζ_t for super- and hypergiants. One of the results found so far is that ζ_t increases with increasing stellar luminosity (for stars with the same T_{eff}-values) and that $\zeta_t = s$, the sound velocity, for stars with $L = \frac{1}{4} L_{lim}$, where L_{lim} is the Humphreys-Davidson brightness limit. This observation leads to the conclusion that shock-wave dissipation of microturbulent energy may be a heat source in super- and hypergiant atmospheres. The next question is whether it contributes to stellar mass loss.

2. RELATIONS TO STELLAR MASS LOSS

De Jager et al. (1986, 1987) have shown that the stellar rate of mass loss can be represented as a function of the efficient temperature T_{eff} and luminosity log $(L/L_\odot)$. The one-sigma value of the adaption of log $(-\dot{M})$ to the interpolation formula is $\pm$ 0.46, while the measurements have an intrinsic one-sigma accuracy per measurement of $\pm$ 0.37. Since the two values are not equal there must be other factors, besides T_{eff} and L influencing $\dot{M}$. To that end consider the residual mass loss values

$$D = \log(-\dot{M})_{obs} - \log(-\dot{M})_{model},$$

being the difference between the observed rates of mass loss and the values calculated with the interpolation formula. It appears that D is

H. J. G. L. M. Lamers and C. W. H. de Loore (eds.), Instabilities in Luminous Early Type Stars, 267–268.

positively correlated with ζ_t/s, i.e. the larger the (scaled) microturbulence, the larger the rate of mass loss (if all other parameters remain the same). In actual practice dD/d log ζ_t = 2.4.

We therefore conclude that mass loss is strongly related to photospheric microturbulence.

Let us next assume that the turbulent energy propagates outward with sound velocity s. The stellar microturbulent energy flux $2\pi R^2 p \zeta_t$ s (erg s^{-1}) appears to be of the same order as the stellar wind energy flux $\frac{1}{2} \dot{M} v_\infty^2$. This is another evidence that stellar winds are fed by microturbulence.

REFERENCES

De Jager, C., Nieuwenhuijzen, H. and Van der Hucht, K.A.: 1986, in Proceedings of Porto Heli Symposium.

De Jager, C., Nieuwenhuijzen, H. and Van der Hucht, K.A.: 1987, Astron. Astrophys. submitted.

Pik-Sin Thé, Hans Nieuwenhuijzen (who made most of the photographs) and Cees de Jager

Nonlinear Dynamics of Instabilities in Line-Driven Stellar Winds

S.P. Owocki
University of California at San Diego
La Jolla, CA 92093 USA

J.I. Castor
Lawrence Livermore National Laboratory
Livermore, CA 94550 USA

G.B. Rybicki
Harvard-Smithsonian Center for Astrophysics
60 Garden St.
Cambridge, MA 01238 USA

ABSTRACT. We have been developing a numerical radiation-hydrodynamics program in order to study the nonlinear evolution of instabilities in line-driven winds from luminous, early-type stars. The present, preliminary version of this program assumes a 1-D, spherically symmetric, isothermal stellar wind that is driven radially outward through absorption of a point source of continuum radiation by a fixed ensemble of isolated, pure-absorption lines. Initial tests of the code indicate that the velocity structure of nonlinear pulses in such a wind may be quite different than assumed in previous analyses. For example, the numerically computed structures show much steeper rarefaction and much more gradual compression than the sawtooth periodic shock structures previously assumed by Lucy. If these preliminary results are confirmed by further computations and analysis, they would have important implications for interpreting radio, ultraviolet, and x-ray observations that are thought to be associated with instabilities in hot star winds.

I. INTRODUCTION

The winds from early type (O and B) stars are thought to be driven by the line absorption of the star's continuum radiative momentum flux (Lucy and Solomon 1970; Castor, Abbott, and Klein 1975, hereafter CAK). Linear stability analyses show, however, that such line-driven winds are highly unstable to short scale, radial velocity perturbations, with derived growth rates 50-100 times a typical wind expansion rate (MacGregor, Hartmann, and Raymond 1979; Carlberg 1980; Owocki and Rybicki 1984, 1985, 1986; hereafter papers I, II, and III). Furthermore, there are

H. J. G. L. M. Lamers and C. W. H. de Loore (eds.), Instabilities in Luminous Early Type Stars, 269–272.

many observed wind characteristics that are not explained by steady, smooth wind theories such as CAK. Among these are the narrow absorption features (Lamers, Gathier, and Snow 1982), the black absorption troughs (Lucy 1982a), and the high ionization states (Lamers and Morton 1976) characteristic of many UV line profiles, as well as the detection of soft X-rays (Harnden et al. 1979; Seward et al. 1979) and nonthermal radio emission (Abbott, Bieging, and Churchwell 1981, 1984). It has been widely supposed that the line-driven wind instability will ultimately lead to a nonmonotonic flow with recurring shocks that may result in many or all of these observed characteristics (Lucy and White 1980; Lucy 1982b; Cassinelli and Swank 1983; Krolik and Raymond 1985; White 1985); but there has until now been no quantitative dynamical calculation following the growth of initially small perturbations into the nonlinear regime. Thus it is not known, for example, whether the actual nonlinear structure is likely to resemble the periodic sawtooth shocks postulated by Lucy (1982b), or the isolated radiation-driven pancakes of Krolik and Raymond (1985), or perhaps neither of these.

We report here on our recent efforts to carry out such a nonlinear dynamics calculation. Section II describes the radiation-hydrodynamics code we are developing, and section III describes some initial results.

II. Radiation-Hydrodynamics Program

In order to study the nonlinear evolution of this instability, we have recently developed a time-dependent radiation hydrodynamics code that numerically integrates a radiatively driven stellar wind model forward in time. The present version of this code assumes a 1-D, spherically symmetric, *isothermal* stellar wind that is driven radially outward through absorption of a point source of continuum radiation by a fixed ensemble of isolated, *pure-absorption* lines. Thus both scattering and line overlap effects are ignored (cf. Paper II), as are the compressional heating, ionization, and X-ray production that might arise if the flow were not simply assumed to be isothermal. The line force, however, is determined not from the usual Sobolev approximation (Sobolev 1960; Lucy 1971; Castor 1974), but from detailed computation of the height and frequency dependence of the line absorption. Thus, although this current code version is not yet suited for making detailed predictions to compare with available observations in ultraviolet lines or X-rays, it does make it possible to study how perturbations on a scale near or below the Sobolev length grow beyond the linear regime.

We have experimented with two different numerical hydro methods. Originally we used the flux-corrected transport (FCT) algorithm from Boris and Book (1976), but we found this to be numerically unstable once steep fronts formed from the amplification of perturbations by the physical instability. Thus, unfortunately, this FCT method fails just when the physics is becoming most interesting. We therefore modified our code to use a staggered-mesh method (van Leer 1979) that is better suited to problems with steep gradients. We also introduced a pseudo-viscosity term to smooth out steep fronts over several grid points and to control the growth of the physical instability on scales near the grid spacing.

III. Preliminary Results

The results obtained so far with this staggered-mesh code version, which resemble those from the FCT version for as long as the latter could run, are quite interesting. These can be summarized as follows:

1. With the initial state chosen to be the CAK steady-flow solution, the wind develops a strong pulse at its base due to the difference between the Sobolev and absorption force at low velocities. This pulse eventually turns into a strong shock, and then into a dense shell structure, bounded by forward and reverse shocks. This bears some resemblance to the pancake flow structure assumed by Krolik and Raymond (1985), except that this shock passes out of the system, and no similar shocks follow behind.
2. The structures that do persist over an extended period of several wind transit times can perhaps be best described as steep rarefaction waves or fronts. These structures are in some sense the opposite of the saw-tooth shock waves postulated by Lucy (1982b), with steep rather than gradual rarefactions, and gradual rather than steep compressions.
3. These rarefaction waves are slowly carried outward by the flow, and with no further acoustic input, it appears they will eventually die out and the flow will reach a steady-state. This is in accord with the characterization of the instability as being of the "drift" type (see paper III).

It should be emphasized that these results, though intriguing, are still quite preliminary, and that further analysis and calculations will be needed to understand fully their physical basis and to determine the extent of their applicability. For example, it is not yet known whether the steep rarefaction waves encountered here are only a transient response of this particular initial condition, or whether they are likely to a ubiquitous result of almost any type of acoustic disturbance.

To distinguish between these two eventualities, we plan next to investigate the nonlinear response to various types of acoustic disturbances at the wind base. We will also explore alternative treatments of the spectral line transfer. As our theoretical calculations become more complete, we are also anxious to apply them to the interpretation of those radio, UV, and X-ray observations that are thought to be associated with this line-driven wind instability.

References

Abbott, D.C., Bieging, J.H., and Churchwell, E. 1981, Ap. J., **250**, 645.
Abbott, D.C., Bieging, J.H., and Churchwell, E. 1984, Ap. J., **280**, 671.
Boris, J.P. and Book, D.L. 1976, Methods of Computational Physics, **16**, Chapter 13, (New York: Academic Press).
Carlberg, R.G. 1980, Ap. J., **241**, 1131.
Cassinelli, J.P. and Swank, J.H. 1983, Ap. J., **271**, 681.
Castor, J.I. 1974, M.N.R.A.S., **169**, 279.
Castor, J.I., Abbott, D.C., and Klein, R.I. 1975, Ap. J., **195**, 157 (CAK).
Harnden, F.R., Branduardi, B., Elvis, M., Gorenstein, P., Grindlay, J., Pye, J.P., Rosner, R., Topka, K., and Vaiana, G.S. 1979, Ap. J. (Letters), **234**, L51.
Krolik, J. and Raymond, J.C. 1985, Ap. J., **298**, 660.
Lamers, H., Gathier, R., and Snow, T.P. 1982, Astr. Ap., **118**, 245.
Lamers, H., and Morton, D.C. 1976, Ap. J. Suppl., **32**, 715.
Lucy, L.B. 1971, Ap. J., **163**, 95.
Lucy, L.B. 1982a, Ap. J., **255**, 275.
Lucy, L.B. 1982b, Ap. J., **255**, 286.
Lucy, L.B. 1984, Ap. J., **284**, 351.
Lucy, L.B. and Solomon, P.M. 1970, Ap. J., **159**, 879.
Lucy, L.B. and White, R.L. 1980, Ap. J., **241**, 300.
MacGregor, K.B., Hartmann, L., and Raymond, J.C. 1979, Ap. J., **231**, 514.
Milne, E.A. 1926, M.N.R.A.S., **86**, 459.
Owocki, S.P. and Rybicki, G.B. 1984, Ap. J., **284**, 354 (Paper I).
Owocki, S.P. and Rybicki, G.B. 1985, Ap. J., **299**, 265 (Paper II).
Owocki, S.P. and Rybicki, G.B. 1986, Ap. J., in press (Paper III).
Seward, F.D., Forman, W.R., Giacconi, R., Griffith, R.B., Harnden, F.R., Jones, C., and Pye, J.P. 1979, Ap. J. (Letters), **234**, L51.
Sobolev, V.V. 1960, Moving Envelopes of Stars (Cambridge: Harvard University Press).
van Leer, B. 1979, J. Comp. Phys., **32**, 101. White, R.L. 1985, Ap. J., **289**, 698.

THE X-RAY EMISSION OF TAU SCO, B0 V, AND THE PROBLEMS POSED FOR EMBEDDED SHOCK MODELS

J.P. Cassinelli
Sterrewacht Sonnenborgh, Utrecht, Nederland and
Washburn Observatory, University of Wisconsin, Madison, Wisconsin, USA

Introduction. Main sequence OB-stars are useful for testing X-ray source models, because the mass loss rates are low as compared to those of the OB supergiants and the Of stars hence the winds are optically thin at nearly all frequencies. We are therefore able to detect the emission from all parts of the outer atmosphere and not be concerned that some source regions are attenuated from our view. The analyses of the X-ray emission and the ultraviolet line profiles are relatively straight-forward.

Tau Sco is a very well studied B0 V star. The discovery of the resonance doublet of O VI (1030A) in its Copernicus satellite spectrum led to the recognition of the now well known "superionization" problem (Rogerson and Lamers, 1975). This led to the prediction that O and B stars would be X-ray sources. Tau Sco is near the lower luminosity limit for stars on the H-R diagram that show easily detectable winds. It is one of only six early-type stars that were observed with the high spectral resolution Solid State Spectrometer on the Einstein satellite. The X-ray data, combined with the results of UV line profiles studies, pose several interesting problems for models attempting to explain the presence of hot gas in the outer atmosphere of the star. Satisfactory explanation of these problems would provide strong evidence in support of a model. Here the focus is on the possibility that the X-ray emission arises from shocks in the winds. Lucy and White(1980) and Lucy(1982) have proposed that instabilities in line driven stellar winds lead to the development of X-ray emitting shocks embedded in the wind. The X-rays from the shocks should then produce the superionization stages through photoionization.

The X-ray spectrum

The SSS spectrum of Tau Sco is shown in Figure 1. Swank(1985) has shown that the spectrum can be explained with a two component model: Most of the X-ray emission arises from a component with a temperature, emission measure, and overlying hydrogen absorption column density of:
$T = 5.3 \pm 0.4 \times 10^6$ K, $EM = 3.6 \pm 0.5 \times 10^{54}$ cm^{-3}
and $N_h = 1.8 \pm 0.5 \times 10^{21}$ cm^{-2}

H. J. G. L. M. Lamers and C. W. H. de Loore (eds.), Instabilities in Luminous Early Type Stars, 273–278.

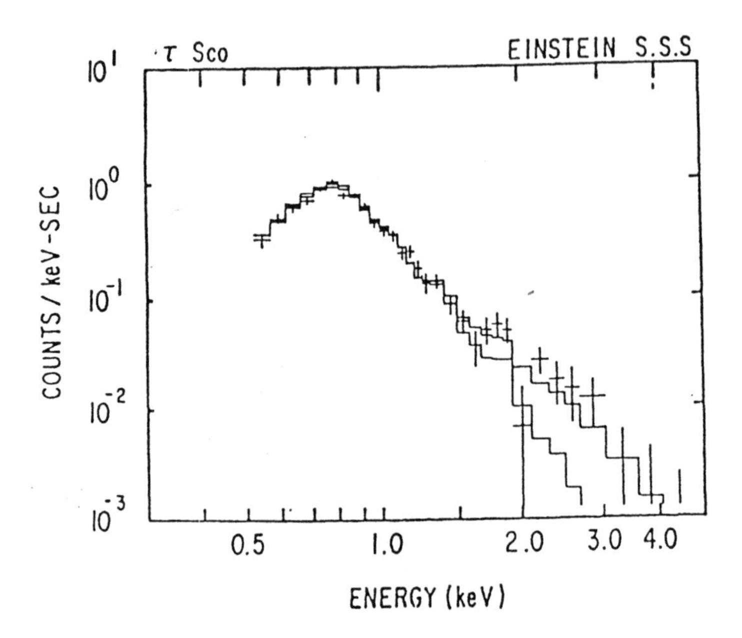

Figure 1. The solid state spectrometer spectrum of Tau Sco. the spectrum below 1.5 keV can be fitted with a source with $T = 5.3 \times 10^6$ K, while at higher energy there is clear evidence for the presence of a second hotter source, with $T > 15 \times 10^6$ K.

The properties of the second or "Hot" component are best described by the 90 percent confidence contours shown in figure 2. The temperature is at least 15 million degrees and the emission measure is of order $10^{53.5}$ cm^{-3}

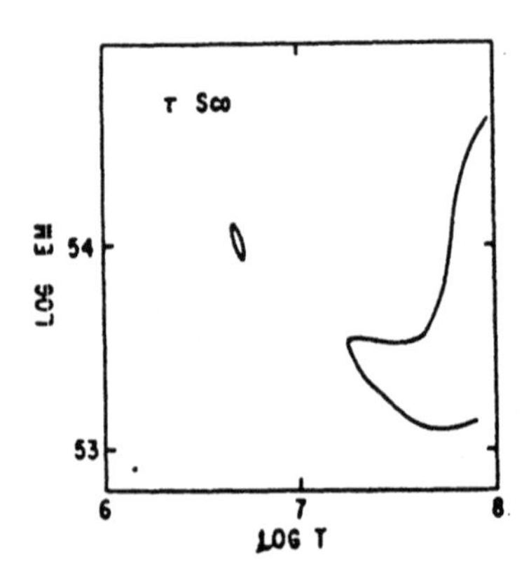

Figure 2. 90 % confidence contours for two temperature fits to the Tau Sco SSS spectrum shown in figure 1. From Swank(1985)

The results of ultraviolet line analyses

The profiles of the O VI and N V lines are shown in figure 3. These and other profiles with shortward displaced absorption have been analyzed to derive the properties of the wind by Lamers and Rogerson (1978), Hamann(1980), Conti and Garmany(1980), Gathier, Lamers, and Snow(1981). The results are as follows: $R = 6.7\ R_o$, $T_{eff} = 32000$ K, v sin i < 10. km/s, v wind = 1800. - 2000. km/s, and the estimates for the Mass loss rate: in M_o per yr are:

Hamann: $10^{-8.9}$

Conti and Garmany: $10^{-8.1}$

Gathier et al: $10^{-7.2}$

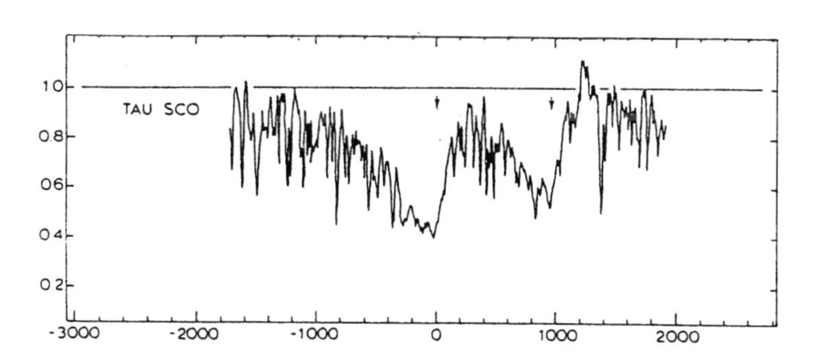

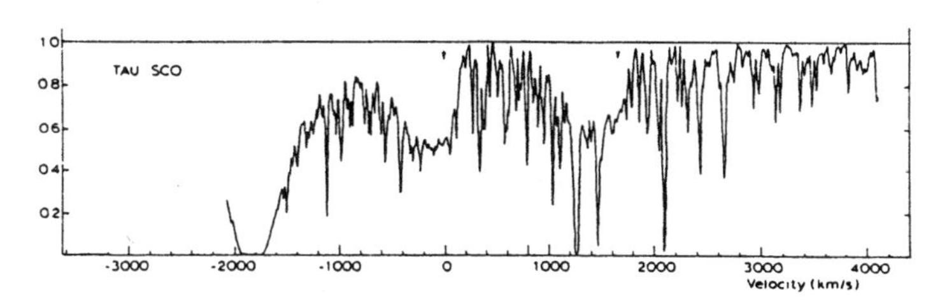

Figure 3. The profiles of the N V 1238, 1242 A lines in Tau Sco are shown in the the top figure, and the profiles of O VI 1032, 1038 A are in the bottom figure. The horizontal axis gives the velocity in the frame of the star. The arrows indicate the laboratory wavelength of the line. There is absorption longward of line centers, and the profile contours indicate that the ions decrease in abundance in the outward direction. From Lamers and Rogerson(1978).

There are two properties of the lines of the superionization stages that are of particular interest: 1) There is absorption by O VI and N V longward of the rest wavelength, indicating that there downflow towards the star with speeds as high as 250 km/s, or that there is "turbulence" in the wind with an rms speed of about 150 km/s (Lamers and Rogerson 1978). 2) The line profile contours show that the fractional abundances of O VI and N V decrease in the outward direction. The corona + cool wind model of Cassinelli and Olson (1979) failed to explain this.

THE PROBLEMS POSED FOR SHOCKED WIND MODELS

I. The x-ray emission measure is very large, comparable with the emission measure of the star's wind.

If the wind had the terminal speed throughout, and $M = 10^{-8}$ $M_\odot$/yr the wind emission measure would be only 4×10^{53} cm^{-3} , an order of magnitude below that required to explain the X-rays. Accounting for a velocity distribution starting at v_s =620 km/s (shock jump needed for $T=5.3\times10^{6}$) increases the emission measure only by a factor about 2. Thus we can say that the mass loss rates of Hamann and of Conti and Garmany must be too small, and that an unexpectedly large fraction of the wind contains very hot gas.

II. The column density required to cool shocked gas at 5 million or 15 millon degrees is as large as the entire wind column density . There cannot be many "periods" in a periodic shock model.

The cooling length derived for a rather general shock by Krolic and Raymond(1985) is $N = 7.7 \times 10^{13}$ x v shock(km/s) squared. For the dominant X-ray component with $T = 5.3 \times 10^{6}$ K, $N = 1.1 \times 10^{21}$ cm^{-2} , For the Hot component (say $T=15 \times 10^{6}$ K so v shock = 1040 km/s), we get $N = 8.8 \times 10^{21}$ cm^{-2}. The full column density of the wind (assuming now that $M = 1 \times 10^{-7}$ $M_\odot$/yr and a contant velocity wind.) $N_{wind} = 2.2 \times 10^{21}$ cm^{-2} ! Thus even with the highest plausible mass loss rate for Tau Sco, the cooling rate is too slow to accommodate the strong shocks that are required to explain the SSS spectra.

III The jump in velocity required to explain the Hot component is more than half of the wind terminal velocity.

A decrease by this amount would lead to a large increase in the density in the flow, effectively forming a shell about the star. There is no evidence for a "shell" in ultraviolet, optical or infrared data for Tau Sco.

IV The spatial dependence of the superionization stages is not explainable by radiative ionization from shocks.

The profiles of the O VI and N V lines indicate that these ions decrease in abundance relative to the electron density in going towards larger radii. The fractional abundance of say O VI is predicted to vary with radius as $q(O\ VI) = J(r) / n_e$ where J is the X-ray mean intensity and n_e is the electron density(cf. Cassinelli and Olson 1979) For the case in which the X-rays are formed in zones distributed throughout the wind, the change in J versus r is small. Therefore $q = 1 / n_e$ and the ion abundance is predicted to increase strongly in the outward direction, contrary to the observed result.

In conclusion: we see that the X-ray emission of this main sequence star presents problems that are especially challanging for the picture that the X-rays arise *only* in shock regions embedded in the wind. These problems are not so apparent in stars that have very large mass loss rates only because of optical depth effects. The presence of the very high temperature gas may indicate the existence of magnetically confined regions at the base of the wind as has been discussed by Cassinelli and Swank(1983).

REFERENCES

Cassinelli, J.P. and Swank, J.H. 1983, Astrophys. J. **271**,681.
Cassinelli, J.P. and Olson, G.L. 1979, Astrophys. J. **229**,304.
Conti,P.S. and Garmany C.D. 1980 Astrophys. J. **238**,190.
Gathier, R.,Lamers, H.J.G.L.M. and Snow, T.P. 1981 Astrophys. J. **247**,173.
Hamann, W.R. 1980, Astron. Astrophys. **84**,342.
Krolick, J.H. and Raymond, J.C. 1985, Astrophys. J. **271**,681.
Lamers, H.J.G.L.M. and Rogerson J.E. 1978 Astron. Astrophys. **66**, 417.
Lucy, L.B. 1982, Astrophys. J. **255**,278.
Lucy, L.B. and White, R.L. 1980, Astrophys. J. **241**,300.
Swank, J.H. 1985 in *The Origin of Nonthermal Heating/Momentum in Hot Stars*, Eds. A.B. Underhill and A.G. Michalitsionos, NASA Publ. 2358.

Joe Cassinelli checking his results

Henny Lamers during his speech at the end of the meeting

"After the ball is over". One of the editors still working hard

NAME INDEX

OBJECT INDEX

Printed in the United States
by Baker & Taylor Publisher Services